CITIES, POVERTY, AND DEVELOPMENT

To Jacqueline,
Daniel and Suzie
To Janine,
Saskia and Rachel

CITIES, POVERTY AND DEVELOPMENT

Urbanization in the Third World

Second edition

ALAN GILBERT
and
JOSEF GUGLER

OXFORD UNIVERSITY PRESS

Oxford University Press, Walton Street, Oxford OX2 6DP
Oxford New York Toronto
Delhi Bombay Calcutta Madras Karachi
Kuala Lumpur Singapore Hong Kong Tokyo
Nairobi Dar es Salaam Cape Town
Melbourne Auckland Madrid
and associated companies in
Berlin Ibadan

Oxford is a trade mark of Oxford University Press

Published in the United States
by Oxford University Press, New York

First published 1992
Paperback edition reprinted 1993, 1994

British Library Cataloguing in Publication Data
Data available

Library of Congress Cataloging in Publication Data
Gilbert, Alan, 1944–
Cities, poverty, and development: urbanization in the Third World
Alan Gilbert and Josef Gugler.—2nd ed.
Includes bibliographical references and indexes.
1. Urbanization—Developing countries. *2.* Rural–urban migration—Developing countries. *3.* Developing countries—Economic conditions—Regional disparities. *4.* Urban poor—Housing—Developing countries. *5.* Community organization—Developing countries. *6.* Social conflict—Developing countries. *7.* Regional planning—Developing countries. *I.* Gugler, Josef. *II.* Title.
HT169.5.G53 1991 307.76'09172'4—dc20 31–32816
ISBN 0–19–874160–X
ISBN 0–19–874161–8 (Pbk.)

Printed in Great Britain
on acid-free paper by
Biddles Ltd, Guildford & King's Lynn

Preface

The twentieth century may come to be seen as the age of urbanization. Urban settlements were first established more than five thousand years ago, but as recently as 1900 only one in eight people lived in urban areas. Before the end of this century half of the world's population will be urbanites, and of these three billion people two-thirds will live in the Third World.

In this extensively revised edition we offer a comprehensive account of Third World urbanization, one which spans three continents and straddles several academic disciplines. Our collaboration has brought together backgrounds in Economics, Geography, and Sociology, extensive research experience in Latin America and in Africa, and sustained work on urban issues. We have drawn on a body of research and writing on Third World urbanization that has expanded rapidly over the past two decades. Along the way we have been assisted by friends, colleagues, and students. Most easily traced are the contributions made by Floyd Dotson, David Drakakis-Smith, Susan Eckstein, Myra Marx Ferree, William G. Flanagan, Howard Handelman, Peter Kilby, Gudrun Ludwar-Ene, Tony O'Connor, Henry Rempel, and Peter Ward, all of whom made helpful comments on individual chapters. Alan Gilbert would like to thank University College and the Institute of Latin American Studies, London, for institutional support. He also thanks the Center for US-Mexican Studies, University of California, San Diego, for six months free of teaching which gave him time to read extensively and to make substantial revisions to his original chapters. Josef Gugler would like to acknowledge the support of the American Institute of Indian Studies for a six month's stay in India that allowed him to familiarize himself with urban research in the subcontinent. He is greatly indebted to library staff at the University of Connecticut and Universität Bayreuth, to C. Saskia Gugler who dramatically improved his prose, and to Debra Crary and Heidemarie Reichert who processed more drafts than any one of us cares to remember.

We recognize that despite all these people's efforts, including those of the many reviewers of the first edition of the book, shortcomings remain. Our only excuses are, first, that the so-called Third World is highly complex and contains a multitude of very different kinds of towns and cities; and, second, that in drawing on a wide range of research we have faced important differences in approach, objectives, and ideology. We shall not please everyone in this new edition, we only hope that we will satisfy most readers.

Alan Gilbert
Josef Gugler

London and Storrs
March 1991

Contents

Tables

Figures

Introduction

The literature on urban development is vast and growing daily. The very volume of this literature, added to its often specialized nature, makes much of the writing inaccessible even to the expert. Much useful work in sociology is unknown to economists, political scientists, planners, or geographers; much interesting Latin American work is unknown to Asian specialists. To help overcome this problem is one of our principal reasons for writing this book. Our main objective is to review current theoretical ideas about urbanization and poverty and to apply those ideas to the diverse conditions found in the Third World. To date, interesting ideas have often been developed and tested only within a particular country or region. For example, 'dependency' theory was developed largely in the context of Latin America, the role and influence of urban ethnic associations tested mainly in Peru and West Africa, and the concept of circular migration examined only in the Pacific region and in Southern Africa. Too few ideas or phenomena have been considered comparatively across different areas of the Third World. Our aim, therefore, is to explain the more interesting theoretical concepts and to examine their relevance to African, Asian, and Latin American experience.

Clearly, this is an ambitious task. For a start, urban development is part of the comprehensive process of societal change. The main objects of our study, the city, the poor, and the so-called Third World, all form part of a complex reality that requires some consideration of factors which, at first sight, seem to be divorced from our main theme. Since urban growth in poor countries is linked integrally to rural change and since the development or underdevelopment of the Third World is linked to the process of development in the First World, we are forced to discuss, albeit superficially, giant issues such as the broad sweep of world history over the past five centuries, the current world economic situation, and the meaning and nature of 'development'. In this respect, recent changes in the social sciences have strongly influenced the form of this book. During the 1970s, studies of the less-developed world, and indeed of urbanization generally, increasingly shifted their attention towards what is commonly known as a 'political economy' approach. This approach was consolidated and diffused widely through the academic community during the 1980s. It was a move towards a more holistic and class-based view of society, away from a functionalist, positivist, and consensual view. In the field of development studies the rise of 'dependency' and 'neo-Marxist' schools of thought shifted the scale of analysis away from the individual. Instead of considering how individual peasants responded to rural–urban income differentials, for example, academic work focused much more on the causes of those differentials. Poverty and its manifestations were no longer perceived as something attributable to an individual person, city, or country, or remediable by national governments using technical planning processes. Rather poverty, or more accurately the state of relative poverty, was seen to be a consequence of a historical process of incorporation into the world capitalist system. To use Dos Santos's (1970) much quoted term, the economies and societies of poor countries had been 'conditioned' by their relationship with colonial and neo-colonial powers.

In urban studies a parallel shift towards a radical approach occurred; Castells (1977) and others reformulating the urban question. They attacked the belief that urban form emerged through a neutral process of individual decision making. For Castells and Harvey (1973), urban areas could be understood only as a result of the conflicts between classes which were a direct outcome of the operation of the capitalist mode of production; urban form, urban issues, urban government, urban ideology could be understood only in terms of the dynamic of the capitalist system. Space was socially determined: the outcome of conflicts between different social classes. Urban 'problems' arose not by chance or through mismanagement but because the interests of one or other social-class fraction had been served by the emergence of such problems. Squalid back-to-back terraced housing in British cities was built for low-paid workers; it was not in the interest of dominant groups to replace it especially as that would increase the costs of labour reproduction. 'Urban disorder was not in fact disorder at all; it represented the spatial organization created by the market, and derived from the absence of social control of the industrial activity' (Castells, 1977: 14–15). State planning might reduce such disorder but could not remove it, for resources were allocated on the basis of a struggle between competing groups. The process of planning might pretend to allocate resources fairly between groups: in practice it did not operate that way. Those who wielded political power influenced planning decisions against the interests of the powerless. Planning did not serve the public interest because there was no such interest.

Of course, some analysis of these issues had always been included in urban and development studies, but the paradigm shift raised them from the second or third rank of importance to top place. The first edition of this book was strongly influenced by that paradigm shift and this second edition continues to adopt certain of its main arguments. Rather than pose the issue of what governments should do to remedy the obvious problems facing the Third World urban poor and how individuals should act in a particular urban situation, we ask why particular phenomena occurred in the first place and indeed why specific government reactions evolved in the way they did. We are concerned with the issue of power and the subsequent allocation of resources between nations and between classes. Why are decisions made in the way they are, who benefits and who suffers from those decisions, and what is the relationship between urban development, society, and the state?

We follow the 1970s paradigm in so far as we accept that conflicts between social groups have an important bearing on the structure of Third World societies and therefore on the process of urbanization. We agree with much Marxist writing about the ideological nature of many of the earlier studies of urban development and planning and in much of the book we follow a political economy approach to the analysis of urban issues. At the same time, it is clear from the happenings of the 1980s that writing from the left does not contain all of the answers that we need. Certainly, the idea that socialism, or at least Marxism, can produce perfect societies has suffered something of a battering in recent years. The revelations emerging from China, the Soviet Union, and Eastern Europe in the last couple of years have shown that socialist revolution is no panacea for the needs of the poor. Without for one moment suggesting that privatization and free-market ideology as propagated by the right provide the correct answers either, we are even more certain than when we wrote the first edition, that a social revolution alone is no answer. To this extent we can only agree with Lefebvre (1970: 220) when he claims that 'the same problems [of urbanism] may be found under socialism and under capitalism with the same absence of response'.

In this new edition, we have also been influenced by the changes in thinking about society which occurred during the 1980s. Within sociology, 'structuralism' has given way to the idea of 'structuration' (Giddens, 1984). Put crudely, the latter approach accepts the possibility of change within broad structural constraints. Human agency can modify society even when the odds are against it. Hence social movements may spring up against dictatorships and transform the way that the state acts. Small countries may transform their economic situation despite the unfavourable workings of the world economic system. In short, there is no automatic route which Third World countries are bound to follow. Their futures are not determined by the inevitable forces of the world economic system and by the societies constructed in response to that system. Nor is there any ideal path along which all societies should seek to proceed. The events of the 1980s, show that neither left nor right has satisfactory answers. The theory of structuration implies that different paths to change can be followed even by societies facing similar internal and external constraints. Unsatisfactory though it may seem, such an approach strongly suggests that there can no longer be a master plan for social change.

Similarly, it is clear that the 1980s have seen major changes in the world economy. Many parts of the Third World have become more firmly incorporated into the world capitalist system as transnational capital has continued to spread its tentacles in search of cheaper labour, new resources, and larger markets. At the same time, the debt crisis has brought a profound decline in living standards in most parts of Africa and Latin America. There has also been a profound change in political ethos at the centre. Social democracy and socialism gave way in many countries to conservatism and neo-liberal thought. The economic and social consequences within the Third World of a decade of Reagan-Thatcherism have been profound. Urban processes have been greatly influenced by that epoch.

While our thinking continues to change in response to the world around us, one unfortunate and undeniable fact of life remains. The world continues to be organized in a profoundly unfair way. Dominant patterns of world trade, investment, and co-operation operate to the detriment of the great majority of humankind. To a significant degree, the Third World is poor because the First World is rich. Admittedly, few indigenous societies were Utopian before Europeans set sail to reach them, but there can be little doubt that the arrival of Cortés, Cook, Vasco da Gama, and much later, of Henry Ford's machines and Rockefeller's oil-rigs, had a traumatic effect on many of those societies. If the idea conveyed in Haley's *Roots* of merrie indigenous Africa (and by implication of merrie Asia, Latin America, and Melanesia) is erroneous, the superimposition of colonialism and capitalism hardly created the cheerful cosmopolitan world of the Coca Cola advertisements. The most that can be said is that poor, unequal, somewhat isolated Africa eventually became somewhat less poor, much more unequal, distinctly more populous, and decidedly more integrated into the world economy. Change there was, but little progress. We would not deny that economic growth occurred in many parts of the Third World, and that inequalities were reduced in some instances, but there can be little doubt that such development opportunities as arose were only partially realized. Third World countries are characterized by patterns of economic growth and income distribution which are little geared to increasing the welfare of the masses.

When we wrote the first edition of this book, we were hardly sanguine about the future facing the Third World. Nevertheless, we did at least expect economic growth to continue in most countries. With hindsight, this was an error. The 1980s have seen a major economic crisis hit many parts of the Third World. Most Africans and Latin

Americans were poorer in 1990 than they had been in 1980; in some cases incomes had fallen back to the levels of the 1960s (World Bank, 1989). Eventually, many of the economies of Africa and Latin America will again expand, but in the increasingly diversified Third World it can no longer be assumed that every country will prosper.

We also have less faith in the willingness of the developed countries to redress the yawning gap that separates the rich from the poor. We had little expectation at the time of the first edition that much would be done. We noted that the Brandt Commission Report had hit the media headlines but felt that it was likely to go the same way as the Pearson Report and the two Development Decades; it would attract some media attention but little governmental action would be prompted by it. At the current time, the whole political philosophy of the First World is hardly conducive to helping poor countries. The right-wing economic policies employed by governments in the United States, Britain, France, Japan, and West Germany, and their obvious support among many voters in those countries, promises a continuingly poor future for most Africans, Asians, and Latin Americans. Indeed, the whole history of the debt crisis and the recent economic decline of Africa and Latin America suggest that the rules governing the world economic order have shifted further in favour of the haves.

What is perhaps still more worrying is that the rich countries do not even perceive this reality. Much of the political dialogue in the developed countries is concerned with the problems of those countries: inflation, unemployment, the urban crisis, farm prices, etc. Many of these problems are being blamed on the few succcessful poorer countries. Newly industrializing countries (NICs) such as Korea, Taiwan, and even the Philippines and Mexico, are being blamed for the lost jobs of the declining industrial regions of the developed world. Rather than blaming the transnational corporations (TNCs) for shifting their production facilities to certain NICs, the poorer countries themselves are often criticized. The truth today is that capital has become truly international and is often highly mobile. There are reasons for concern in the developed countries; but the source of change is still more worrying for most Third World countries. Similarly, the recent recognition that the world's environment is being damaged by current patterns of resource use, is likely to rebound more upon the poor than the rich. It is the excesses of deforestation in Brazil and the fear of what future increases in the numbers of refrigerators in China and India will do to the ozone layer that seem to dominate policy in many developed countries. As usual, the United States seems reluctant to admit that the way its own economy and society is organized can possibly be in error.

It is absolutely obvious from the events of the 1980s that we were right to emphasize these kinds of issues in the first edition. Urbanization simply cannot be understood unless placed in a broader societal context. The future of the Third World city cannot be understood without questioning the basis of the world division of labour and the interests of governments ostensibly managing the city.

Within this broad perspective we seek in this book to compare the processes of urban change in different Third World countries. We are trying to generalize, but are aware that differences in the ways in which poor countries were integrated into the world economy created many distinctive urban features. More women migrate to cities in Latin America, but more men migrate in most Asian and African countries; many families invade land in Latin American cities, but organized squatting is less common in most African and Asian cities; urban areas house most Argentines and Mexicans, but a minority of Indians and Ghanaians. Why did these differences emerge and to what extent are they tending to disappear as the international economy tightens its embrace? How far can we generalize meaningfully about urban conditions and about the prospects for change? By comparing the literature on African, Asian, and Latin American cities and societies, we hope to demonstrate both the strengths and the weaknesses of recent generalizations emerging from social science writing. In preparing this book, we are seeking to unravel some of the main issues that will determine our future. Understanding may not bring change, but sensible change rarely occurs without it.

The Third World

Throughout the book we use the much abused term the 'Third World' to refer to the world's poor countries. We are well aware of the deficiencies of this term. It may be that there are now more accurately six worlds (O'Connor, 1976), or four worlds (Wolf-Phillips, 1979). It may be that older terms, such as less-developed countries, developing nations, poor countries, etc. are equally appropriate descriptions. It may be that there is no meaningful distinction between the First, Second, and Third Worlds. It may be that the Third World has come to an end (Harris, 1987), or even that it never existed.

Our use of the term 'Third World' is merely a convenient shorthand. In this book it includes all countries with a per capita national income below $3,000 in 1978 or a life expectancy of less than 70 years (World Bank, 1980: 148–9). This arbitrary, if still meaningful, dividing line means that all the nations of Asia, Africa, Latin America, and the Caribbean are included with the exceptions of Israel, Japan, Hong Kong, and Singapore. Unlike some authors, therefore, we include several poor countries often described as part of the Second World: socialist, centrally planned economies such as Cuba, the People's Republic of China, Vietnam, Laos, Kampuchea, and the Democratic Republic of Korea. We have included them because of their poverty and because of their recent colonial experience. Perhaps more controversially, we exclude the nations of underdeveloped Europe (Seers, 1979); nations which in terms of per capita income and life expectancy are little different from some of the 'poor' countries we have included. While Albania, Greece, Portugal, Romania, and Yugoslavia all fall below our statistical dividing line and might justifiably be included in the 'Third World' we have excluded them partly on the grounds of convention and partly because of our own lack of knowledge of Southern Europe.

Current Patterns of Urbanization

Even if there are clear signs of counter-urbanization in the United States and Western Europe (Berry, 1976; Hall, 1982; Vining, 1982), the world's population is becoming increasingly urban. In less-developed countries, the pace of change since 1950 has been startling. Table 1.1 shows that between 1950 and 1985 the proportion of the population living in urban areas exactly doubled. Fuelled by changes in the countryside, high rates of fertility, falling death rates and rapid cityward migration, most Third World countries have been transformed from rural to urban societies in two or three decades. The larger cities have been expanding rapidly, often doubling in size every fifteen years.

Table 1.1 shows that in 1950 the populations of virtually every part of the Third World lived mainly in the rural areas. In China, South and Southeast Asia, Melanesia, and most of sub-Saharan Africa less than one person in six lived in a town or a city. Indeed in the whole of Africa, Asia, and Latin America, only Japan, temperate South America, and Southern Africa contained urban majorities. While parts of the Third World had begun to experience rapid urban growth, most societies remained predominantly rural in nature.

Since 1950 urbanization has become a world-wide phenomenon. While the pace of change has varied considerably between countries and regions, virtually every country of the Third World has been urbanizing rapidly. Only in a few socialist countries such as China, Cuba, and Kampuchea has government policy temporarily slowed the urban transformation. Most regions have seen their people leaving the countryside in ever-increasing numbers. In places, the movement was so rapid that migration soon ceased to be the predominant cause of urban growth. By the 1970s, for example, growth in many Latin American cities was sustained principally by natural increase; former migrants were now bearing children in the cities.

Table 1.1

Percentage of total population in urban areas by region, 1950–1985

Major region and subregion	1950	1970	1985
AFRICA	13.2	22.5	29.7
Eastern Africa	5.3	10.3	18.1
Middle Africa	8.1	24.8	35.6
Northern Africa	23.2	36.5	42.1
Southern Africa	36.5	44.1	52.5
Western Africa	9.5	17.6	24.9
LATIN AMERICA	40.9	57.4	69.0
Caribbean	33.0	45.8	56.5
Mexico and Central America	39.5	53.9	63.3
Temperate South America	62.8	77.9	84.3
Tropical South America	36.5	56.1	70.4
UNITED STATES AND CANADA	63.6	73.8	74.1
ASIA	15.9	23.9	28.1
China	11.1	20.1	20.6
Japan	50.3	71.2	76.5
Other East Asia	23.2	47.4	66.8
Southeastern Asia	13.4	20.2	26.3
South Asia	15.6	19.5	25.2
Western Asia	23.3	43.2	55.0
EUROPE	54.8	66.7	71.6
Eastern Europe	42.2	53.5	61.5
Northern Europe	70.8	82.4	86.1
Southern Europe	44.9	56.2	62.5
Western Europe	63.2	76.4	79.6
OCEANIA	64.5	70.8	71.1
Australia and New Zealand	78.7	84.4	85.2
Melanesia	2.0	15.1	20.2
Micronesia and Polynesia	20.6	32.3	41.6
USSR	39.4	56.7	65.6
More-developed regions[a]	53.4	66.6	71.5
Less-developed regions[b]	15.6	25.4	31.2

Note: Definitions of urban areas are based on national criteria.

[a] Includes Europe, United States and Canada, Japan, and Australia and New Zealand.

[b] Includes all other regions.

Sources: UN (1982); UN (1988).

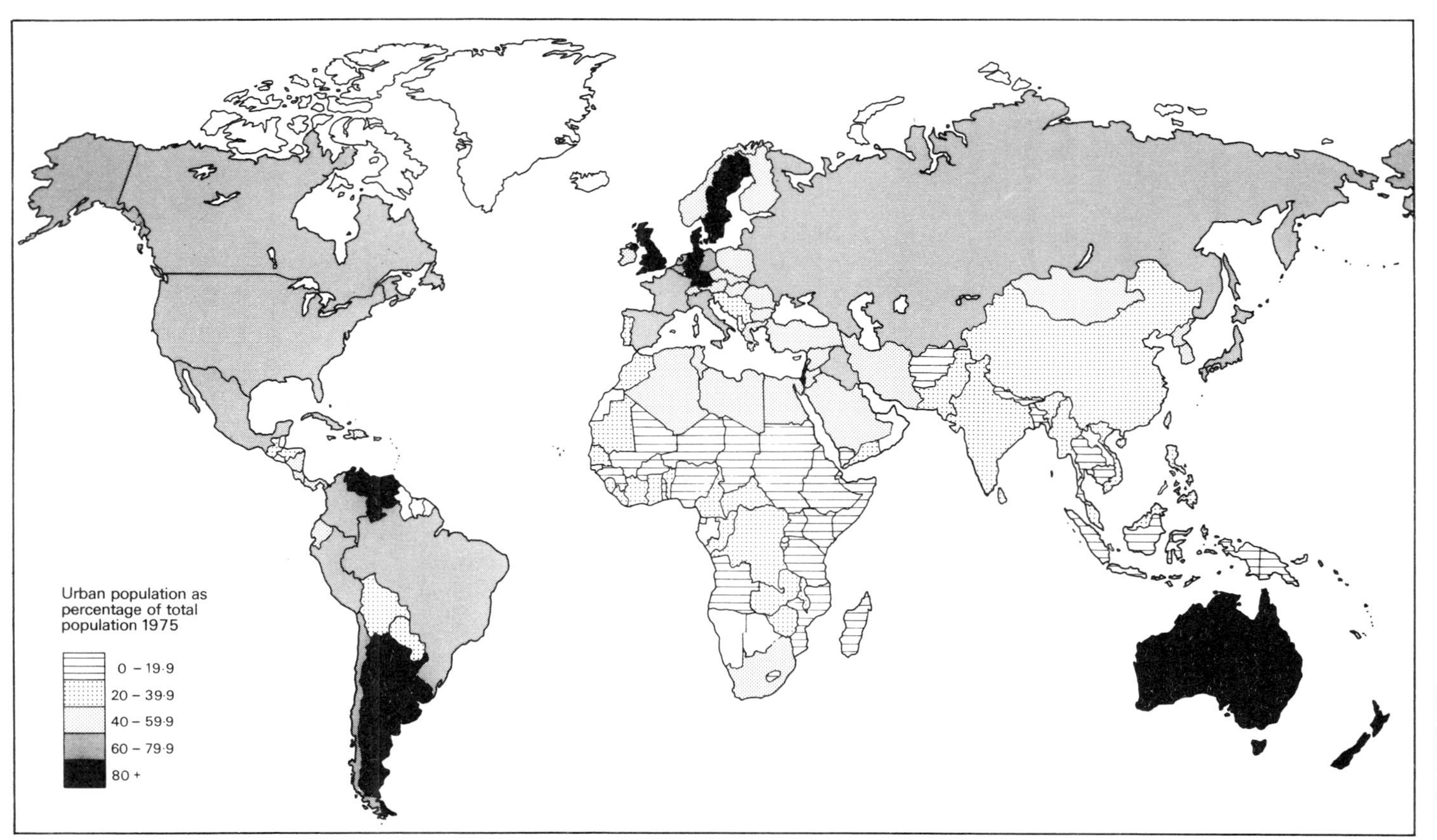

FIG. 1.1 Urban population as percentage of total population, 1985

Table 1.2

Urban population growth and development indicators for larger Third World nations

Country[a]	Urban population as % of total population[bc]		Annual rate of urban growth[bc]			Per capita income[c] ($US)	Annual growth in p.c. income	Life expectancy at birth[c]	Total population[c] (millions)
	1987	1960	1980–7	1970–80	1960–70	1987	1965–87	1987	1987
Uganda	10	5	5.0	7.0	6.3	260	−2.7	48	15.7
Ethiopia	12	6	4.6	6.9	6.1	130	0.1	47	44.8
Malawi	13	4	8.6	6.2	6.6	160	1.4	46	7.9
Mali	19	11	3.4	5.5	5.4	210	n. a.	47	7.8
Sudan	21	10	4.2	6.8	6.9	330	0.5	50	23.1
Kenya	22	7	8.6	6.8	6.6	330	1.9	58	22.1
Mozambique	23	4	10.7	6.8	6.6	170	n. a.	48	14.6
Madagascar	23	11	6.4	5.2	5.1	210	−1.8	54	10.9
Zimbabwe	26	13	6.3	6.4	6.8	580	0.9	58	9.0
Tanzania	29	5	11.3	8.3	6.3	180	−0.4	53	23.9
Ghana	32	23	4.1	5.2	4.6	390	−1.6	54	13.6
Nigeria	33	13	6.3	4.9	4.7	370	1.1	51	106.6
Senegal	37	23	3.3	3.3	2.9	520	−0.6	48	7.0
Zaïre	38	22	4.6	7.2	5.2	150	2.4	52	32.6
Cameroon	46	14	7.4	7.5	5.6	970	3.8	56	10.9
Ivory Coast	52	19	6.9	8.2	7.3	740	1.0	52	11.1
Zambia	53	23	6.6	5.4	5.4	250	−2.1	53	7.2
South Africa	57	47	3.3	3.1	2.8	1,890	0.6	60	33.1
Sub-Saharan Africa[d]	29	14	6.4	6.2	5.7	421	0.0	52	409.8
Nepal	9	3	7.8	4.7	4.3	160	0.5	51	17.6
Bangladesh	13	5	5.8	6.8	6.7	160	0.3	51	106.1
Thailand	21	12	4.9	3.5	3.6	850	3.9	64	53.6
Sri Lanka	21	18	1.2	3.7	4.3	400	3.0	70	16.4
Burma	24	19	2.3	4.0	3.9	n. a.	n. a.	60	39.3
Indonesia	27	15	5.0	3.6	3.7	450	4.5	60	171.4
India	27	18	4.1	3.3	3.3	300	1.8	58	797.5

Pakistan	31	22	4.5	4.3	4.0	350	2.5	55	102.5
China, PR	38	19	11.0	3.1	3.4	290	5.2	69	1,068.5
Malaysia	40	25	5.0	3.5	3.5	1,810	4.1	70	16.5
Philippines	41	30	3.8	3.6	3.8	590	1.7	63	58.4
Korea, R. of	69	28	4.2	4.8	6.4	2,690	6.4	69	42.1
Low-income Asia[d]	30	18	5.0	4.2	2.1	732	2.1	62	2,489.9
Yemen Arab R.	23	3	8.4	7.3	7.5	590	n. a.	51	8.5
Algeria	44	30	3.9	6.4	6.1	2,680	3.2	63	23.1
Morocco	47	27	4.5	4.5	4.3	610	1.8	61	23.3
Turkey	47	30	3.4	4.6	5.1	1,210	2.6	64	52.6
Egypt	48	38	3.7	3.0	3.4	680	3.5	61	50.1
Syrian Arab R.	51	37	4.5	4.7	4.8	1,640	3.3	65	11.2
Iran	53	34	4.2	4.9	4.7	n. a.	n. a.	63	47.0
Tunisia	54	36	2.9	3.8	3.8	1,180	3.6	65	7.6
Iraq	72	43	4.9	5.4	6.2	n. a.	n. a.	64	17.1
Middle East and N. Africa[d]	49	31	4.5	5.0	5.1	1,227	3.0	62	240.5
Guatemala	33	33	2.9	3.7	3.6	950	1.2	62	8.4
Ecuador	55	34	5.0	4.5	4.4	1,040	3.2	65	9.9
Peru	69	46	3.2	4.4	5.0	1,470	0.2	61	20.2
Colombia	69	48	2.9	3.9	5.2	1,240	2.7	66	29.5
Mexico	71	51	3.2	4.5	4.8	1,830	2.5	69	81.9
Brazil	75	46	3.7	4.3	4.8	2,020	4.1	65	141.4
Venezuela	83	67	2.6	4.2	4.7	3,230	−0.9	70	18.3
Chile	85	68	2.3	2.4	3.1	1,310	0.2	72	12.5
Argentina	85	74	1.9	1.8	2.0	2,390	0.1	71	31.1
Latin America[d]	69	52	3.1	3.7	4.2	1,720	1.5	67	353.2

[a] Only countries with populations of more than 7 millions are included.
[b] World Bank (1980: 148–9).
[c] World Bank (1989: 164–5, 224–5).
[d] Unweighted averages.

n. a. Not available.

Table 1.2 shows that by 1985 urbanites had become a majority in the more affluent parts of the Third World. Most people in Latin America, the Middle East, and the Far East now live in cities. Even in the poorer areas, urban growth is proceeding apace. Between 1970 and 1985 there was a dramatic jump in the proportions of urban people in Middle and Eastern Africa, the Far East and, Western Asia. Table 1.2 shows that by 1987 numerous countries had joined the ranks of the predominantly urban societies: in Africa, the Ivory Coast, South Africa, Tunisia, and Zambia; in Asia, Iran, Syria, and South Korea; in Latin America, Brazil, Ecuador, and Peru. By the end of the decade, it is likely that only the poorest countries of Africa and Asia will still contain majorities of country folk. Even there, life will have been dramatically changed by the urban revolution.

Patterns of Inequality

The patterns and processes of urban development that we describe in this book are those of an unequal world. There is no way in which the process of urban development can be understood in isolation from the processes that generate that inequality. We have already referred to the inequalities which exist at the world level but there are also important inequalities within Third World countries. Table 1.3 indicates the extent of this inequality. The poorest quintile received as little as 2.4 per cent of household income in the Ivory Coast in the middle 1980s compared to the top 10 per cent's share of 43.7 per cent. In Zambia the degree of concentration at the top was still greater with the richest 10 per cent of households receiving 46.4 per cent of income in 1976.

Unweighted averages of these figures demonstrate that the distribution of income in Third World countries is generally much less equal than that in most industrialized nations, the top decile of households in each group receiving 36.4 per cent and 25.4 per cent of income respectively. Admittedly the difference is less marked at the other end of the scale, although it has to be remembered that the poor in the Third World countries were receiving a low share of a very small household income. Clearly, most Third World households are both absolutely and relatively poor.

These inequalities within Third World countries have contributed to the urban forms which have emerged. In the absence of such wide inequalities in access to resources, the phenomena we describe would be very different. While shanty towns, unemployment, petty services, and pavement dwellers are in part an outcome of national poverty and the unequal world distribution of income, their form and severity would be less marked if Third World societies were more equal.

Table 1.3
Distribution of income in selected countries: percentage share by household group

Country	Year	Lowest 20%	Lowest 40%	Top 10%
Bangladesh	1981–2	6.6	17.3	29.5
India	1975–6	7.0	16.2	33.6
Kenya	1976	2.6	8.9	45.8
Sri Lanka	1980–1	5.8	15.9	34.7
Zambia	1976	3.4	10.8	46.4
Indonesia	1976	6.6	14.4	34.0
Philippines	1985	5.2	14.1	37.0
Ivory Coast	1985–6	2.4	8.6	43.7
Thailand	1975–6	5.6	15.2	34.1
El Salvador	1976–7	5.5	15.5	29.5
Malaysia	1973	3.5	7.7	39.8
Mexico	1977	2.9	9.9	40.6
Korea	1976	5.7	16.9	27.5
Average	—	4.8	13.2	36.6
Spain	1980–1	6.9	19.4	24.5
Italy	1977	6.2	17.5	28.1
New Zealand	1981–2	5.1	15.9	28.7
United Kingdom	1979	7.0	18.5	23.4
France	1975	5.5	17.0	26.4
Germany	1978	7.9	20.4	24.0
Japan	1979	8.7	21.9	22.4
Sweden	1981	7.4	20.5	28.1
United States	1980	5.3	17.2	23.3
Average	—	6.7	16.8	25.4

Averages are unweighted.

Source: World Bank (1989). Detailed sources are provided on p. 280 of that publication, which also notes that although '. . . the estimates shown are considered the best available, they . . . should be interpreted with extreme caution.'

1

Urban Development in a World System

Urban form is a direct outcome of the ways in which different societies are organized. In an unequal world, therefore, it is not surprising that urban societies should also be unequal. Indeed, it has been argued that cities have served élite groups throughout history and that it is only in élitist societies that cities can actually develop. Wholly egalitarian societies based on what Polanyi calls 'balanced reciprocity' cannot produce cities. Without a central power and a mechanism to generate a surplus over consumption and to concentrate it into urban areas, cities cannot grow. It is only in societies based on 'redistribution', where a surplus over consumption can be appropriated by a particular group, that urbanization is possible. But, the fact that urbanization depends upon some mechanism to concentrate wealth does not lead us very far. Through time, the world has contained a wide range of urban societies with fundamental differences between them. The theocratic societies of ancient Mexico or China, for example, supported urban areas very different from most contemporary cities. The ancient cities contained administrative and religious élites who were supported by the agricultural surplus extracted through a combination of military force and moral pressure. In so far as the élites in these societies posed as the representatives of God and even as gods themselves, their rural subjects had little choice but to offer tribute. By contrast, modern industrial societies rely less upon force, although European colonialism and contemporary military dictatorships suggest that force is scarcely alien to capitalism, and more upon a complicated web of interlinked interests. Capitalist societies contain élites as powerful as those of theocratic societies, but the mechanisms for generating and accumulating a surplus are clearly very different. Rather than justifying tribute in terms of religious or magical functions, the ideology of capitalism legitimizes urbanization in terms of its contribution to the growth of the gross national product. Whether all capitalist cities perform that role is a debatable issue. Certainly, a case can be made that urban areas in developed countries are 'generative' in Hoselitz's (1957) sense; the services of the banks and finance houses of London have long generated income for the rest of Britain, as have the factories of Birmingham and Manchester and the ports of Liverpool and Southampton. But they are also as 'parasitic' as the cities of the Third World with respect to the natural resources and the primary products emerging from the less developed countries. Throughout the world cities extract surplus, whether it be from local agricultural areas or from half-way across the globe. Today large areas of the world are integrated into a single economy. Rural areas and mining centres produce for distant populations and consume products manufactured far away. Within this world economy, individual cities perform specialized functions and their individual prosperity depends greatly upon their position in this economic system. Indeed, it is one of the major arguments of this book that the size, role, and characteristics of individual cities reflect the world roles of the societies of which they form part. Thus the evolution of a city in the American Middle West or in southern India cannot be understood solely in terms of its local or even its national functions; its development is dependent on the way it is linked into the wider system. Needless to say, the strength of the linkages between different cities and the world system vary greatly. Nevertheless, in so far as the development of the world economy has created an interlinked economic system, different cities perform the roles allocated to them within that system.

In this respect, the modern world is different only in degree from the ancient world. In the past, regions were allocated roles by an imperial power and the cities within those regions reflected those roles. In the Roman Empire major communication centres, such as Alexandria, and cities that controlled rich agricultural regions, such as Damascus, acted as administrative centres of the empire and grew in size and prosperity; at the apex of the urban hierarchy was the imperial capital. The magnificence of Rome was generated by the tribute or surplus available to it. In this sense, it would seem to support Sjoberg's (1963: 220) proposition about urban growth in the pre-industrial world: 'at any given stage of technological development, the grander the empire, the grander the size and number of its cities.'

But should this proposition be confined to the ancient or pre-industrial worlds? Perhaps its value is nowhere better demonstrated than in the development of British cities up to 1900. London, for example, has always had national importance, but its growth since 1500 has been stimulated by its position as one of the centres of world power. Spanish silver and gold stolen by pirates such as Drake, Hawkins, and Vernon sustained sixteenth-century and seventeenth-century England. The slave trade and other forms of international commerce maintained momentum in the seventeenth and eighteenth centuries along with the expansion of colonies in the Americas. The activities of the East India Company extracted huge revenues for Britain during the eighteenth century in India (Chamberlain, 1974; Dutt, 1968). Rapid urban expansion, however, awaited the industrial revolution of the eighteenth century.

During the nineteeth century the growth of the textile, steel, and engineering industries led to the unparalleled expansion of Manchester, Glasgow, Liverpool, Birmingham, and the mill towns of Lancashire and Yorkshire. Such industrial and urban expansion, of course, did not benefit all of Britain's trading partners; not every nation agreed with the principles of liberal political economy. And when other less enlightened nations objected to the logic of free trade, or other European powers attempted to limit British penetration through political means, Africa, Asia, and the Caribbean were formally enlisted into the British Empire. Latin America was not incorporated into the formal empire, but British capital built railways, extracted nitrates, stimulated the expansion of cattle and cereal production, and generally accelerated export-based growth. A combination of military force, advanced industrial technology, and improved communications created and sustained both a formal and an informal empire. The great Crystal Palace Exhibition of 1851 is apt enough comment on Britain's reliance on the rest of the world. Victorian Britain's wealth and self-confidence, and therefore the greatness of its cities, were based on its foreign 'empire'. British cities expanded and the British urban system evolved on the basis of this new world division of labour.

British was exceptional but not atypical. Most of the great cities of Europe evolved and prospered on the basis of foreign empires: Paris, Madrid, Amsterdam, Vienna are notable examples. The grandeur of European, and indeed North American, cities at the turn of this century can only be explained in terms of the international division of labour and the evolution of a world urban system.

Urban Expansion in the Third World

The cities of Europe and the United States would have emerged in less grandiose form without the benefits of empire but urbanization in the Third World would have been very different. Karl Marx once described modern history as the 'urbanization of the countryside'; it is a peculiarly accurate aphorism for most parts of the Third World. Indeed, it is possible to argue that without the intrusion of industrial capitalism and imperialism some Third World societies would still lack major cities. In major parts of America and Africa urban development was superimposed by capitalism on essentially rural societies. In Peru, Mexico, India, the Middle East, and China, where indigenous urban civilizations had already developed, urban forms were radically altered. The impact of European expansion from the sixteenth century onwards transformed urban structures in the Third World. The functions and forms of contemporary Third World cities cannot be understood without a consideration of this process. As Hoselitz (1953: 204) has pointed out 'the cities of contemporary underdeveloped countries are hybrid institutions, formed in part as a response of a indigenously developing division of social labour and in part as a response to the impacts made upon less advanced countries by their integration into the world economy.' The next few pages will describe this process of European capitalist penetration and its effects on Third World urbanization.

The most obvious effects of European and later United States expansion in the Third World were the creation of new cities, the generation of new urban forms, and sometimes the destruction of existing urban cultures. In Latin America the Spanish removed most signs of the conquered civilizations. The great cities of the Incas and the Aztecs were demolished and replaced by new Iberian architectural forms. Thus Tenochtitlán, the Aztec capital, was destroyed and Mexico City erected on the same site: The sacrificial temples of Moctezuma and his conquered civilization were replaced by a Catholic cathedral and the new city's main square, the Zócalo. Similarly in Peru, Cuzco, the 'city of the sun', was obliterated as the Spanish superimposed their own urban forms on Inca society. During the next three centuries the Spanish created a wholly new urban form and settlement system in the Americas. Small towns based on a central square, a church, and a gridiron street pattern were established throughout the Spanish realm (Hardoy, 1975; Morse, 1971). Many cities were constructed in previously non-urban areas; most of today's major cities were built by the Spanish: Lima, Buenos Aires, Bogotá, Caracas, Veracruz, and La Paz were all founded by the new Spanish masters. Elsewhere in South America, the Portuguese assumed a similar role and established most of Brazil's contemporary major cities including Rio de Janeiro, São Paulo, and Salvador de Bahia.

The process of urban growth came earlier and was more marked in the Spanish and Portuguese empires than elsewhere, but in other parts of the 'traditional' world the arrival of Europeans was eventually to lead to the foundation of cities which were to dominate the emerging settlement systems. Some of the fort settlements established by the British, French, and Portuguese along the shores of West Africa were to become the 'primate' cities of the twentieth century; only where urban patterns were long established, in the Sudanic belt and among the Yoruba, were European cities less influential. In the rest of sub-Saharan Africa, the story was similar: Johannesburg, Cape Town, Nairobi, and Salisbury were all established by Europeans.[1] Similarly, in Asia, Europeans built new urban forms which complemented the existing interior cities of the Indian and Chinese civilizations. Europe built coastal cities to act as foci for trade, cities which were to become future metropolitan centres. Thus in India, while the ancient cities of Delhi and Hyderabad survived in

Towns & cities that were conquered were converted into society & similar to the homeland country. They were also imbued w/ the culture of their conquerers' homeland (country)

modified form, the British founded Calcutta, Bombay, and Madras. Only in China, North Africa, and parts of the Middle East do the indigenous centres still dominate the urban systems, albeit with certain foreign additions and considerable modifications.

Throughout the 'traditional world' Europeans destroyed, transformed, or distorted indigenous civilizations. The newly founded cities reflected the new power structures and exercised functions relevant to the interests of Europe. They were beginning to become part of a world economic and social system. They were moving towards an equivalent, international position to that which Sjoberg (1963: 223) describes for the cities of traditional empires.

> The lower ranks of the ruling élite have often resided in the lesser cities scattered about the realm so that the hierarchy of cities has reflected, albeit crudely, the status system in the society as a whole. These lesser cities have been situated astride the principal communication routes; thus the local or provincial rulers could maintain contact with the control centre—the capital—and the key political function.

The new cities clearly reflected the aims of European expansion and the social system being superimposed on the new territories. The Spanish aimed at the conquest and subjugation of the indigenous populations in America and the Philippines. New territories were to provide wealth for the crown and their populations were to be 'civilized' and converted to Catholicism. The city was the instrument of conquest. New cities were established throughout the empire and each urban area granted jurisdiction over a large tract of land and its dependent population. The first of the newcomers tended to pre-empt land surrounding a new town and to incorporate municipal commons into private ownership. These newcomers

> maintained discipline, accorded civil and criminal justice, and distributed the dividends of enterprise-as-conquest: the assignment of native labor in encomienda for estates, mines and public works, the disposition of tribute and services owed by the new Amerindian vassals incorporated or in process of incorporation into the new order. At the local level, the colonial conquerors, transformed rapidly into mine operators, large-scale agricultural proprietors, and cattle ranchers, consolidated their position in the town councils whose members they chose. They were the local aristocracy no matter what their metropolitan social origins. (Stein and Stein, 1970: 70–1)

City and country became inextricably linked with most of the conquerors living in the city and most of the conquered in the countryside. The city became the centre of the local aristocracy and the main link in the chain of political control held by Spain.

In South-East Asia cities were also the vanguards of the new colonial and trading empires. At first, the modest trading aims of the Westerners were represented in small maritime towns which were scarcely rivals of the great inland metropolitan capitals. They competed only with the towns of the coastal Asian traders. But, as Western technology exerted its superiority and Europeans began to dominate regional patterns of trade, the towns came to reflect this dominance. By the nineteenth century the European cities were the main centres of power. It was only where existing economic and cultural systems were strong that the impact on indigenous urban form was reduced. This condition was satisfied in China where European influence was long confined to Shanghai and a few other coastal cities. It was also true of many Arab countries where 'the strength, pervasiveness and cohesion of Islam as a total social system and the fact that the Muslim city was a vigorous, strongly functioning system with long traditions and roots' allowed urban society to limit the effects of European economic and cultural pressures (Fisher, 1983: 95).

In general, however, European cities prospered throughout the Third World. In Latin America they benefited from control over land and labour. Elsewhere cities grew more on the basis of trade, although most gained from their location between the imperial capital and the outer edges of the empire. But such a position also limited their autonomy; their roles were limited by the nature of their relationship with the 'mother country'. In the Caribbean, and indeed in colonial America generally, the functions of cities reflected the limited freedom given to the colonies by the colonizing power. As Clarke (1974: 224) points out, 'historically, Caribbean towns were ports, administrative centres, retail outlets but never locations for manufacturers. They were pre-industrial by predilection and prescription.' When the colonies became politically independent there was little change in their position in the world system. Most Latin American societies had gained their independence by 1830, but by then they were already strongly integrated into the world trading system, and they were to become still more enmeshed over the next fifty years.[2] In Asia and Africa political independence came much later and therefore gave the cities in these societies still less scope for an independent existence. In many cases the only change independence brought in Asian and African cities was to substitute local citizens for European administrators and élites.

Dependent Urbanization

There are close parallels between our present argument and the school of thought that emerged in Latin America known loosely as 'dependency' theory. The critical argument in this approach is that Latin American development was 'conditioned' by the region's incorporation into the capitalist mode of production (Dos Santos, 1970). The social and economic structures which have emerged in the region represent the outcome of a historical process of interaction with Iberian, then British, and later North American expansion. This interaction explains the economic and social-class formations which have emerged in Latin America, the urban structures contingent on those formations, and indeed the structure of trade, technology, and investment between the developed world and Latin America. Later authors (Wallerstein, 1974; Amin, 1974; Emmanuel, 1972) have applied this argument to other parts of the Third World.

Dependency formulations gained wide circulation largely as a result of the work of André Gunder Frank (1967, 1969). Given the popularity of Frank's work, it is perhaps necessary to qualify some of the ideas presented in his early writings. Frank argued that Iberian conquest absorbed pre-Columbian America into the world system of capitalism. As a result of this incorporation, economic surplus was extracted by Spain (mainly in the form of silver) and Portugal (mainly through cotton, sugar, and gold) and later, when Spain and Portugal became economic dependencies of newly industrialized Britain, the surplus found its way to Britain. Latin America was therefore linked into a chain of surplus extraction which both 'developed' Britain and 'underdeveloped' Latin America; development and underdevelopment were not to be viewed as separate processes but as different outcomes of the same process. The chain of surplus extraction created centres and peripheries both on a world scale and within Latin America. At a world scale, Latin America came to supply the developed countries with raw materials and in return imported manufactured products. At a regional level the major cities acted as the centres, with the provinces and rural areas relegated to peripheral status. The Latin American city acted as a crucial link in the chain of surplus extraction. Within the city local élites benefited from foreign domination and from the periphery's underdevelopment. Once political independence had been won, the interests of those dominant national élites were to perpetuate the process of dependent development and to impede structural change. Rural underdevelopment and poverty for the masses were the inevitable outcomes of this historical process.

This is now seen to be a crude and somewhat inaccurate account (Palma, 1978; Oxaal, Barnett, and Booth (eds.), 1975; Roxborough, 1979; Kay, 1989). First, Frank was clearly wrong in asserting that the arrival of Cortés and Pizarro led to the immediate incorporation of Latin America into the capitalist system. As Laclau (1971) indicates, a distinction must be made between capitalism as a mode of exchange and capitalism as a mode of production. The exchange of products on a market basis is a necessary, but not a sufficient, condition of capitalism. What is required is a shift in the ownership of the means of production and the emergence of classes; a bourgeoisie and a proletariat. Even today some parts of Latin America remain to be fully incorporated into the capitalist mode of production.

Second, the aftermath of the Iberian conquest modified existing productive systems in a multitude of ways. In some places subsistence agriculturalists farmed much as they had always done, merely giving the surplus to the Spanish rather than to the Aztecs or the Incas. Elsewhere, *haciendas* emerged whereby the Indian communities were forced to work on the land of the *hacendado* in return for the right to work their own plot of land. In still other places plantations were established as the mechanisms of surplus extraction. Clearly the process was similar in the sense that surplus was generated for the benefit of local and foreign élites. On the other hand, recognition of the fact that pre-capitalist forms of enterprise survived and in some cases were re-created by the new élites introduces a whole new level of sophistication into the argument (Pearse, 1975).

Third, it was implicit in the Frankian argument that the process of surplus extraction had impoverished Latin America, even though the facts suggested a different interpretation. For example, the investment of British capital in Argentina and Uruguay had created railways, urban infrastructure, and the basis of cattle, wool, mutton, and cereal production. The incorporation of many areas into the capitalist mode of production had generated wealth. It was first pointed out by Cardoso (1972) and Warren (1973) that a necessary distinction should be made between 'dependent' development and impoverishment. While impoverishment might be the outcome of capitalist expansion in certain areas, in others it might increase material prosperity. In such areas, surplus would be extracted by dominant class interests but from a larger product. Poor groups would not become poorer, the process was one of relative, not absolute, exploitation.

Fourth, the local impact of export generation differed considerably. According to the nature of the export product, regions might prosper or decline. As Furtado (1971) has argued, the size and the form of the multiplier linkages were critical factors in determining the regional effect. Bolivian tin, employing few miners and offering little opportunity for local processing, created few multiplier effects. By contrast, Colombian coffee and Argentinian wheat established large numbers of producers, stimulated the construction of railway networks, and encouraged industrial development, thereby offering an opportunity for local expansion (McGreevey, 1971*b*; Balán, 1976).

Current interpretations of dependency theory all accept these arguments, but differ widely on many issues (Palma, 1978). Indeed, differing strands of Marxist and neo-Marxist thought have created a position where there is no such thing as dependency theory, probably no longer a single dependency approach, and conceivably not even a recognizable dependency school. While this debate is critical at the theoretical level and for political prescription, it is less important for the current theme. What is critical here is to underline that the developed and the underdeveloped countries did not emerge independently; the development of the one was integrally related to that of the other. The dominant social formations and productive systems of the Third World have emerged in response to colonial and capitalist development. If social and economic formations in the Third World have been 'conditioned' by the expansion of capitalism, urban forms have clearly been affected by a similar process. It is this argument that has led to Castells' (1977) invention of the term 'dependent urbanization' and to Harvey's (1973: 232) statement that 'global metropolitanism is embedded in the circulation patterns of a global economy . . . different city forms are contained within that economy.' It is an approach which has currently been employed to explain urban and regional change in the developed world (Massey, 1984; Cooke, 1987; Gregory and Urry, 1986).

The following discussion stems from this argument. The present forms of urban development in the Third World can be understood only as an outcome of the historical process of expansion by capitalist powers. At the same time, the effect of that expansion cannot be understood except in terms of the nature of raw-material production and the forms of the indigenous societies that were incorporated. In the next few pages I shall consider the process of urbanization in the light of these two different sets of factors. I shall first examine the nature of indigenous societies, their urban institutions, their economic and military strength, and their response to European contact; and second the timing and form of European contact: to what extent did Spanish, Dutch, Portuguese, British, Turkish, Belgian, or US colonialism establish different forms of urban structure and to what extent did different forms of export product establish different kinds of social and economic structure and thereby influence the form of urban society that eventually evolved in the Third World?

Society and Urban Tradition before European Contact

Europe encountered a diverse range of societies in its expansion after 1500; primitive hunters and gatherers, agricultural societies, theocratic military societies, industrial economies. The India of the eighteenth century was a great manufacturing as well as a great agricultural country; 'Indian methods of production and of industrial and commercial organization could stand comparison with those in vogue in any other part of the world' (Anstey, 1936: 5). By contrast, European observers of some African peoples were anything but impressed. Elements of the 'noble savage' come across in some accounts, but many evoke the image of Darkest Africa, full of cannibals. Burton (1856: 65), for example, observed a Danikil caravan in Eastern Africa and commented that the 'men were wild as orang-outangs and the women only fit to flog cattle.' Such accounts are as much a comment on the observers as on the observed and there can be little doubt that the vanguard of European expansion did not constitute the cream of European society. As Kiernan (1972: 25) points out, 'these Europeans were pirates, traders, grabbers and settlers by turn . . . and it gave the world a picture of Western civilization very much like the picture of Islam that the Arab slave dealers gave.' This vanguard did anything but persuade the great Eastern civilizations that they had anything to learn from Europe. Proud China despised Western culture and its agrarian Confucian ideology conservatively refused to adopt Western ideas or ideals. The reaction in the Islamic kingdoms of the Middle East and India was much the same.

The diversity of the newly discovered world was due more to its vast geographical area than a function of isolation. Certain regions had remained relatively autonomous for many centuries, particularly in the Americas. But in most places conquest by other peoples was a recurrent fact of life. In what is now contemporary Mexico the Aztecs had come to dominate many other civilizations compelling them to provide the human sacrifical offerings that fed the gods. In most parts of Asia, Europeans encountered societies which were under the domination of foreign or semi-foreign élites; large areas of India had been invaded over the centuries by Aryans, Arabs, and Mongols; China was controlled by the Manchu dynasty. In Africa major population movements frequently led to conflict. In the West Islamic traders and teachers from across the Sahara had superimposed a new religion and urban form on existing societies and forced some peoples towards the coast. In Southern Africa Europe came into contact with African civilizations at the time of the *Mfecane* (great smashing): the series of conflicts among indigenous societies which 'scattered African chieftains in fragments across half the continent of Africa' (Davenport, 1977: 10). The conflicts between peoples was not only characteristic of the world Europe 'discovered' but also a great aid in its incorporation and subjugation. Thus, slavery in West Africa was facilitated by inter-group antipathy and the existing slaving tradition. The Spanish conquest of Mexico was eased by the way Cortés could enlist the support of Indian peoples previously conquered by the Aztecs. In India the East India Company made alliances with one kingdom while in conflict with others.

The diverse, and in certain places rapidly changing, societies encountered by Europeans clearly possessed widely differing urban traditions. In the Pacific and in major parts of Africa urbanization was non-existent, however loosely the term is defined. By contrast, urbanization in the Americas, the Middle East, India, and China often made European cities look infants in comparison. Indeed, what Wheatley (1970) has described as the areas of 'primary urban generation' were confined to what is now the Third World.[3] The great urban achievements of the 'traditional' world greatly impressed the European who encountered them. Cortés found himself unable to describe the beauty and greatness of the buildings he found in Moctezuma's capital. Further south, the Spaniards marvelled at the magnificence of the Inca cities (Burland, 1967). In other parts of the world it was the size of the cities as much as their architecture that impressed in Europeans. In China before 1900 there were more large cities than in Europe, not only in absolute numbers, but even in proportionate terms. In the Middle East ancient Alexandria and Baghdad are both estimated to have had 1 million inhabitants. In 1803 Delhi's palace, fort, great mosque, and most of its population were encircled by a wall 9 kilometres long. In sub-Saharan Africa similar urban magnificence was absent and a strong urban tradition had evolved only in a few regions. Major towns existed in the area under Islamic influence, both in the Sudanic belt and on the East African coast. In the Yoruba territories it was common for agricultural peoples to live in towns for social, religious, and defensive reasons; and after the nineteeth-century wars the traditional Yoruba towns, as well as new cities, became swollen with refugees (Mabogunje, 1962).

Undeniably great though many of these cities were, they were not the cities of modern Europe. They were more akin to what Sjoberg (1960) has described as the pre-industrial city. These cities were reliant on simple inanimate forms of technology and dependent on religion. In Inca Peru power derived ultimately from Inti, the sun god; the construction of the city was the highest expression of Inca art and organization of the state. Its culmination occurred in Cuzco, 'the navel of the universe and the seat of the sun's earthly descendant, the Sapa Inca' (Burland, 1967: 40). In traditional Asia cities 'were intended as cosmic creations substantive and symbolic pinnacles of, and resplendent thrones for, the great tradition, enshriners as well as administrators of the relatively homogeneous and particularistic culture, to which the market towns and the peasant villages of the little tradition also belonged' (Murphey, 1969: 68). The sacred determined their form and their very existence. Throughout the 'traditional' world, in fact, Wheatley (1967: 9) has suggested that this 'cosmo-magical symbolism . . . informed the ideal type traditional city in both the old and the new worlds . . . brought it into being, sustained it and was imprinted on its physiognomy.'

These great cities were supported by, and dictated orders to, the rest of society. In ancient China each city received its orders from Peking and passed them down the urban hierarchy. In Inca Peru town officials formed part of an efficient civil service hierarchy. Such cities engaged in commerce and manufacturing but these functions were subordinate to religious and state needs. Even in the Yoruba cities, inhabited by agriculturalists and traders, the main function was as the seat of an *oba*—'the visible symbol of the deity . . . the High Priest of his kingdom' (Ojo, 1966: 75).

These urban civilizations were to affect profoundly the form of modern cities. In both their locations and their morphology, 'traditional' cities were to help to mould European urban influence. And, in so far as these urban civilizations differed from one another, they made different kinds of contribution to the evolution of modern cities.

European Contact: Colonialism and Neo-Colonialism

Diversity was not confined to the colonized cultures; it was just as characteristic of the colonizing powers. Indeed, European expansion into the Third World assumed a marked variety of forms. Such variety was partly the outcome of the range of European nations engaged in colonial and commercial expansion; Spanish, Portuguese, Dutch, British, French, German, Belgian, and Italian styles of expansion were in many respects very different. The nature of European contact was also influenced by the kind of territory and civilization discovered. Thus the effects of the British in Australia and North America were very different from those in Africa. European influence also changed through time: the styles of the British in India changed markedly from the days of the East India Company to the Viceroyalty; the effect of Spanish colonialism in the sixteenth century was very different from that in the early nineteenth. Indeed, the form of capitalism had changed from mercantile capitalism to industrial capitalism and later to monopoly capitalism (Baran, 1957; Barratt-Brown, 1974). In addition, while many Third World countries had experienced contact with one European nation, others had come into contact with several. In the Caribbean, islands were constantly changing hands; Cuba was at different times under the rule of Britain, Spain, and the United States.

Different colonial experiences obviously had diverse economic and social consequences throughout the colonized world. In North America, Australia, New Zealand, and Argentina the unfettered expansion of capitalist agriculture in temperate latitudes eventually led to high levels of per capita income. In areas where an indigenous population was numerically superior, the local economic and social consequences were more variable. European expansion rarely helped indigenous commerce and agriculture but nor were indigenous institutions and enterprise automatically destroyed. Frequently Europeans did not even try to transform precapitalist agriculture and commerce into capitalism. As Foster-Carter (1978: 51) has pointed out, 'capitalism neither evolves mechanically from what precedes it, nor does it necessarily dissolve it. Indeed so far from banishing precapitalist forms, it not only coexists with them but buttresses them, and even on occasion conjures them up *ex nihilo*.' Local variation conditioned by the European presence is the only adequate description of the economic impact of colonialism. Similarly, its effect on the demographic structure was equally variable. In North America, Argentina, and Australia extensive European settlement allied with the extermination of the indigenes created a new population structure; in the Caribbean the local population was destroyed and replaced with African slaves and indentured labour; in Spanish America both extermination of the indigenous population and racial mixing occurred; in most of Asia and Africa Europeans came and went with little effect; in Southern Africa Europeans coexisted uneasily with African populations. The urban consequences were similarly diverse.

It is probable, however, that the variations in European impact became less marked in the nineteenth and twentieth centuries. During the stage of mercantile capitalism Europe was interested mainly in trade and its effect on Africa and Asia was more limited. Under the phase of industrial capitalism Europe's effect on the rest of the world was greater. Whether Karl Marx was right in arguing that colonialism and imperialism were ordained by the expansionist logic of capital accumulation and the falling rate of profit remains a matter for debate, but industrial and monopoly capital certainly opened up the world (Barratt-Brown, 1974: 184). European powers colonized and/or established production facilities in most Third World countries. British and, later, French, Belgian, Dutch, and US capital developed mines and plantations, railways and ports, factories and cities. Through direct political control, through investment and with the ever-present threat of the gunboat, the world came to be dominated by Europe and the United States. From this time on, the world began to look more similar. The cities associated with these developments, which began to expand in the late nineteenth century, resembled one another. The tram and the railway, the suburban house, and the occasional dash of town planning created passable imitations of European cities. The so-called modern cities contained all the advantages and extravagances of European urbanization together with the additional disadvantages of general poverty. The new cities were frequently insanitary, badly located, and tasteless. As Fisher (1976: 112) has noted, both in Europe and in the United States, the industrial revolution rapidly bred an immense self-confidence and arrogance. Nature, it was automatically assumed, had been conquered and man was in full control. Such men felt no need to harmonize their activities with those of nature and, so far from being symbols of reconciliation, the upstart industrial cities of Victorian Britain were built in flagrant contempt of the natural order: this flagrant contempt of the natural order carried over into the Third World.

Urban functions and settlement systems naturally reflected the general orientation of the economy: cities were concerned with international trade. What Murphey (1969: 72) has said of Asia is valid for most parts of the Third World. By the time of independence 'national life had come to centre on Western-developed ports to an irreversible degree . . . As each Asian country had responded to Western stimuli and altered its own outlook, its world had been refocused on its seaward gates, originally the funnels for export and the vestibules for Western manipulators, but ultimately also the breeding grounds and the apexes of a new Asia.' These apexes were to develop into the primate cities which would dominate national settlement patterns. Into these cities would eventually move large numbers of rural migrants: their land alienated by the intrusion of capitalist enterprise, their numbers swollen by lower mortality rates. The cities we recognize today as being of the Third World had begun to emerge. Urbanization characterized by

> an urban population unrelated to the productive level of the system; an absence of a direct relation between industrial employment and urbanization, but a link between industrial production and urban growth; a strong imbalance in the urban network in favour of one predominating area; increasing acceleration of the process of urbanization; a lack of jobs and services for the new urban masses and, consequently, a reinforcement of the ecological segregation of the social classes and a polarization of the system of stratification as far as consumption is concerned. (Castells, 1977: 57)

Recurrent Patterns

Social scientists seek to generalize and urbanization has long been the object of such generalization. Pirenne (1925), Sjoberg (1960), Mumford (1975), Friedmann (1961), Schnore (1965), Castells (1977), and Harvey (1973) have all in their distinctive ways sought to generalize about urban change in different parts of the world. One of the enduring elements in these efforts has been the proposition that cities have assumed a more homogenous form through time. Whether one accepts the validity of Schnore's (1965) argument about urban form in Latin America converging towards the North American model, or Davis and Hertz's (1954) hypotheses about urbanization increasing lineally with levels of development, or Castells's (1977) theory of the capitalist logic underlying urban development, the assumption underlying all these arguments is that urban form and development have become more universal. Not infrequently, normative statements have been implicit in this assumption. Underlying the approach of the 'modernization school' is the belief that the emergence of the modern cities and log-linear city-size systems improves conditions for the populations of poor countries.[4] It is divergence from this pattern in the form of *favelas*, primate cities, or inflated service sectors that slows the process of economic development and lowers the living standards of Third World populations.[5] A contrary view is presented by the Marxist literature, where generalization has attempted to show that conditions of poverty and inequality are not confined to the Third World but are entrenched even in the most affluent of capitalist cities; what planners designate as 'problems' are in fact the inevitable urban outcomes of capitalist development. One of the objectives of this book is to examine in what respects urban forms are becoming more universal and to discuss the implications for urban development of that tendency.

At first sight the suggestion of a common universal pattern of urban development seems absurd. It is true that Paris recently had its *bidonvilles*, but there is no way that poverty in Paris can be compared with that to be found in Calcutta or Jakarta, either in the severity of that poverty or in the relative and absolute numbers of people involved. Urban poverty in the Third World is on a scale quite different from that in the developed countries; it is a poverty, moreover, that is likely to persist for many years to come. In other respects, however, more credence can be given to the notion of universality.

Technology has always imposed certain similarities on city form. It is Sjoberg's (1960) contention, of course, that technology is the common factor which determined the form of the pre-industrial city. Technology created the social order comprising a small élite and much larger lower-class and outcast groups. The élite dominated the feudal city by controlling the main religious, political, administrative, and social functions. The spatial manifestation of this control was the wealthy, exclusive central core surrounded by an extensive area of poor settlement. Similarly, ideology and beliefs have always influenced urban form. Wheatley's (1969) concept of 'cosmo-magical symbolism' offers a means by which the structures of Inca, Yoruba, ancient Cambodian, Indian, and Chinese cities can be related. But pre-industrial cities, for all their similarities were very diverse. The absence of inanimate energy may have limited communications in the cities of both the Incas and the ancient Chinese, but at least the Chinese had the horse and the wheel. Ancient cities may have shared a belief in God which articulated their societies and legitimized their élites, but in some societies this gave rise to human sacrifice, in others to slavery, and in others to harmonious social relationships.

By comparison, technology in the modern world is more universally accessible, at least to the élites and to the subsidiaries of transnational corporations. Ancient cities were similar to one another in the sense that their inhabitants lacked the ability to travel rapidly or to build tall buildings cheaply. The modern city adopts a similar form throughout the world because the ability to employ the latest technology has become universal. Given the ability to pay, any country can build an underground railway, a television system, or a chain of supermarkets. As a result of this transferability of technology, and its critical associate, an international order of production, modern cities have grown to look and feel more and more alike. Few large cities in the world now lack skyscrapers, motorized transportation, municipal fountains, shopping malls, and public housing projects. Know-how has spread and an international business of architecture, planning, engineering, and construction has developed to implement it. Colonialism first led to the international diffusion of urban form, although differences between the colonial powers led to major local variations. Modern technology has completed the internationalization process so that few cities contain distinctive-looking buildings. Ignore the lettering on the buildings and the skyscrapers of Tokyo or London look like those of Nairobi or São Paulo.

Major components of the new technology have been the car, the television, and the concrete block. The introduction of the car brought with it air pollution, traffic congestion, commuting, and suburban development in increasing numbers of cities. The diffusion of the television set and the radio, helped to internationalize consumer tastes and even urban values (Felix, 1983; Armstrong and McGee, 1985). There are few Latin Americans who have not heard of Dallas, few urban Africans who have not seen European fashions, few Asian city-dwellers who do not aspire to the 'norm' of a two-car, suburban life-style. Increasingly, the clothing and the manners of the middle classes look the same. Increasingly, even the urban poor of different continents are beginning to resemble one another. What is true of clothes, is gradually true of patterns of food consumption. Macdonald's and Kentucky Fried Chicken, Sushi bars and Parisian bistros are gradually spreading into Third World cities; many middle-class families are giving up using traditional grains in favour of wheat.

It is not only patterns of consumption but also patterns of production which have become more international. The subsidiaries of international corporations have sprung up in most cities. Car production takes place in most Latin American countries, in India and the Far East. Large plants turn out textiles and electrical goods in increasing numbers of cities. In order to maintain and support modern industry, producer services have spread across the globe. The names of hotel chains, accounting firms, and banks are now truly international; there are few cities where they are absent. This process of internationalization has further encouraged the homogenization of culture; office employees and factory workers tend to adopt similar life-styles across the globe. Managers all tend to live in large residential homes in well-serviced, car-dependent suburbs.

Of course, there are important variations. The poorest cities have been able to adapt the new technology much less readily, and the poor of those cities have often continued to eat and dress in traditional ways. Some societies have rejected certain aspects of the internationalization process. Iran has recently turned against the international life-style and Muslim fundamentalism threatens to raise the barriers against Western values in other Middle-Eastern areas. Socialist societies may contain elements of the new technology and changing consumer tastes, but the influence of the car and suburban living is far less marked. As a result, population-density and land-value gradients, seemingly so ubiquitous in the 'capitalist' city (Ingram and Carroll, 1981; Mills and Tan, 1980; Mills, 1972), are seemingly absent from the 'socialist' city (French and Hamilton, 1979). Similarly, it is quite clear that the diffusion of production techniques and consumer tastes is occurring at very different rates. In Africa, the appearance, if not the size, of many cities has changed relatively little through time, whereas the form of Latin American and Far Eastern cities has been almost totally transformed. The so-called New International Division of Labour has not touched every society equally (Armstrong and McGee, 1985). Some countries have entered the ranks of the newly industrialized economies while others have stagnated. Even within countries, there are major differences between the largest cities and many provincial centres. As Walker and Storper (1981) have commented the new world economy is 'a mosaic of unevenness in a constant state of flux'.

If the city has undoubtedly become more international, most cities have only been partially affected. The proportions of middle-class people, the numbers of transnational subsidiaries, access to television sets, let alone electricity and drinkable water, all differ widely. Knowledge of the existence of new technology is not the same as adoption; a desire for a suburban house and a car is not the same as possession. This is a major reason why many writers are today much less optimistic about the benefits of urbanization. The know-how to produce good sanitation and infrastructure, rapid communication and transportation certainly exists. In this sense, the optimism that characterized the work of architects and urban planners such as Le Corbusier, Doxiadis, and others, was justified. As Mumford (1975: 651) put it 'the oppressive conditions that limited the development of cities throughout history have begun to disappear'. Unfortunately, even if the internationalized economy is arguably reducing inequality and improving living conditions in the cities of the developed world, it is only too easy to argue that most Third World cities occupy an underprivileged position in an unequal world economy. It is only too easy to argue that in Third World cities deeply penetrated by international economic forces, social polarization is increasing (Kowarick and Campanario, 1986; Evers, 1975; Walton, 1977).

Internationalization has the potential for improvement but whether new forms of technology are used to emancipate the poor or whether to subjugate them depends upon the developmental priorities selected. In the past, most decisions about urban development were national decisions. Colonialism introduced international control, and today the integrated world economy means that many of the critical decisions about technology, employment, and economic growth are made in the offices of the transnational corporations in Tokyo, New York, and Frankfurt. Increasingly, the World Bank makes decisions about the pricing of infrastructure provision, not local governments (Linn, 1983; Anderson, 1989).

Whether the shifting locus of decision making matters to the poor depends greatly upon the nature of local decision-makers and the extent to which the poor were previously integrated into the market economy. Many would argue that the logic of capitalist development is inherently undesirable because it contains an integral force towards inequality. Certainly, Harvey's (1973) description of the logic of capitalist expansion in Baltimore, with the parallel development of slum ghettos and affluent suburbs is applicable, to a greater or lesser extent, throughout the world. In the Third World city, of course, the relative poverty of the majority is accentuated by absolute deprivation. If the Baltimore slum-dweller feels poor, the poor in Indian cities, are both absolutely and relatively poor. It is only too easy to paint Dickensian pictures of squalor, poverty, and crime in many Third World cities.

At the same time, it is an open question whether the urbanization process is accentuating poverty and increasing inequality. While many argue that it is, and during the 1980s this was easy to demonstrate for most parts of Africa and Latin America, it is doubtful whether urbanization is creating poverty as opposed to merely reflecting wider societal changes. For example, rising levels of unemployment in Latin American urban areas during the 1980s are a function of the debt crisis more than urban growth. Similarly, the proliferation of shanty towns is more a function of rural distress and high rates of demographic growth than of urban processes *per se*. If there is all too much evidence of worsening urban conditions, the picture is highly variable across cities.

As such the vulgar statement that urban inequality and poverty are worsening, is a reaction of the heart and not of the head. Capitalist urbanization does not create equality, but it not infrequently improves urban living conditions even for the poor (Wells, 1983; Dixon, 1987).

The truth is that however similar Third World cities look, their position in the world economy is highly diverse. As such their prospects for growth and the nature of their urban development differ profoundly. As such, while most prophets of Third World urban catastrophe are too pessimistic and most technocratic planners and city mayors too complacent, both are right some of the time. Unfortunately, what seems eminently clear, is that urban poverty will continue to plague every Third World city in the years to come. There is no obvious reason, given the present distribution of world resources, for contemporary descriptions of Third World cities to be redundant in fifty years time. The grandchildren of the urban poor will continue to suffer from social polarization, material deprivation, and political repression. If urban conditions may well have improved in Seoul or São Paulo, they may well have deteriorated in Calcutta or Addis Ababa. The detailed future of the Third World city is in doubt, but there is little reason to believe that the urban poor are about to disappear.

2

Urban Agglomeration and Regional Disparities

Economic development occurs unevenly across space and time. In terms of regional disparities, therefore, most Third World countries exhibit wide differences in levels of income and economic activity between urban and rural areas, between large cities and small, and between central regions and the periphery. In many countries, especially those attempting to follow an accelerated capitalist path to development, such disparities are often widening. There seems little evidence of any levelling of regional, or urban/rural welfare differences in most poor countries. In response, national populations are on the move; migrating from poor regions to areas of economic growth. In the process, big cities are growing ever larger.

The welfare considerations of this process are intriguing and have been debated at length in the literature. Many have argued that increasing spatial concentration is undesirable; some the opposite. The point that must be made here is that we must always be careful to distinguish between the welfare of people and the welfare of regions. If we do not then we may fail to recognize that greater regional prosperity does not guarantee greater prosperity for all. Nor, since most poor rural areas contain rich landlords and every affluent city contains many poor people, can we assume that growing spatial concentration will automatically harm the poor. As Chapter 8 will show, regional policy has often foundered on this misplaced assumption. For the moment, however, policy is not the issue. This chapter is concerned with describing current patterns of spatial concentration and the reasons why there is a continuing tendency in most Third World countries for growth to occur in very restricted geographical areas. In order to explain these trends, I consider three theoretical models of spatial change. Policy discussion is delayed until Chapter 8.

Contemporary Spatial Disparities

Economic and social change has been associated with the emergence of wide geographical disparities throughout the Third World. These disparities are linked to the nature of the economic model that underpins development in most of these countries and to the acute personal income disparities which have emerged. In this section it is sufficient to detail three broad patterns of spatial concentration and inequality: economic and social differences between urban and rural areas; regional disparities; and the degree to which one city dominates the national urban structure (urban 'primacy').

Rural–urban disparities

Major differences are apparent between the standards of living of urban and rural populations (see Chapter 3). With the partial exception of a few socialist nations, where some effort has been made to reduce differentials, most Third World rural areas contain a higher proportion of very poor people, offer fewer and less adequate services, and have more limited opportunities for well-paid work. Indeed, in those developing countries which have achieved some measure of industrial and commercial development, urban–rural disparities are often very marked. In Indonesia, almost half of all rural families were living in poverty in 1980 compared to only one-fifth of urban families (Hamer, 1986: 24). In Brazil, 42 per cent of rural families lived in poverty in 1974–5, compared with 10 per cent of those living in smaller urban centres (Pfefferman and Webb, 1975).

Similar differences are apparent in terms of infant mortality. In India, for example, rural babies are twice as likely to die than the urban-born; in the Philippines urban babies have a 50 per cent higher chance of surviving than rural babies (Pryer and Crook, 1988: 9). Similar kinds of figures can be replicated for most kinds of health measure in the Third World (Bryant, 1969).

Regional disparities

However measured, regional disparities in Third World countries are extreme. In terms of almost any kind of socio-economic indicator, medical provision, schooling, industrial activity or whatever, certain regions demonstrate a marked superiority over the rest of the country. In 1985, for example, more than 50 per cent of all Kenyan industrial enterprises, 52 per cent of all waged employees, and 60 per cent of all formal sector incomes were concentrated in Nairobi compared to a mere 7 per cent of the national population (ISHOC, 1987; Kenya, 1986). In Thailand, the Bangkok metropolitan area generated 54 per cent of all manufacturing value added in 1984 while containing only 10 per cent of the national population (Thailand, 1986). In the Philippines, the Manila metro region contains 60 per cent of manufacturing firms and generates

40 per cent of national output, even though it contains no more than 12 per cent of the nation's people (Philippines, 1986).

In general, regional disparities in less-developed nations are far wider than those in developed countries. Williamson (1965) shows that the average differential in per capita income, as measured by the coefficient of variation, was only 0.19 for 11 developed countries in the 1950s compared to 0.42 for the poorest 9 countries in his survey. More recent data for a number of poor countries support the idea of major spatial imbalances (table 2.1). In Peru there was a fourteenfold difference between the per capita regional product of the richest and poorest provinces in 1981, in Brazil an eightfold difference in 1980, and in the Philippines a sixfold difference in 1983. The average of the twelve coefficients of variation (V_w) listed in table 2.1 is 0.46.

Regional income disparities have two other major characteristics. The

Table 2.1

Regional inequality in selected less-developed nations

Nation	GNP per capita 1981	V_w	V_{uw}	Date	Number of regions
Chile	2,560	0.34	0.65	1980	13
Mexico	2,250	0.43	0.34	1980	32
Brazil	2,220	0.50	0.49	1980	25
Malaysia	1,840	0.38	0.38	1985	14
Korea	1,700	0.25	0.24	1978	11
Colombia	1,380	0.30	0.33	1977	23
Peru	1,170	0.67	0.82	1981	23
Philippines	790	0.63	0.54	1983	13
Thailand	770	0.76	0.67	1985	6
Indonesia	530	0.74	0.51	1980	6
India	260	0.32	0.40	1981/2	20
Bangladesh	140	0.28	0.58	1984/5	20

Sources: Respective national censuses and statistical yearbooks. World Bank (1983), *World Development Report 1982*, table 1.

Note: Williamson (1965) uses two main measures of inequality (V_w and V_{uw}), both based on the coefficient of variation.

$$V_w = \frac{\sqrt{\sum_i (y_1 - \bar{y}^2 (n_1/n)}}{\bar{y}}$$

where n_1 = population in region i, n = national population, y_1 = income per capita of region i, and $\bar{y}$ = national income per capita.

$$V_{uw} = \frac{\sqrt{\sum_i (y_1 - \bar{y})^2/N}}{\bar{y}}$$

where N = number of regions.

first is that there seems to be little tendency towards greater equality in most Third World countries; in the period since 1970 the limited data reveal that three countries have seen the level of regional disparities increase (India, Indonesia, and Thailand), four have seen a decline (Brazil, Korea, Mexico, and Peru), and a further three have experienced no change (Colombia, Malaysia, and the Philippines). A similar characteristic was observed for the 1960s (Gilbert and Goodman, 1976).

The second characteristic is that, despite the argument in the literature that there is a clear relationship between rising per capita income and levels of regional disparity, no clear pattern is obvious in table 2.1. The lack of a clear relationship is certainly due to the fact that regional disparities are profoundly influenced by the form of economic and social organization, the nature of the political regime, the geography of natural resources and economic activity, and the pattern of population distribution.

Urban primacy

In many Third World countries, most large-scale modern activities, most forms of social infrastructure, and most centres of decision making are found in a single major city. This concentration is mirrored in the way one city dominates all others. Thus Lima-Callao has 10 times the population of Arequipa, Peru's next largest city; Montevideo, 17 times the population of Salto, Uruguay's second city; Buenos Aires, 10 times the population of Córdoba; Tehran, 7 times the population of Esfahan; and Bangkok 50 times that of Thailand's second city, Khon Khaen.

This phenomenon is usually known as urban 'primacy'. It is often measured by the 'four-city index'; calculated as a ratio of the population of the largest city to the sum of the populations of the four largest cities.[1] Table 2.2 shows how common primacy is in Third World countries. Of the 75 countries with a per capita income lower than $1,800 in 1973, 55 exhibited primacy or high primacy. It is important to note, however, that 20 poor countries did not have primate urban-size distributions, for example, India, South Africa, the Yemen Arab Republic, and Zambia. In addition, there were several countries demonstrating dual primacy; two cities dominate the urban-size distributions of Brazil (São Paulo and Rio de Janeiro), Ecuador (Guayaquil and Quito), Syria (Damascus and Aleppo), and Pakistan (Karachi and Lahore).

Table 2.2 also shows that while primacy is more prevalent in Third World nations it is by no means confined to them. Rather, 11 of the 28 high-income countries were primate or high-primate, with the dominance of Paris, Vienna, or Copenhagen serving to remind us that high primacy is not exclusively a Third World phenomenon. Indeed, various efforts to relate the degree of primacy to levels of urban development or per capita

Table 2.2
Degrees of primacy and levels of per capita income

Gross national product per capita, 1973	Numbers of countries with	
	Low and non-primate distributions*	Primate and high-primate distributions*
Developed countries (more than $1,800)	17	11
Less-poor Third World countries ($601–1,800)	5	17
Poorest Third World countries ($0–600)	15	38
Total	37	66

* For definitions see n. 1, p. 264 below.
Source: *World Bank Atlas*, 1977; and United Nations *Demographic Yearbooks*.

income have proved inconclusive (Berry, 1961; Mills, 1972; Mehta, 1964 El-Shakhs, 1972; Richardson and Schwartz, 1988). If a relationship exists it is of a much more complex form.

In sum, therefore, while primacy is not limited to Third World nations, it is highly characteristic of them. And if high primacy is not limited to countries with any single characteristic, it is especially common in those which are small, highly centralized, and of medium income.[2] Whether such a phenomenon constitutes the social and economic problems that so many claim, is a theme I explore in Chapter 8.

Models of Regional Change

Two broad theoretical approaches offer useful insights into the process of urban and spatial concentration and the related phenomena of regional welfare and income disparities. The first is John Friedmann's 'Centre–periphery' model; the second the work emerging from neo-Marxist writing.

The centre–periphery model

Friedmann's (1966) work had the major virtue of drawing together a wide range of ideas to create an evolutionary model of economic development and spatial change. It incorporated the ideas of Myrdal (1957) and Hirchmann (1958) about how market forces accentuated regional inequalities, general models of economic development (Rostow, 1960, Prebisch, 1950), and regional planning strategies (Isard, 1960;

Rodwin, 1973), to produce a simple, normative model of spatial development in less-developed countries. The author himself no longer accepts many of the assumptions he made (Friedmann and Weaver, 1979; Friedmann and Douglass, 1976), but it is interesting to study his model because it incorporates many of the dominant ideas about spatial and economic development held in the 1960s and still held by many today. By explaining and criticizing the model we shall understand better some of the processes that have led to the emergence of urban primacy and regional disparities in so many Third World nations.

The model consists of four stages which trace the evolution of a spatial system from a sparsely populated and newly colonized nation to a fully integrated urban and regional system in a developed country (Figure 2.1). The first stage conceives of an unexploited region with scattered rural settlement which is populated by a colonial power. The settlers establish mining and agricultural activities and urban centres evolve to service and administer these enterprises. Those cities which handle foreign trade or are located in rich agricultural regions prosper, but the urban system generally consists of regional centres with little interdependence.

The second stage is marked by the beginnings of industrialization and the growing concentration of investment into one or two main cities. Regional disparities increase as resources continue to flow to the more productive regions away from those which are now 'locationally obsolete'. Gradually, a dualistic spatial structure emerges '. . . comprising a "centre" of rapid, intensive development and a "periphery" whose economy, imperfectly related to this centre, is either stagnant or declining' (Friedmann, 1966: 9). The major urban outcome is the emergence of a 'primate' city.

The third stage is marked by increasing industrial maturity and by rising political consciousness in the periphery opposed to the spatial concentration of wealth. Such opposition evokes a response from the national government in the form of a regional development policy. Such a policy is essential because 'on the whole, the unrestrained forces of a dynamic market economy appear to be working against a convergence of centre and periphery' (p. 18). Eventually the combination of national government intervention, provincial initiative, and the effects of rising national income reduce regional disparities and bring about greater spatial balance. Provincial cities now contain dynamic economic activities and help to stimulate agriculture in the surrounding regions. Poverty is not abolished, but is now confined to limited areas rather than being endemic in the whole periphery.

During the final stage a fully integrated space economy emerges which combines 'national integration, efficiency in the location of individual firms, maximum potential for further growth, and minimum

1. Independent local centres, no hierarchy

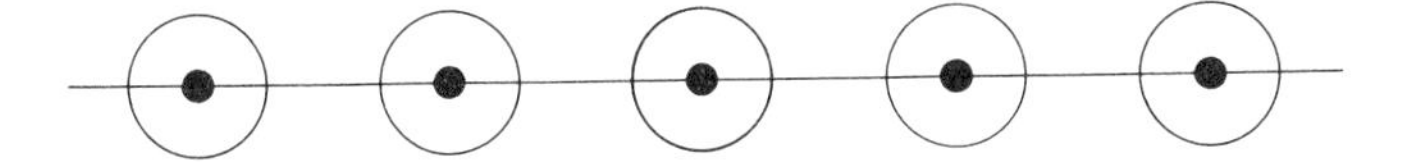

2. A single strong centre

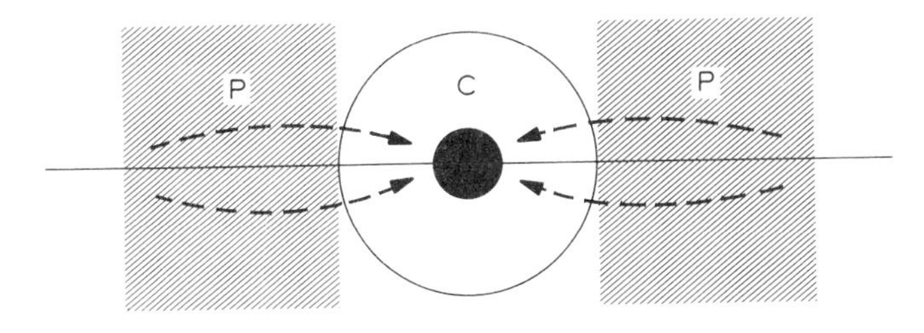

3. A single national centre strong peripheral subcentres

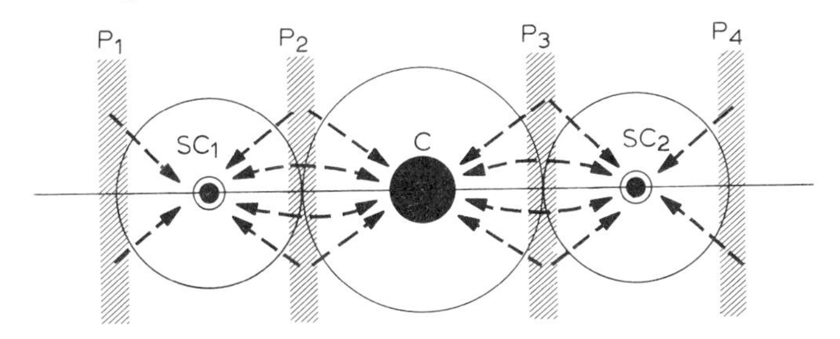

4. A functionally interdependent system of cities

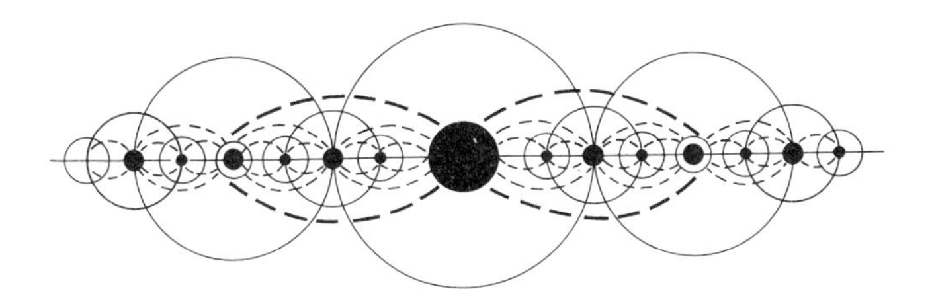

(After Friedmann)

FIG. 2.1 A sequence of stages in spatial organization
Source: Friedman (1972)

essential interregional imbalances' (p. 37). In these conditions a balanced urban-size distribution emerges with regional centres competing effectively for resources with the national capital.

Friedmann's model is both descriptive and normative; it sketches the evolution of the space economy in less developed countries as well as prescribing the form that future regional policies should take. This model is the basis of a series of regional development strategies focusing on the concept of the 'growth centre' which are still being employed today (UNCRD, 1976; Appalraju and Safier, 1976; Kuklinski, 1972; Rondinelli and Ruddle, 1976). It is important, therefore, to consider the weaknesses of the model both as a description of the past and as a guide to government policy.

First, the model was made less relevant to most Third World societies by the assumption that original European colonization took place in a sparsely populated region. While Australia, North America, South-West Africa, and lowland South America conform in good measure to this assumption, most parts of the Third World do not fit at all. In most places European contact came face to face with large established populations (Horvath, 1972). Friedmann's model implies that the colonizing population settled and developed the area's resources and best lands without conflict and with the spirit and goodwill of a Daniel Boone. In fact colonial settlement was much more varied and warlike; European settlement in Asia was confined to the coasts for many years by the power of the existing civilizations; the Spanish conquistadores of Latin America located their settlements close to the main mines and cities of the conquered indigenous populations; the British did not just settle Australia but stole Aboriginal land. The lack of a *tabula rasa* created conflict and unequal societies, not merely an unintegrated urban and regional system.

Second, Friedmann considerably underplays the role of foreign influence. While external demand is seen to be an important stimulus to export production and foreign technology a major source of innovation, it is assumed throughout that most important political decisions are taken within the country and that foreign influence is favourable to high rates of economic growth. The historical circumstances facing most less developed countries hardly accord with this conception. Major economic decisions were always influenced by colonial or neo-colonial powers and sometimes resolved directly through political control or military threat. Similarly, the decisions of the élites who emerged in most independent Third World countries were conditioned by, and linked to, the interests of major foreign business groups. Industrial activity was owned and controlled by foreign corporations, exports were channelled through foreign-dominated cartels, and political decisions were subject to pressure from London, Paris, and Washington. The model errs in so far as it 'assumed that the problems of the dependent countries could be understood by analysing phenomena occurring within their geographical borders' (Stuckey, 1975: 97).

The third criticism is that the model is largely apolitical. It assumes that national governments introduce regional policies because it is in the public interest to do so. Beyond stating that regional politicians press for favourable treatment to compensate for past neglect, it fails to examine the nature of political influence and its spatial repercussions.[3] Clearly, different political forces evoke different kinds of national response. The kind of regional policy demanded by provincial élites politically aligned to national parties controlled from the centre will be very different from the response evoked through popular protest in the periphery. Because he believes government to be based on consensus rather than on conflict, Friedmann ignores the political feuds and rivalries that go to make up most government policies. Most importantly, he neglects the possibility that the political will to help the poor regions emerges less out of a sense of public spirit than as a response to political pressure generated by social protests (Castells, 1977).

Fourth, the phenomenon of poverty is treated as a regional rather than as a social phenomenon. This bias underlies the argument that the introduction of regional programmes to raise agricultural productivity and to improve transportation will eliminate poverty in the periphery outside certain isolated pockets. Higher per capita income will lead to the integration of centre and periphery and remove poverty. In reality, of course, poverty is maintained both in the periphery and increasingly in the centre as rural and small-town migrants move in search of better work opportunities. The model fails to treat the issue of poverty in the cities or indeed the personal disparities in income which exist in most Third World societies. Because poverty is treated in purely spatial terms, it is easy for Friedmann to slip into the belief that regional development will eliminate its worst manifestations.

Finally the model assumes that most poor countries will eventually become developed. In this assumption, Friedmann tends to follow the treatise of Rostow (1960: 166) and his belief that the 'tricks of growth are not difficult'. Unfortunately, even if high rates of economic growth can be achieved in the Third World, there is little evidence to suggest that most of these nations are developing highly educated, well-housed, and more equitable societies. The assumption that was so common in the 1950s, that there was a unilineal development path along which most countries would someday pass, is today much less widely accepted. Greater awareness of the relationships between developed and less developed nations and of the social-class interests which dominate political decision making in so many poor nations strongly point to the rejection of a unilineal development path. As such, the pattern of regional and urban change posited in Friedmann's model is unlikely to be replicated in most less developed countries.

These criticisms are not meant to remove all faith in the model, for there are numerous elements which still repay attention. The description of how the centre grows at the expense of the periphery and the critique of the spatial equilibrium model are masterpieces of clear exposition. In addition, some policy recommendations still offer good advice to the regional planner. However, what the model lacks is sufficient scepticism of the benefits offered by economic growth and of the social consciences of most Third World governments. Both these ingredients are offered, in good measure, by the following alternative model of regional development.

The model of peripheral urbanization

Various strands of writing, principally from the left, have tried to link the urban process in the Third World to the political economy school of development. This general approach may be labelled as a model of peripheral urbanization, even if there are important differences between individual writers and even if the word model is too limiting for such a broad range of thought. This approach derives from the more historical approach to studying development associated with Cardoso and Faletto (1969), Frank (1969), and Wallerstein (1974). It is linked to the strengthening of Marxist scholarship in Britain and the United States both in urban studies (Harvey, 1973; Castells, 1977) and in development studies generally (Warren, 1973; Portes and Walton, 1981; Oxaal *et al.*, 1975).

This approach differs from the early Friedmann model in adopting a more explicitly historical interpretation of urban change, in linking urbanization to global changes in production and trade, and in adopting a broadly Marxian interpretation of the dynamics of capitalist expansion to the peripheral countries of the world. The basic ingredient, therefore, is the attempt to explain urban change in the Third World in terms of the worldwide process of capitalist development. Whereas the Friedmann model was mainly concerned with analysis at the national and regional levels, the peripheral urbanization model is concerned with the impact of international capitalism on national urban systems.

The neo-Marxist models view the internal structure of Third World countries as part of the world system of production and consumption (Wallerstein, 1974). The establishment of agricultural and mining activities by capitalist enterprises in the periphery gradually incorporates manpower into wage–labour relationships and thereby generates surplus value. The surplus value generated is transferred abroad either through the operations of foreign corporations or through unequal trade relationships. In either case, the extracted surplus is unavailable for investment in the regions of production. The surplus value that remains in the Third World nations is extracted by élite groups, mainly concentrated in the major cities, and therefore rarely benefits the periphery.

This simple model must be modified in several ways in order to explain the differing spatial and social structures of different Third World countries. The mechanism of capital accumulation varies according to the form of capitalism, the dominant mode of production, and its relationship with other modes of production within individual Third World countries. The articulation of dominant and subordinate modes of production determines social formations in the Third World society and the amount of surplus extracted. In turn, the degree of social-class conflict and the needs of the dominant mode of production determine the relationships between the main actors in the capitalist Third World economy: the foreign or multinational corporations, national entrepreneurial groups, the state, and the poor. Most decisions are made by one or more of the first three sets of actors. The specific interests of these fractions of the dominant class may well diverge, but their basic concern, to maintain the capitalist system which benefits them all, is shared. The critical role of the state is to arbitrate between these conflicting class fractions and to legitimize the system to the poor. While the state will frequently favour the dominant-class fractions through the instrumentalist mechanisms described by Miliband (1969, 1977), still more essential is its maintenance of stability in the system (Poulantzas, 1969, 1973, 1976).

The relationships between different fractions of the dominant class and between that class, the proletariat, the peasantry, and pre-capitalist populations depend upon the stage of capitalism developed in the periphery. Marxian writers have distinguished a series of stages of peripheral capitalism, but a commonly cited typology is that described by Castells (1977). World capitalism has been characterized first by colonialism, then by capitalist commercial domination, and finally by monopoly capitalism; these stages at times having overlapped. The essential difference between the stages is that the dominant mode of production changes its form through time so that the method of surplus extraction and the consequent social formation change also. Thus, under colonialism, surplus is extracted through foreign political control over trade, investment, and domestic economic policy. Under capitalist commercial domination the Third World nation will be politically independent, but the surplus will be extracted through unequal trade arrangements. In the stage of monopoly industrial and financial domination, the peripheral country will have developed its own manufacturing and agribusiness sectors, but multinational corporations

Need to know what are the causes of migrants leaving rural areas!

will control these sectors through subsidiary or finance-holding companies. Surplus will then be extracted through profit repatriation, royalty, and patency arrangements, and through accounting mechanisms within the corporation. Increasingly, a new sub-stage of monopoly capitalism is being identified: the New International Division of Labour (Fröbel *et al.*, 1981; Palloix, 1973; Walton, 1982).

The expansion of peripheral capitalism generates a strong process of urbanization. First, cityward migration increases as precapitalist forms of agriculture are disrupted by the penetration of commercial agriculture or by the introduction of monetary taxes on the rural population. Similarly urban migration is generated as the viability of craft industry in the periphery is undermined by cheap imports and later by the products of national manufacturers. Second, surplus generated in the periphery is extracted by national bourgeois groups and representatives of foreign capitalist interests based in the main urban centres. The process of extraction leads to the expansion of the main transportation and market centres and to the rapid growth of the national capital and main ports. Third, the growth of manufacturing within the Third World economy concentrates production still further in the largest cities, stimulates the growth of a national state bureaucracy to encourage the process of industrialization, and leads to the concentration of higher-income groups in the major centres where surplus is accumulated. Fourth, labour moves to the largest cities to look for work and produces surplus in the national centres both through wage labour and through petty-commodity production which supports the expansion of the capitalist sector. Fifth, the state acts to support industrial expansion by providing infrastructure in the main urban centres and by legitimizing the continued functioning of the system through the provision of social services to selected groups. The capitalist systems requires a strong state, both to subsidize the process of surplus accumulation in the private sector (through the development of communications and the provision of education and health services, etc.) and to maintain the system thereby ensuring both the state's and the system's reproduction. Sixth, as metropolitan development accelerates, private capital begins to deconcentrate to areas within the metropolitan region but outside the main city in order to avoid rising land prices, labour costs, and traffic congestion (Rofman, 1974). The state may encourage the process of deconcentration or introduce measures to encourage decentralization.

The essential element in the process is the need for the capitalist system to accumulate capital. The spatial formation is an outcome of the particular needs of the capitalist system at any historical conjuncture and of the conflicting interests between the distinct social classes to control the surplus. Normally, dominant class interests, both directly and through their control of the state apparatus, accumulate surplus which is recirculated within the largest urban areas at the expense of the periphery. The spatial effect is to increase regional disparities and to undermine the autonomy of the subordinate regions. The system may impoverish the poorest regions by a very high rate of surplus extraction (absolute exploitation) but equally it may lead to higher income levels as well as a high rate of surplus extraction through the raising of labour productivity (relative exploitation). The latter process is more typical during the stage of monopoly capitalism.

The model of peripheral urbanization is a generalized description of the relationship between global capitalism and local urban expansion and change. In so far as it underlines the key roles that colonial and capitalist expansion have played in the urban process in most Third World countries, it is a necessary and important advance in thinking. As with any political economy approach, it has the major strength of linking urbanization to the dynamics of social-class conflict. It has thereby managed to demystify some of the technocratic discussion of urban development, revealing the important political dimensions of urban change.

Nevertheless, certain problems are appearing in this model as the literature expands and develops. First, there is a strong measure of determinism in much of the writing. It sometimes appears that particular problems associated with urban development, for example increasing urban primacy or deteriorating urban conditions, are deemed to be inevitable outcomes of the diffusion of international capitalism. As Walton (1982: 128) argues, many interpretations 'allow little room for social organization and cultural tradition'. They fail to recognize the diversity of response to outside forces; for example, how shanty dwellers in large cities react innovatively to poverty and organize politically to ameliorate their problems. As such, some accounts fail to recognize the potential for social protest as labour and neighbourhood organizations mobilize support in peripheral urban formations.

Second, many writers who have adopted this approach have assumed that the model of peripheral urbanization is so different from previous approaches that nothing is worth considering from the earlier literature. As such, some writers have been busy reinventing the wheel, failing to note that their conclusions are often similar, sometimes identical, to those of other writers using a different language (Timberlake, 1985; Slater, 1975). The defence of ideological purity or evangelistic belief in something new has sometimes produced poor scholarship. It has also given rise to much unthinking verification of the model; something that was also common during the heyday of Friedmann's centre–periphery model.

Third, as the interpretations of the more intelligent advocates of this approach become more sophisticated, a new danger is appearing.

Ironically, this is the complete opposite of the early determinism; the danger is now the excessive flexibility of the interpretation. In so far as the variations apparent in the urbanization process in different parts of the Third World are being incorporated, the model is becoming less easy to describe and synthesize. The incorporation of real situations is demonstrating that certain tenets of the model are often incorrect. For example, although cityward migration seems to be a key, universal element in the model, there is increasing evidence of circular paterns of migration linked to capitalist expansion (Chapman and Prothero, 1985; Long and Roberts, 1984). Ecuadorian and Peruvian workers move from the highlands to the coast and back (Peek and Standing, 1982; Matos-Mar and Mejía, 1982); urban workers in São Paulo sometimes commute out to the rural periphery so as to increase their incomes (Goodman and Redclift, 1981). Urban unemployment in peripheral cities does not increase consistently, it ebbs and flows, rises and falls. Employment patterns also change, for example the process of proletarianization occurs in one city but is reversed in another (Portes, 1985). Similarly, political reactions within Third World cities are highly variable; rapid cityward migration rarely produces either an apolitical lumpenproletariat or a revolutionary social movement. In one city, social protest is widespread, in the next it is strongly and effectively repressed by the state. While these variable reactions can be accommodated within the model, incorporation of the highly varied local responses weakens the model's simplicity. As is true of all models, the early simplification of reality is useful to focus attention on a particular facet of the complex whole. More detailed analysis, however, gradually undermines many, often most, of the key assumptions. This is beginning to occur with the model of peripheral urbanization.

The Process of Urban and Regional Concentration

In the absence of any single wholly satisfactory model of spatial evolution I shall resort here to a more descriptive account of urban and regional development in Third World countries. This account takes ideas from both Marxian and the non-Marxian literature and is premissed on the perception that both intellectual traditions demonstrate certain important similarities of interpretation about spatial change. Essentially, I am seeking to explain why urban 'primacy' has emerged in so many Third World nations together with its corollary, the limited development of provincial centres and the emergence of major regional disparities.

At the same time, I am endeavouring to reconcile generalization with local variation. In particular, to demonstrate how outside intervention and the penetration of capitalism has brought major changes in urban organization while differing local circumstances and responses have produced variable urban patterns.

The emergence of distorted urban and regional patterns can be linked to three broad phases of change in the world system. The first is the development of Third World economies as sources of raw materials and markets for the industries of the developed countries. The second is the emergence of inward-oriented industrialization and modernization policies, generally known as the import-substitution process. Finally, there is the newly emerging effort to encourage the development of exports, both manufactures and raw materials, as part of a so-called New International Division of Labour.

The development of raw-material producers and markets for the industrialized countries

Throughout the Third World the growth of major cities was linked to the growth of export production and international trade; São Paulo grew on the basis of coffee, Accra on cocoa, Calcutta on jute, cotton, and textiles, and Buenos Aires on mutton, wool, and cereals. Whether or not export-linked metropolitan development led to urban primacy was dependent upon the degree to which one or several centres controlled the flow of international trade. In those few cases where control over the production, transportation, and profits was spread among a variety of centres, non-primacy was the result. More typically, the national capital controlled the flow of exports, the revenues deriving from those exports, and the importation of goods financed by the export flow. Primacy in Third World countries, therefore, can be explained in terms of the geographical location of export production, the transport networks which emerged to ship those exports and, most fundamental of all, the location of the main recipients of the profits generated by international trade.

It is no coincidence that so many primate cities are major ports. The coastal cities generally benefited because they were located close to the centres of export production or commanded the main channels of trade.

> The great port cities in Asia, as elsewhere, arose where navigable water and productive land met and where internal lines of access were concentrated. The major river mouths were obvious places of this sort; in the island countries, strategically located coastal sites could develop similar access to the most productive areas by coastal shipping . . . Given this sea-oriented pattern, Singapore could easily serve the whole of the tin and rubber belt along the west coast of Malaya, as Colombo could for the quite compact plantation areas of Ceylon, Manila for the Philippines as a whole, and Batavia for Indonesia, where commercial production was heavily concentrated in Java and along the east coast of Sumatra. (Murphey, 1969: 79)

In Latin America and Africa, rivers were usually less adequate export channels and many railways were built by British, French, and US capital to serve this role. Some railways did serve other functions: in India the British built lines to sensitive frontier areas and linked the major cities for strategic reasons, but the typical function was to move exports to the main ports. Even today the orientation of the main railway systems of Africa and Latin America reflects that dominant pattern. Both the mass of lines in Argentina spreading out from Buenos Aires like the spokes of a wheel and the isolated tracks of the Ecuadorian, Peruvian, Angolan, Kenyan or Tanzanian systems had the common purpose of linking the mineral and agricultural export areas with the main ports.

The major ports often developed into primate cities as a consequence of their control over international trade. But primacy was not always linked directly to the location of export production. In Latin America several major capitals emerged in areas physically distant from the main production areas and export routes. In Venezuela petroleum is still produced in, and shipped from, areas hundreds of kilometres away from the national capital, Caracas; in Chile most of the copper is produced in the far north of the country. But these paradoxical situations serve only to emphasize the point made earlier that the critical issue in the evolution of primacy is control over the surplus of profits generated by international trade. In Venezuela the foreign revenues derived from petroleum poured into the government exchequer after 1935 and served to swell the government bureaucracy in the national capital (Lieuwen, 1961). This flow of funds led to the concentration of high-income groups in the city and to the development of a lucrative import business. In Peru, almost a century earlier, the growth of the guano trade had led to Lima's rapid expansion. Within a few years a wholesale market and slaughter-house had been built, a new water system constructed, and a telegraph system opened. New avenues, squares, and parks were opened, the population and geographical area of the city expanded, and a luxurious life-style established for the rich of the city (Romero, 1949). In Peru, as elsewhere, control by the state bureaucracy and private interests, located in the national capital, over the surplus generated by international trade was the critical factor in the domination of the primate city over the provincial cities.

Clearly, however, the degree of control exercised by the élites in the capital city varied greatly between countries. In several Third World nations the benefits of export expansion contributed to the growth of more than one large city. More balanced regional development was normally associated either with a division of functions between cities or with control over different export products. Thus, in Colombia the national capital was neither a port nor the main centre of export production. While Bogotá was a major centre of international commerce and political power, control over the coffee sector lay also in the cities of Medellín, Manizales, and Pereira to the west. Similarly, the lack of a dominant port led to competition among four main centres for the nation's international trade. With mountainous terrain limiting further the control exercised by any one city over the domestic economy a rank-size distribution emerged. Bogotá undoubtedly benefited from export expansion, but so did a whole series of provincial cities. In countries with a more recent colonial past, local agents of Empire controlled the surplus less than they channelled it to the Imperial capital. Thus, in India surplus extraction benefited London more than it did the main Indian cities. Nevertheless major urban centres developed precisely because they were the channels of the export trade; jute, tea, and indigo went through Calcutta, tobacco, and cotton through Madras, textiles from Bombay, and cotton and tobacco through Karachi. The combination of geographical size and the development of non-contiguous export areas led to the emergence of several major port cities. Delhi, the interior viceregal capital, played little part in the export traffic and consequently failed to dominate the urban system.

By contrast, the overwhelming primacy that developed in Argentina was linked to the monopoly of one city over most administrative, commercial, and industrial functions. Buenos Aires combined physical proximity to the export areas of the Pampas, a stranglehold over imports, the ability to attract many of the massive flow of foreign immigrants, and most critically, a monopoly over political decision-making. In 1880 Buenos Aires was voted the capital of a new federal nation. But

> [if] symbolically, the provinces gained a capital, in reality, Buenos Aires, seconded by other urban centres, continued to draw upon the resources, talents, and ambitions of the country and left the other provinces and rural areas drained and depressed. Politicians might come from the interior or the countryside, but in Buenos Aires they quickly forgot their origins and adopted the life and attitudes of the porteño city. (Scobie, 1964: 105)

The political status of the nation seems to have had little influence beyond affecting the rate and form of surplus extraction. Under foreign rule most administration was concentrated in the colonial capital; after independence new forces occupied the same positions and maintained control of the primate centres. Of course there are exceptions. In nineteenth-century Latin America, independence was often achieved by rural power groups whose concern was to limit the dominance of the national capital over their regional fiefdoms. Such a political situation led to a decline in urban primacy in the first few years of independence in a number of countries (Morse, 1971). In Africa and Asia, on the other hand, freedom from colonial rule tended to accentuate primacy as new functions and institutions were created in the new national capital. Frequently independence led to tighter control over the export surplus and to the bureaucratization of government. Not infrequently it also led to the espousal of an inward-oriented development strategy, the consequences of which we examine in the next section.

Urban primacy and the emergence of major cities was an outcome of export expansion and the channelling of the benefits arising from international trade. In this sense the urban-regional structure was a result of the superimposition of a foreign trading system onto an indigenous system. In turn the size and prosperity of the major city or cities was related to the revenues created by export production and to the proportion of that income which remained in the country of origin. Similarly, urban development in the periphery of Third World nations was linked to the process of export expansion. According to the form of this expansion, urban development was either dynamic or truncated.

At the local level the urban effects of foreign-owned plantations or mines differed greatly from those of small-scale peasant production. In fact one can posit that while certain forms of export production stimulated balanced regional growth and the evolution of active market towns, other forces limited such developments. Thus plantation systems tended to restrict urban growth by paying workers low wages, by reducing outside opportunities for small-scale agriculture through the control of land, and by monopolizing the services normally provided through small towns (Cross, 1979; Morse, 1975). Specialized export production generated little in the way of multiplier effects since the harvest was shipped directly to the ports and the profits from the plantations flowed to national urban centres or out of the country altogether. Admittedly, São Paulo's coffee plantations represent an exception to this generalization, but even there fifty years' expansion was necessary before urban development really got under way. Urban growth after 1900 can be explained by the fact that most coffee planters lived in the state and began to diversify their investments. Encouraged by the availability of a cheap and literate immigrant workforce, and helped by federal government provision of infrastructure, they created a major industrial metropolis. With this one exception, and noting differences in local plantation system, the effect of plantation agriculture on local urban development was generally depressive. Whether we consider the tea plantations of India or Sri Lanka, the banana and sugar estates of the Caribbean, or the rubber, tobacco, and tea estates of Sumatra and Java, the urban effects were similar.

Mining enterprises were more likely to generate local urban development even though most mining centres were also enclaves with a national economy. Roberts (1978: 59), for example, notes that even 'where enclave production was strong, as in the mining towns of the central region of Peru, miners' incomes and revenues from contracts for food supplies, transport, and construction for the mining company substantially raised regional income levels . . . [which] made possible the diversification of village economies and the expansion of crafts.' Nevertheless, local processing of the raw material was limited to basic sorting and loading, and refining was often carried out in some foreign port. The mining town would attract migrants, either on a permanent or on a semi-permanent basis, but would generate little in the way of rural–urban linkages. Money would be transferred from mines to rural community, but supplies for the town would originate mainly in extra-regional, even foreign, regions. According to the wealth of the mines, the ownership, and the social system operating, mining might create metropolitan cities, for example Johannesburg, or small depressed centres such as the tin towns of Bolivia. In either case mining activity would be linked more closely to the international than to the local economy.

By contrast, small-scale peasant production tended to stimulate local urban growth. The production of coffee in western Colombia, of cocoa and palm oil in West Africa, and of rice in South-East Asia created relatively large and prosperous agricultural groups. A surplus over subsistence stimulated demand for consumer products and for agricultural inputs such as fertilizers, seeds, and tools. This demand was channelled through local markets rather than through the mine or plantation store. Continuing growth of the export economy led to the gradual evolution of a diversified, integrated urban system, and in areas of exceptional prosperity, gave rise to an indigenous process of industrial and commercial expansion.

Urbanization at the local level was strongly affected, therefore, by the form of export production (Balán, 1976). At the same time no simple relationship between product type and urban development can be detected. Since the production methods of the same product often varied, the urban consequences were very different in different regions. Thus wheat production created very distinct urban system in Argentina and the United States. Similarly sugar, which helped to create the Mexican Revolution in Morelos by the way the plantations expanded and took over peasant land, had less marked and possibly more beneficial effects in Northern Peru (Roberts, 1978: 58; Womack, 1969). In some cases different technologies and forms of social organization have varied for the same product even in the same country. In Brazil, for example, Balán (1976: 158) notes that 'coffee has been produced by slave workers, plantation tenants and share croppers, and independent farmers. Although technology has changed, it has not done so in any drastic way.' Thus it is a combination of the linkage effects, together with the methods of production, that determine the local impact of an export product. As a consequence, the local manifestations of dependent export development varied dramatically from product to product and from region to region. Indeed Roberts (1978: 60) has warned that students should 'pay particular attention to the way in which capitalism expands within underdeveloped countries and not simply focus on the mechanisms through which the dominant world economies have extracted a surplus from underdeveloped countries.'

This warning clearly applies also to consideration of those regions beyond the orbit of export production but which were affected by its development. The national consequences of success for export expansion sometimes had important repercussions on such regions.

First, the expansion of export production led to the annexation of land from indigenous groups. Such annexation might be carried out by force, as in Latin America in the seventeenth century, or through the extension of a commercial market for land, as in Mexico in the nineteenth century. Sometimes the alienation of land would occur so as to permit export production, sometimes to supply export areas with food supplies. Export expansion, especially in the nineteenth and twentieth centuries, often went hand in hand with the commercialization of domestic agriculture. In British South and East Africa concessions granted by, or forced upon, local populations constrained indigenous agricultural expansion. Clearly the consequences of land alienation differed according to the total cultivable area available, whether or not indigenous groups were turned into wage-earning labour forces and according to the social and economic systems established in the rural areas. But since a common element in such land alienation was the concentration of wealth in the hands of the expanding élites and the reduction of land available to the displaced groups, the consequences were always likely to truncate urban development.

Second, the expansion of the export sector was likely to weaken the political autonomy and the economy of provincial regions and thereby undermine the growth of major provincial centres. Not infrequently export expansion attracted imported manufactures which undercut the markets of indigenous enterprises based in provincial cities.

The growth of export revenues gave national governments a new source of income which was often many times greater than their previous revenues. While the provinces might be relieved of their former tax bill, they were now forced to request funds from the national government and they lost their previous power to control or withdraw contributions to the exchequer. In Peru guano and nitrate exports, which financed Lima and the national exchequer, left the provinces without any real autonomy (Romero, 1949). Elsewhere control over the national export surplus gave political control to the dominant élites of the capital cities and led to stagnation of the provincial economies. In many cases regional economies have never recovered from this reversal.

Inward-directed industrialization and modernization

The world depression of the 1930s and the effects of the Second World War led to a spontaneous process of industrial expansion in the larger Third World nations, especially those in Latin America. Protected from imports of manufactured goods from the developed countries, first by the fall in the prices of primary exports and therefore a shortage of foreign exchange with which to purchase consumer imports, and later by the shift of industrial production in the developed countries away from export activities towards the war effort, Latin America's industry prospered (Frank, 1967). This experience suggested that a new and more successful strategy of development might be embraced. Rather than continuing to rely on imports of manufactured products, Latin American nations could themselves industrialize. Intellectual justification for such a strategy was provided by Raúl Prebisch (1950). His study of trade between Latin America and Britain between 1870 and 1930 demonstrated that the terms of trade had turned in favour of Britain and against Latin America. The prices of primary exports from Latin America had fallen relative to the prices of manufactured products from Britain. Every ton of coffee or tin exported from Latin America bought less in the way of manufactured products in 1930 than in 1870. It was this shift in the terms of trade that lay at the heart of Latin America's underdevelopment. The solution to the dilemma was for Latin America to industrialize. Questionable though much of the argument and statistical evidence was, the report had an enormous influence in Latin America. Through the United Nations Economic Commission for Latin America, which he directed, the industrialization strategy was propagated throughout the region. By the 1960s no Latin American country had failed to adopt the recommendation. Import tariffs had been raised, quota restrictions introduced, infrastructure was being constructed, and encouragement offered to manufacturing enterprise.

As countries in other parts of the Third World approached and obtained independence the industrialization/import substitution strategy became conventional wisdom among most new administration. India had adopted an import-substitution policy even before independence, as the British encouraged the industrialization of the country. After 1949 this strategy was continued during the first and second national plans and led to the establishment of major steel and engineering complexes. India, of course, was something of an exception both because of her dual adoption of Soviet and 'Western' expertise and practice and because of the size of her population. In the smaller countries of the Third World import-substituting industrialization could never achieve the same credibility. Nevertheless even in the smallest newly independent African or Caribbean state elements of the policy were embraced. If a steel industry was too ambitious, plastics, food-processing, or light engineering were clearly not; most governments went ahead with infrastructure, industrial parks, and incentives to foreign and local investors.

This is not the place to examine in detail the benefits and costs of import-substituting industrialization in different parts of the Third World. Suffice to say that in few places was it an outstanding developmental success. In most places it brought economic growth, but the majority of the population did not participate in the benefits of that growth. In the largest and more affluent Third World nations the process led to the development of a sophisticated and diversified manufacturing sector. Even in Brazil, Argentina, or Mexico however, it failed to resolve the recurrent problems of too few jobs and too much foreign technological and financial dependence (Bergsman, 1970; Furtado, 1971; King, 1970). In the medium-sized and poorer countries it led to the creation of inefficient companies producing, at costs well above competitive world prices, a small, rather privileged industrial work-force, and the promise of an industrial miracle which would never be satisfied. In the majority of Third World countries the results were derisory. Perhaps novelist Naipaul's (1969: 215–16) ex-minister of a newly independent Caribbean state best sums up the worst consequences of import-substituting industrialization:

> We encouraged a local adventurer to tin local fruit. This was a failure. It hadn't occurred to anyone concerned to find out whether local people wanted local fruit tinned; no one else did either. The same man went in later for tinning margarine and was a success. The margarine was imported, the tins were imported. Our effort was to operate a machine that turned the flattened tins into cylinders. We capped one end, filled the cylinder with the imported margarine, and capped the other end. I remember the process well. I opened the factory. Our margarine was slightly more expensive than imported tinned margarine, and had to be protected. I believe the factory employed five black ladies, whom we photographed looking grave and technical in white coats. Industrialization, in territories like ours, seems to be a process of filling imported tubes and tins with various imported substances.

The locational consequences of industrialization were clear-cut. Throughout the Third World industrial development occurred most rapidly in the largest cities and encouraged the accentuation of metropolitan and primate city development. The tendency to concentration was both cause and effect of other centralization processes and of the efforts to promote rapid economic growth. In Mexico it led to the national capital increasing its share of manufacturing employment from

35 per cent in 1950 to 47 per cent in 1975; in Brazil to the state of São Paulo increasing its employment share between 1950 and 1970 from 39 per cent to 49 per cent. The advantages of a central location are legion in most capitalist countries, but in less developed countries the benefits are possibly greater still (Alonso, 1971; Gilbert, 1974*b*). Underlying these advantages is the fact that the export-orientation phase of development tended to create a highly concentrated urban complex which contained the bulk of the higher-income groups, the greater part of the social and economic infrastructure, the termini of the transport system, and the national government bureaucracy. In choosing their locations, few foreign and national enterprises were reluctant to eshew the advantages of these concentrations; advantages which were magnified further by the forms of industry being established.

The initial stage of industrialization saw the establishment of companies producing directly for the consumer market. Textiles, foodstuffs, and clothing were followed by electrical, plastics, and light-engineering companies making vacuum cleaners, refrigerators, and other consumer-durable products. Many of these products were limited to the higher-income groups in society, most of whom lived in the major cities. Consequently, most market-oriented companies tended to concentrate in the largest cities. In turn the establishment of modern plants had a detrimental effect on existing manufacturing and artisan industry. Small companies amalgamated with the new, local artisan activity failed to compete with the modern plants. The consequence was that the dynamism of labour-intensive industry in the smaller urban centres was undermined.

Primate city location was favoured further by the dependence of so many new industries on imported parts, machines, and fuel. Indeed, one of the greatest ironies of import-substituting industrialization was its failure to cut imports. Rather than importing the finished refrigerator, the various parts that would make up the finished product were imported and assembled locally. On occasion the various parts would cost more than the final item would have cost ready assembled. In locational terms the dependence on imports emphasized the advantages of coastal cities. Thus, those primate cities that were ports or were closely linked to ports gained a further stimulus to their industrial expansion.

Many of the new industrial enterprises were associated with, or financed by, foreign corporations. Quijano (1971) and Castells (1973) have argued that such links encouraged the 'very unbalanced' pattern of urban development which characterizes Latin America. Certainly there is evidence to support the contention that foreign-owned industry is more concentrated in metropolitan centres that in other cities, whether the country be Argentina (Rofman, 1974) or, for that matter, the United

Kingdom (Holland, 1976). But numerous exceptions suggest that the link between foreign ownership and spatial concentration is not so straightforward. In Colombia, for example, foreign investment during the fifties led to the expansion of neither of the main industrial cities, but to the industrialization of Cali and Barranquilla (Gilbert, 1970). It was only foreign investment that was sufficiently mobile to take advantage of the locational advantages offered in those cities. Much the most plausible argument, therefore, is that foreign investment locates in the most advantageous location; this is usually the primate or capital city. In this sense foreign investment responds to existing centralization more than creating it.

In fact all industrial companies, irrespective of the source of their capital, seek access to the national government bureaucracy. For throughout the Third World national governments wield increasing economic influence. In most Third World countries the state manipulates exchange rates, import tariffs and licences, public utility charges, infrastructural provision, wages, and industrial prices. As a consequence, managers are engaged in constant dialogue with the government. Such contact would be maintained by telephone, cable, or by regular visits to the capital city; however, the nature of business relationships in Third World countries dictates that such contacts are made personally with the minister or director of a government institute. In such circumstances it is not surprising that access to the government machinery is a much quoted rationale of industrial location in Third World countries (Gilbert, 1970; Lavell, 1971).

In many countries, indeed, it can be argued that it is the location of government and the paraphernalia of modernization rather than industrial growth *per se* that is the principal source of urban and regional concentration. In most African and Caribbean countries, where industrial growth is limited, expansion of the government bureaucracy has been a major stimulus to urban concentration. Well-paid government bureaucrats constitute an important market for imported manufactured products, and for the shops which sell them, for the construction industry and for domestic services. One of the incidental outcomes of 'modern' administration and efforts of planning economic development has been the accentuated growth of urban complexes. Whatever the level of industrial development, national governments have sought to mobilize savings and to centralize decisions over the allocation of investment. Thus the surplus created in rural areas and in the hinterlands of provincial cities tends to be channelled towards the primate city. The growth of banking taps savings from every region and allows investment in projects anywhere in the nation; but most of the viable projects are located in the major cities. Sometimes the mobilization of funds is manipulated through state institutions. The Cocoa Marketing

Board of Nigeria long financed industrialization and other national growth policies through surpluses extracted from peasant producers. In so far as these funds were invested in Lagos and Ibadan, rather than in the production areas, the large cities grew at the expense of lower-order centres (Bauer, 1954). Variations of this transfer of funds have been described in numerous countries (Baer, 1964; Fitzgerald, 1976), invariably with the result that the metropolitan centres have grown at the expense of provincial cities.

Once under way industrial concentration is difficult to stop. Commercial, wholesale, and transport businesses emerge to service the manufacturing companies, and indeed the emerging government bureaucracy. Some urban diseconomies develop but are less significant for most companies than the advantages to be derived from the central location. As traffic congestion, air pollution, increasing travel costs, and rising housing prices become major problems, governments start to intervene to reduce the negative effects. Political stability and the continuation of low industrial wages are assured through subsidies or through controls on food prices and fares. Middle-income groups receive subsidized public housing and health services. Political lobbying by powerful urban interests helps reduce the negative effects of large city growth.

Nevertheless, a point is eventually reached when these diseconomies begin to slow the pace of urban growth. This may occur as a result of changing government policies; the costs of subsidizing transport or the increasing severity of water shortages or air pollution may recommend some modification in urban-growth patterns. The private sector may in fact anticipate such a change; and begin to spontaneously deconcentrate certain kinds of industrial activity. Companies wishing to increase production and constrained by the cost of land in the metropolitan area may relocate to nearby towns and cities. This point seems to have been reached in a number of the Third World's largest cities. There is evidence to suggest that the growth rates of the metropolitan areas of Buenos Aires, Mexico City, São Paulo, and Seoul have all recently fallen (Richardson, 1989; Townroe and Keen, 1984). Many are now growing more slowly than the larger provincial cities although the more typical pattern is for metropolitan growth to spill over into nearby urban areas. During the 1970s, in the state of São Paulo, for example, 10 municipalities within 150 kilometres of the metropolitan area were attracting industrial enterprise and growing rapidly (Hamer, 1985). Indeed, population growth in urban areas outside the metropolitan area exceeded that in greater São Paulo. This kind of deconcentration occurred long ago in many developed countries and has been under way for some years in several Latin American countries (Gilbert, 1974*a*; Rofman, 1974). But, if it demonstrates the ability of free enterprise to adapt to changing economic and political realities, it is clearly not the kind of deconcentration that most provincial leaders would like to see (Gwynne, 1985). The growth of Greater São Paulo may be slowing but the benefits are not being felt in Recife, Salvador, or Belém. Indeed, the process of deconcentration seems almost to have strengthened industrial growth in the central region by reducing avoidable costs arising from urban diseconomies while retaining many of the benefits from metropolitan location.

The New International Division of Labour

It is currently fashionable to argue that the international economy has entered a new phase. A new international structure is emerging which is modifying the links between developed and less developed countries. The new stage is linked to profound changes in the world economy which are altering the political economy of development in rich and poor worlds alike. The new stage is accentuating global interdependence, increasing trade and integrating national economies more fully into a global production system. The key features of this new stage of development are the growing mobility of international capital (Jenkins, 1984), the internationalization of currencies and producer services (Thrift, 1986), the growth of manufacturing production in certain parts of the Third World for export to markets in developed countries (Fröbel, Heinrichs, and Kreye, 1980; Corbridge, 1986), and the increasing commercialization and export-orientation of Third World agriculture (Feder, 1977; Sanderson, 1985).

The most commonly noted symbol of the New International Division of Labour is the development of international manufacturing. The emergence of the 'world car', whereby parts are made in many different countries and assembled in strategically located plants, is one widely known example. Linked to this process is the relocation of manufacturing plants from developed to less developed countries. Many transnational corporations have chosen to open plants in Taiwan, the Philippines, or Mexico, rather than increasing their investment in factories 'at home'. This development has led to a large increase of imported manufactured products entering the developed countries. Increasingly, clothes, steel, and electronic products are made in Third World countries. This shift has given rise to anguish among unemployed workers in developed countries, and some measure of euphoria in the favoured 'newly industrializing countries'.

Another important manifestation of the New International Division of Labour is the way that some transnational corporations have encouraged the commercialization of agriculture in certain Third World countries (Goodman and Redclift, 1981; Sanderson, 1985). Domestic production has given way to production for export in countries such as Brazil and Mexico. In Brazil, soya bean production has increasingly replaced that of black beans, a staple of the local diet. The result has been to increase exports but also to generate an increasing demand for imported foodstuffs (Barkin, 1985). Agriculture, like industry, has become more and more part of the international economy.

The causes of these shifts are several. Manufacturing shifts have been facilitated by improved communications technology, the invention of the silicon chip and its incorporation into most electronic products and indeed into most kinds of production process. Such shifts have also been encouraged by the willingness of Third World governments to reduce foreign-exchange, labour, and pollution controls over transnational corporations, a tendency strongly encouraged by the debt crises faced by so many poor countries during the 1980s. The process has been further encouraged by the pressure on profits since the late 1960s and particularly since the oil crisis of the early 1970s. A major stimulus for companies to develop plants in certain Third World countries has been the cutting of labour costs. Reorganization of the production process has allowed a measure of deskilling which has permitted the employment of relatively unskilled Third World workers, particularly women. This development has had the major advantage of undermining the negotiating power of labour both in the developed and less-developed countries (Fernandez-Kelly, 1983).

Clearly, these changes in the world economy are significant even if there is reason to suspect that their importance has been exaggerated. It is certain that the changes have only affected a minority of Third World countries. The newly industrializing countries (NICs) have clearly benefited but, according to the definition used, only Brazil, Colombia, Korea, Malaysia, Mexico, the Phillipines, and Taiwan are included in the Third World as defined in this book. There are major limitations on the ability of most developing countries to participate (Schmitz, 1984; Corbridge, 1986). Indeed, it is all too clear that many African and Asian countries have seen a decline in their export earnings in recent years rather than an increase. Similarly, it is likely that the degree to which integration into the world economy has affected local autonomy may also have been exaggerated. Even in Brazil and Mexico, the terrible effects of the debt crisis are arguably as much a consequence of internal decision making as of their changing role in the world economy (Gilbert, 1987*b*). Clearly, too, the effects on individual countries are highly variable; in some countries the benefits have been considerable, elsewhere negative.

In so far as the New International Division of Labour is producing new patterns of growth, how are these changes affecting urban development? One starting point is to consider the effect of export manufacturing growth on the distribution of industrial activity (Meyer, 1986). In certain countries, totally new areas of manufacturing activity have emerged. These have often been linked to the creation of free-trade and enterprise zones by Third World governments. In these zones, transnational corporations are permitted to import duty-free inputs and export the finished product. Since many of the new zones have been established in areas without a manufacturing tradition, some measure of urban deconcentration has been encouraged. In China, four new zones were created in 1980, all in the south-east of the country close to Hong Kong, Macao, and Taiwan. The aim of the zones is to attract foreign investment bringing with it new forms of technology and different styles of management (Phillips and Yeh, 1983; Sklair, 1985).

Similarly, in Mexico, most of the new export plants are located on or near the border. Of the 1,655 *maquiladora* plants in 1989, 80 per cent were located in border cities such as Juárez, Tijuana, and Mexicali (INEGI, 91: 193–4). Only 14 such plants had been established in Mexico City and 16 in Guadalajara, Mexico's second city. Given the acute level of geographical concentration in Mexico, this process is a significant shift in manufacturing location. At the same time, the change should not be exaggerated. In 1980, the border zone *maquiladoras* only accounted for 4.1 per cent of Mexican industrial employment (Urquidi and Carrillo, 1985). Even if that proportion has increased markedly during the decade, it still represented only one-tenth of Mexican manufacturing employment in 1986 (Mexico, 1989).

The other countries where export-oriented manufacturing has developed on a large scale, are Korea, Malaysia, and Taiwan. While there has been some degree of decentralization in Korea during the past two decades, some of this has been encouraged by firm government controls over location in Seoul and by the construction of free-trade zones (Renaud, 1981). In any event, while cities such as Taegu, Daejeon, and Gunsan have increased significantly, more than half of the new industry has been located in the vicinity of Seoul and Pusan (Rondinelli, 1982). In addition, it is important to point out that Korea is geographically small and has very good communications and infrastructure systems. Whether this is the principal motivation underlying deconcentration or the locational requirements of export-oriented industry is consequently difficult to evaluate.

Likewise, it is difficult to generalize about the urban effects of recent developments in export agriculture. In Brazil, the dramatic programme of development in the Amazon region has generated considerable urban growth. Similarly, the development of fruit and vegetable production in north-west Mexico has helped stimulate urban growth in Baja California. But, if these are relatively successful examples, the experience is hardly replicable in most parts of the world; few parts of Asia have any agricultural frontiers left. In so far as export agriculture will develop further it will occur in existing areas of production. Consequently, the potential for radical changes in the pattern of urban development will be limited.

In sum, therefore, the effects of the New International Division of Labour on urban development in Third World countries is very uneven. In some countries, the growth of export manufacturing and agriculture has been dramatic; elsewhere it has been very limited. In some countries, new concentrations of industrial production have emerged; elsewhere the opposite has occurred. As a result, we can only wait to see whether the global economy will embrace the Third World still more closely and observe carefully the urban effects in the countries fortunate enough to be selected.

3

The Urban–Rural Interface and Migration

Until the last century, many rural populations had little connection with urban centres. They lived in quite self-centred societies. They operated subsistence economies maintaining only limited external contacts. As explained in Chapter 1, the expansion of the capitalist system, however, under way for half a millennium and accelerated by the Industrial Revolution, incorporated ever more outlying regions into the emerging world economy (Wallerstein, 1974, 1980, 1989; Chase-Dunn, 1989). Existing cities were integrated into the new system, their functions transformed. New cities were established to exercise political control and to channel resources to the metropolitan centres. And rural populations all over the world have been drawn into the urban nexus.

The process of incorporation into the world system has spread across the entire globe. The self-centred society that had only limited contact with the outside world has virtually disappeared. Rural populations have become subject to political control exerted from urban centres. Coerced to provide labour, conscripted into the army, and subjected to taxation, they became part of colonial or national polities—even while they remained by and large disenfranchised, an issue to which I will return in Chapter 7.

Along with political incorporation has come economic incorporation. Rural populations began to produce for urban markets. Or they went to work on plantations, in mines, or indeed in the cities. Whether they sell their products on the market or their own labour, they are part of a far-reaching economic system, a system which is beyond their control. They experience the vagaries of the world economic system: their earnings from a crop are depressed by a sudden drop in its price on the world market, they lose their jobs during a recession. At the same time, their culture is transformed as they are indoctrinated by foreign missionaries, taught in schools according to curricula mandated by urban élites, and exposed to radio programmes, films, and most recently television series, produced in distant cities, some of them half-way across the globe.[1]

For many peoples the incorporation into the world capitalist system was traumatic. American Indians were forced into the *encomienda*. Africans were enslaved and shipped to the Americas. Indians and Chinese were sent around the globe as indentured labour. Colonial governments conscripted labour. Eventually, the imposition of taxes provided a more subtle means of coercion: unless they grew cash crops or sold some of their cattle, people were forced to earn wages to pay their taxes.

Nearly everywhere incorporation created new desires in the rural areas that only money could satisfy. The high degree of self-sufficiency of traditional farmers thus was ever more compromised as they became dependent on goods and services in the market. Today rural populations need cash to settle taxes, purchase manufactured goods, and pay school fees. Some continue to farm their own lands. Many no longer own their land; they have become proletarians: some are tenant farmers or share-croppers at the mercy of landowners, others have become wage-earners on plantations, in mines, or in the cities.

Some rural populations were exploited to such an extent that their living conditions declined. Elsewhere, specific groups experienced pauperization. For most rural dwellers, however, living conditions improved; in terms of better health and longer life the change was usually dramatic. However, these same improvements accelerated population growth, and population pressure on the available land became severe in many areas.

Incorporation into the world system brought considerable differentiation to what had frequently been quite egalitarian societies. Certainly, many had known severe inequalities in the past. Captured enemies were held as slaves, entire people were subjugated and forced to provide goods and services for their masters. Still, a measure of equality founded on general access to land was the more common pattern. In any case, as these societies became part of larger societies, new elements of differentiation came to the fore. The first to be converted by missionaries and to attend their schools had a head start in employment as teachers, government officials, or commercial clerks. The first to accumulate a little capital in employment, or from the sale of their crops, established themselves as traders or transporters. And the first to become agents for the colonial government or the independent state expanded their control over land or derived the benefits that flow from wielding patronage. Rural populations thus came to experience relative deprivation. As incorporation proceeded, they recognized their own poverty: they saw a few in their midst rise to levels of affluence undreamt of in the past, and they came face to face with the life-style of outsiders—missionaries, traders, government officials, foreign experts, and tourists.

With the perception of a better life enjoyed by a few came an awareness of the means towards such an end. Throughout the world today few are the rural dwellers who have not sold and bought in markets or shops, who have not seen what a school certificate can do for the future of a child, who have not listened to a firsthand account of work in the city. Some improve their condition while staying where they were born, or moving to other rural areas as farmers, traders, or artisans. But rural prospects appear dim to many, the urban scene more promising.

The Urban–Rural Gap

Cities are centres of power and privilege. This is true throughout the Third World today. Certainly, many urban dwellers live in desperate conditions. In a survey of various street occupations in Jakarta in 1972, average daily earnings ranged from 120 to 365 *Rupiah* (equivalent to $US 0.30 to 0.90) in different trades. Most could barely feed their families; the very poor could manage only because they were single. However, even those in the poorest trades said that they were better off than they had been in the rural areas. There was an increase of nearly two-thirds in reported income, in terms of the unweighted average for the different trades (Papanek, 1975). In a survey of squatter settlements in Delhi in 1973–4, 54 per cent of household heads were reported to be employed as casual labourers. They found work for only 240 to 260 days a year on average, but this was more than twice the working days they had in the village. And while their wages were low, their average earnings over the years were two and a half times what they could earn in the rural areas (T. K. Majumdar, 1978). Rural–urban migration in Third World countries can be fully understood only if we grasp the condition of the rural masses.

Even socialist countries, while reducing income inequality within the urban and the rural sector respectively, find it difficult to deal with inter-sector inequality. In China, Mao Zedong persistently drew attention to problems associated with urban–rural and industrial–agricultural inequality. Moreover, he promoted social movements and implemented far-reaching state programmes whose explicit goals included reducing and eliminating urban–rural inequalities. In fact, however, most of the extreme suffering that ensued from the disastrous failure of the so-called Great Leap Forward in 1958–61 fell on the peasantry, and per capita rural income and food consumption did not regain the levels achieved at the start of collectivization in 1955 until the late 1970s. The state imposed the organization of production through collectivization, mandated the crops to be grown for compulsory sale to the state, and controlled prices through a virtual state monopoly over the bulk of marketing activity. As a result, the rural sector fell far behind the urban sector. Estimates of the urban–rural income ratio in the late 1970s range from 2.2 to 1 to as high as 5.9 to 1. The high estimates include subsidies that went largely to the urban sector. By virtually every measure, whether that of relative income, status, mobility, educational opportunity, or even cultural patterns, the urban–rural gap widened in the course of two decades of mobilizational collectivism to the detriment of the countryside and rural people. As a consequence of the sweeping reform measures enacted in the late 1970s, the absolute and relative position of the rural sector improved dramatically in the 1980s (Selden, 1988: 153–80). Even then the urban–rural income ratio remained as high as in other poor countries in Asia (Parish, 1987).

Comparisons of urban and rural incomes are notoriously problematic.[2] The usual basis for comparison is an index of urban wage rates and a crude index of agricultural incomes, such as the prices which cash crops fetch. These measures signal sudden shifts in urban or rural incomes, for example the rapid rise in urban wages in most African countries around the time independence was attained. But they have serious shortcomings. Adjustments for the difference in the cost of living between urban and rural areas are difficult to make. And agricultural income is a function not only of prices paid to producers, but also of output: growing population pressure on land will tend to decrease output per capita; improvements in farming techniques will raise it.

The foremost difficulty for any comparison of urban and rural incomes to explain migratory behaviour is a need for disaggregation. Average urban wage rates have little relevance for the unskilled migrant. And rural opportunities vary according to the endowment of the region of origin, and the migrant's local position, i.e. access to land and capital inputs. The point is well illustrated by Shaw's (1976: 74–105) analysis of rural out-migration in Latin American countries. Variations in the land tenure system are significantly related to the rate of rural out-migration from different provinces in both Chile and Peru. Comparing 16 countries, Shaw shows that it is not absolute population pressure, but access of labour to land as mediated by land ownership, that affects rural out-migration (table 3.1). The average rate of out-migration was highest for those countries—Mexico, Peru, Venezuela—in which more than half the land was held by *latifundios* and where more than half the farms were *minifundios*.

An evaluation of the urban–rural balance of opportunities has to consider collective consumption as well as individual incomes. The urban areas, and especially the major cities, invariably offer more and better facilities than their rural hinterlands, and provide superior education and training, for the migrant's children in particular. Expert medical care and drugs can be found. Piped water assures clean water and releases women from the drudgery of fetching water over long distances. Electricity supersedes the kerosene lamp and the open fire. Some migrants eventually find subsidized housing. Here again, there is the serious problem of disaggregation. Migrants frequently experience

Table 3.1

Average yearly rates of rural out-migration for sixteen Latin American countries classified by patterns of farm size, 1950–1960[a]

Distribution of farms	Distribution of land	
	Less than half held by *latifundios*[b]	More than half held by *latifundios*[b]
Less than half are *minifundios*[c]	0.56	1.60
More than half are *minifundios*[c]	1.48	2.33

[a] These rates present the arithmetic yearly averages of net rural–urban migration during census periods around the 1950s as a percentage of the rural population at the beginning of the census period. The estimates are based on vital statistics using average rates of natural increase from United Nations *Demographic Yearbooks* and population data from national censuses (Shaw 1976: 5).
[b] A *latifundio* is defined as a farm in excess of 500 hectares.
[c] A *minifundio* is defined as a farm less than 5 hectares.
Source: Shaw (1976: 103).

severe discrimination in access to these urban amenities. Indeed, for some, housing and sanitary conditions are worse than where they came from. Still, on balance most enjoy more amenities than those who stayed behind.

Urban–rural differences are best assessed with an indicator that directly measures well-being. Mortality data are an obvious choice. Health and life are universal values, thus providing measures with cross-cultural validity. Increasingly reliable estimates of infant and child mortality have become available for a large number of Third World countries. They show a quite consistent pattern: infants and children have a better chance of survival in urban than in rural areas in all but a few countries. In many countries the gap between urban and rural mortality rates is huge (table 3.2).[3] Such shortcomings of the data as remain tend to overstate urban and to understate rural mortality.[4] Still, the problem of disaggregation remains. At the urban end in particular, mortality rates among migrants presumably are above average. The difference should not be exaggerated though: migrants usually constitute a large proportion of the urban population, and hence strongly affect urban averages. In most countries the urban–rural differential is sufficiently large to suggest a real improvement in living conditions for migrants.[5]

Why People Move

According to the substantial body of research on rural–urban migration which has accumulated over the last three decades the evidence is

Table 3.2

Infant and child mortality rates in urban and rural areas, 1980s

Country (date of survey)	Infant deaths[a]		Child deaths[b]	
	Urban	Rural	Urban	Rural
Botswana (1988)	39.4	39.5	18.0	16.8
Brazil (1986)	75.4	107.5	12.4	15.9
Burundi (1987)	84.6	87.2	85.6	108.6
Colombia (1986)	38.5	40.8	9.9	17.5
Dominican Republic (1986)	71.4	69.1	23.0	24.8
Ecuador (1987)	51.3	77.8	14.3	37.6
Ghana (1988)	66.9	86.8	68.8	82.9
Guatemala (1987)	66.0	84.9	35.5	50.2
Indonesia (1987)	50.9	84.1	28.4	43.1
Liberia (1986)	138.4	160.7	90.5	93.3
Mali (1987)	92.0	144.0	122.5	185.9
Morocco (1987)	66.1	91.0	16.4	50.9
Nigeria (1990)	75.4	95.8	58.9	123.8
Peru (1986)	57.9	104.1	21.9	60.4
Senegal (1986)	69.1	101.3	71.7	165.2
Sri Lanka (1987)	34.6	32.1	5.7	11.0
Thailand (1987)	26.3	40.5	8.3	11.6
Togo (1988)	72.8	86.9	62.7	89.4
Trinidad and Tobago (1987)	36.3	27.5	4.8	2.5
Tunisia (1988)	49.9	64.1	12.9	25.5
Uganda (1988/89)	104.7	106.0	67.6	93.0
Zimbabwe (1988/89)	37.4	64.0	16.6	35.0

[a] Probability of dying between birth and the first birthday, for the ten-year period before the survey, per thousand.
[b] Probability of dying between the first and the fifth birthday, for the ten-year period before the survey, per thousand.

Source: Institute for Resource Development, Columbia, Md., from the birth histories of women aged 15–49 at the time of the survey.

overwhelming: most people move for economic reasons.[6] When people are asked why they moved, they usually cite the better prospects in the urban economy as their chief reason. Also, migration streams between regions have been shown to correspond to income differentials between those regions.[7] And over time, as economic conditions at alternative destinations change, migration streams alternate accordingly.[8]

Material considerations are of prime importance to most people. Certainly, poor people who ignore their material circumstances are rapidly threatened in their very survival. Migration entails costs, economic and frequently psychological, as well as risks. Those migrants who are not motivated by the prospect of material rewards are a minority. The 'bright lights' theory of rural–urban migration has enjoyed a certain vogue, but the simple fact is that most new arrivals do not have the means to spend much time in bars, dance halls, or movie theatres. Indeed, many people, when the rural environment where they have grown up offers a similar standard of living and equivalent prospects for their children, prefer to stay rather than move to the city. As Hemalata Dandekar (1986: 219–20) reports:

> I asked [a textile worker] if he didn't prefer to live in Bombay. Wouldn't he miss the excitement if he went back to live in Sugao, I wondered. 'What kind of question is that?' he said. 'There is no question about it. Of course I would live at home if I could make enough money there.' Waving his arm to encompass the dirty pavement and the roaring pedestrian and vehicular traffic outside the door of the Sugao *talim* (exercise place) where we were sitting, he said, 'There's no excitement here, the air smells of the mills, the food is bad, and there is nowhere to go for a walk that isn't as crowded as this.' He was right. It did smell of the mills. We were in the heart of the textile area of the city, and the nearest beaches, which are literally the lungs for congested Bombay, would be packed with the city's population strolling shoulder to shoulder. 'In Sugao,' he continued, 'at least the fields are open and the breeze from the hills is fresh, the *bhakari* made in one's own home tastes so much sweeter than what the *khanawal bai* (the woman who prepares meals for pay) throws on your plate here.' Bombay, the queen of India's cities, with its thriving commerce, cosmopolitan and heterogenous population, more open society, and huge entertainment industry, has little to offer one in his economic class. Working long days in the textile mill, he earns barely enough to maintain himself in the city and his family in the village in a very modest lifestyle. He has little surplus income with which to splurge on the city's luxuries and allurements.

The sight of severe and widespread poverty in Third World cities easily leads to the assumption that migrants do not know what to expect, that illusions about the prospects lying ahead bring them to an urban environment in which they find themselves trapped. This happens occasionally. Grindal's (1973) account of a group of northern Ghanaian immigrants in Accra suggests that they had been misled by returning migrants who described the South as a land of great wealth, where the buildings are many storeys high, where the people ride in cars or on bicycles, where the 'social life' abounds, and where one can earn money for things such as bicycles, clothing, and finery. Such myths were perpetuated by returning migrants who wished to build up their image and their exploits. They underplayed the problems they had encountered in the cities of the South. For successor migrants whose expectations had thus been raised, the first contact with southern urban life was an unexpected and often shattering experience. Their pride forced many to remain in the South in order to spare themselves the humiliation of coming home in poverty.

Studies not limited to a small ethnic group tell a different story.

Caldwell (1969: 122) surveyed predominantly migrant areas in the four principal cities of Ghana. Nearly two-thirds of the migrants said that life in the town approximated what they thought it would be. Among those whose experiences did not match their expectations, almost half had been overly apprehensive about urban conditions. Only about one in six migrants had been disappointed. The unexpected disappointments of the town divided almost evenly into less economic opportunity than anticipated and greater problems. Studies throughout the Third World similarly report, time and again, that most migrants consider that they have improved their condition. They are satisfied with their move.[9]

The relative success of most migration is due in large part to the fact that it is embedded in social relations. Migrating is not a solitary affair. The days when elders disapproved of young men 'running' away and left them no alternative but to abscond at night (Banton, 1957: 48–59; Rouch, 1956; Elliott Skinner, 1965: 67) are long past. Going to town became the thing to do. Patterns of migration were established. The urban experience took on positive connotations. Thus, in the 1950s, young men in many parts of the Sudanic belt were expected to spend one or several spells of seasonal migration in Ghana (Rouch, 1956). Cultural norms exalted the challenge to the young to prove themselves, the experience to be gained. Such norms can become so generally established that individuals are swept along even when they do not share the economic rationale for going to the city.

Today, rural communities virtually everywhere accept the out-migration of young adults. Communities have developed migration strategies. Their strategies are informed by the experience of migrants who have kept in touch, who return to the village on visits or to stay, and by villagers who have visited kin and friends in the city. These strategies are modified over time as experience dictates. Potential migrants are thus presented with quite well-defined options.

The decision to migrate in turn is rarely an individual one, rather it is usually a family decision. Much rural–urban migration of individuals is part of a family strategy to ensure the viability of the rural household (van Velsen, 1960; Arizpe, 1981). Parry (1979: 44–5) characterizes the situation in a region in northernmost India in these terms:

> Kangra . . . is a district with a high population density and insufficient land to meet even the barest subsistence requirements of its people. A large proportion of the adult men are forced to look for work outside, but because even the smallest holdings offer a degree of security and even the most factious kin groups will preserve their own from total destitution, few abandon their villages altogether. These social and demographic conditions allow the city access to a vast pool of cheap labour; while—to a significant degree—the urban economy supplies the material base for the perpetuation of the rural social structure . . . Provided that it is understood that the local community only maintains itself as a

community to the extent that it retains its peasant basis, we might characterize the Kangra economy as a remittance economy backed up by subsistence agriculture.

The family character of the migration decision is obvious where young women are sent out to supplement the family income from their urban wages—Trager (1988) gives an account of such a strategy from the Philippines; Kate Young (1982) traces the 'expulsion' of young daughters from rural Oaxaca, Mexico, to the requirements of their parental households. The migration decision appears as more of an individual decision where migrants are family heads who take family decisions largely on their own.

Finally, migrants typically receive considerable assistance in the move, in adapting to the urban environment, in securing a foothold in the urban economy. In a sample of blue- and white-collar workers who had moved to Bombay, over three-quarters had one or more relatives living in the city. More than half gave this as an important factor for choosing Bombay over another city. Nine out of ten reported that they had been assisted by relatives or friends on their arrival: about two-thirds received free accommodation and food, and two-thirds of the blue-collar workers and over one-third of the white-collar workers acknowledged help in finding a job (M. S. Gore, 1971: 48–52, 62–7). Similar accounts abound. Indeed, in some cases potential migrants wait in their village until their urban contacts signal a job opportunity.[10]

Kin groups can mobilize greater resources than nuclear families. Frequently a wide range of relatives can be drawn on to help pay for an education for the future migrant, provide a home for children who are sent to town to go to school, offer the newly arrived migrant shelter and food for a while, take care of parents and assist wife and children who stay behind. The extended family thus acts as an agent of urbanization (Flanagan, 1977; Eames, 1967).

The push from rural areas and the pull of urban areas are often distinguished in discussions of migration. Indicating push or pull stresses a particular motive in the decision to migrate. Refugees, for example, may be said to be pushed out of their rural homes. During the civil strife that followed the Partition of India in 1947, about 16 million people fled across the newly established boundaries. Most of those uprooted from rural areas sought a new beginning in cities. War, the man-made calamity, frequently makes rural areas insecure so that peasants pack up and leave for the relative security of cities. During the many years of the blood-bath in Indo-China, peasants sought shelter in the cities. Many others were relocated by force. In Indonesia, the independence struggle, as well as regional rebellions following independence, led to a mass exodus from the affected rural areas. Civil wars in Malaysia, the Sudan, Zaïre, Ethiopia, Chad, Angola, and Mozambique made peasants abandon their ancestral lands. In Colombia *la violencia*, the violent conflict in the countryside which lasted for over a decade, was a major force in rural–urban migration.

Elsewhere droughts, earthquakes, cyclones, volcanoes, or floods have brought not only immediate physical danger, but threatened hunger and disease as well. Such disasters thus frequently force rural dwellers to abandon their homes and seek relief in urban areas. They are commonly referred to as 'natural', but they are man-made to the extent that political action, or inaction, increases the severity of their impact. Thus the famines of the 1970s and 1980s in sub-Saharan Africa were not simply the outcome of an act of nature. They must be traced first and foremost to government policies. Policies that did little to develop rural areas, leaving much of the peasantry in a precarious condition, resourceless to deal with a natural calamity. The peasantry was disadvantaged in general, and few specific efforts were made to render marginal lands less prone to drought, for example by supporting well construction or reafforestation. Indeed, many rural areas lack the roads to carry relief supplies: the hungry have no alternative but to move to urban areas or camps. Urban decision-makers' disregard for the rural masses was exposed when some governments, refusing to acknowledge famine conditions, delayed relief operations. There is also the culpability of rich countries: in an age of highly productive agriculture in major parts of the world, and efficient global transport, there is no justification for hunger anywhere.[11]

When men instigate wars, when they are unable to control the elements, or unwilling to help fellow men in their struggle with nature, entire populations become refugees. There are also less noticeable refugees, individuals who, having fallen foul of the rural community, or the locally powerful, seek refuge in the city.

The push from rural areas is dramatized in the case of refugees. Even in such extreme cases, however, it can be seen that a comparison is at stake: the refugees move to more secure settings. The decision to migrate involves choosing among locations. This is also true where the pull looms large. To take the archetypal case, joining the Gold Rush implied a perception of more limited opportunities at home.

Discussions of rural–urban migration, and of changes over time, tend to focus on the urban labour market—the topic of our next chapter. The rural areas, however, are experiencing major transformations that have profound effects on migration. First of all, population pressure on land increases as the rural population continues to grow in most Third World countries: natural population growth in the rural areas exceeds out-migration in all but some already highly urbanized countries.[12] Second, access to land is transformed: communal land tenure breaks down, land becomes concentrated in the hands of the rural wealthy and absentee landlords, even while elsewhere land is redistributed under land reform programmes.[13] Third, agricultural labour is made redundant in some countries through a shift to either extensive farming such as cattle ranching or to capital-intensive farming such as the introduction of tractors. Fourth, as rural areas become more fully incorporated into national and international markets, they become increasingly subject to the price dictates of markets, some of them free, many government-regulated.

Migrating to Join the Unemployed?

Rural–urban migration continues unabated throughout the Third World. 'Why do so many come', the question usually goes, 'when urban unemployment is widespread and underemployment common?' To which a peasant might respond with the counter-question: 'Why do so few go, when the rural–urban gap is so unmistakable?' Two interpretations explain migratory behaviour under these circumstances. Both establish that the decision to migrate is a rational response to economic conditions. Variations in the structure of urban labour markets account for the difference between the two interpretations.

In Tropical Africa, analysis focused on migrants coming in search of jobs that offered wages and working conditions regulated by legislation and/or collective bargaining. They would spend several months trying to secure such a job but, if unsuccessful, eventually return to the village. Thus in Kampala, Uganda, Hutton (1973: 61–2) found a clearly established pattern in the middle 1960s. Of the unemployed men she interviewed, three-quarters intended to leave if they could not find work, typically within less than six months. More than three-quarters of these intended to return to their rural home. Going home, however, was only a temporary measure: only 11 per cent of the unemployed surveyed said that they would stay there.

In the 1950s and 1960s much urban unemployment in Tropical Africa conformed to this pattern. With independence, urban wages rose substantially in many countries—the advent of independence raised expectations that governments judged had to be met at least to some extent. Rural–urban migration surged, the labour shortages that had plagued colonial governments vanished, and urban unemployment appeared. Much labour migration had been short-term, and new immigrants faced little competition from entrenched urban workers and their descendants. Moreover, independence was frequently accompanied by a significant rise in urban employment. The system of recruiting unskilled labour approximated a random process. Since minimum wages were high, relative to rural incomes, even an extended job search was a promising strategy. Joining the urban unemployed, rural–urban migrants tried their luck at the urban job lottery (Gugler, 1969).

The probability of securing urban employment, along with the rural–urban real-income differential, was incorporated into an econometric model of rural–urban migration and job search by Harris and Todaro (1968, 1970).[14] The probability of obtaining urban employment was defined as the proportion of the urban labour-force actually employed. The assumptions underlying this definition were problematic even in the early stages of urban unemployment in Tropical Africa (Gugler, 1976). The most comprehensive test of the basic proposition was based on a survey of 5,500 households in seven Tanzanian towns in 1971. An analysis of the propensity to migrate according to level of education provided evidence for the significant role of both rural–urban income differentials and urban employment probabilities. In contrast to Harris and Todaro, however, the employment probability was defined as the ratio of the net number of jobs created over a four-month period (the estimated average time spent in the job search) to the number of unemployed (Sabot, 1979: 120–7).

In retrospect it is clear that the urban job lottery pattern occurred in exceptional circumstances. More commonly, labour turnover is low, job creation slow, and recruitment anything but random. A more widely applicable interpretation of rural–urban migration has to focus on the fact that urban labour markets, like many markets, are fragmented in a variety of ways, i.e. different categories of people enjoy differential access to earning opportunities. As we shall see in Chapter 4, access is usually largely a function of three criteria: education and training, patronage, and gender. Differential access in turn shapes the composition of the migrant stream. The role of formal education as a prerequisite for access to the more privileged strata motivates parents in rural areas and small towns to relocate with their children or to send their children away to better or more prestigious schools. For those who have climbed the educational ladder, the most attractive career opportunities are in the city. Others, while not so fortunate, have the right connections and come with reasonable assurance that the assistance of their kinsman, fellow villager, or patron will get them a job. Women, finally, are disadvantaged in the urban labour markets of the Third World, as everywhere else, but their migratory response varies across major Third World regions as we shall see in a moment.

The Characteristics of the Migrants

Given the rural–urban gap, wholesale emigration from the disadvantaged rural areas might be expected. But the cities are less than hospitable to new immigrants; only the highly trained, well connected, and hardy venture there. The stream that appears enormous at the urban end constitutes only a small proportion of the rural population: in spite of rural–urban migration, the rural population continues to grow in nearly all Third World countries. Migration then is selective. Three characteristics stand out in particular: socio-economic background, age, and gender. They determine prospects in the city and hence affect the decision to move or stay.

The socio-economic background of migrants affects their urban prospects. At one end of the spectrum are the many migrants who come from neglected and impoverished regions, such as much of Burkina Faso, or from the lower strata of quite differentiated communities, for example village India. They are poor and ill-prepared for any but the most menial tasks. At the other end of the spectrum are migrants from unusually developed regions, or more typically members of privileged rural minorities. They attend the better schools, and advance their education sufficiently to gain access to promising careers in public administration, with a major company, or as professionals.

Young adults predominate among migrants in search of employment.[15] They are usually unmarried; but even when married, they have less at stake in the rural areas than their elders. They frequently lack control over resources, land in particular, and wield little power in local affairs. To put it into more general terms, they are at a transitional stage between adolescence and adulthood, not yet firmly committed to an adult role in the local setting. In that respect, they enjoy an advantage in the urban economy: less established, they are more adaptable to the urban environment. And if migration can entail accepting marginal earnings with the prospect of eventually bettering them, then the potential rewards are highest for the young starting on a lifetime urban career.

The migration of women has received scant attention in the huge migration literature. When marriage requires migration, it is usually the woman who moves. Thus in India many women migrate at marriage from rural to urban areas. There is also, however, a reverse movement of brides from urban to rural areas. In fact, while women outnumber men among rural–urban migrants, they are outnumbered by men in net rural–urban migration. If we are concerned with the latter, with urbanization, urban sex ratios indicate distinct patterns of sex selectivity of net rural–urban migration for different world regions (table 3.3).[16]

In South Asia, the low position of women translates into high national and even higher urban sex ratios. Discrimination does not only affect the chances of survival, it also shapes patterns of rural–urban migration. Women move usually as dependants. Men outnumber women among the migrants because many are unmarried or leave their wives behind.

Table 3.3
Urban sex ratios in major Third World countries, 1947–1986

Region/country	Year	Sex ratio: males per 1,000 females		
		National	Urban	Adjusted Urban[a]
EAST AND SOUTH-EAST ASIA				
Burma	1953	977	1,040	1,065
	1983	985	991	1,007
China	1982	1,074	1,099	1,024
Indonesia	1961	973	1,001	1,029
	1971	968	1,000	1,033
	1980	988	1,002	1,014
Korea, Rep. of	1960	1,008	998	990
	1966	1,014	1,004	990
	1970	1,008	1,007	999
	1975	1,012	996	984
	1980	1,005	999	994
	1985	1,002	992	990
Malaysia	1970	1,015	1,017	1,002
	1980	1,006	1,012	1,005
Philippines	1970	988	933	944
	1980	1,007	955	949
Thailand	1956	995	1,071	1,077
	1970	991	983	992
	1980	993	963	970
Vietnam	1979	942	1,065	1,131
SOUTH ASIA				
Bangladesh	1961	1,076	1,422	1,322
	1974	1,077	1,294	1,201
	1981	1,064	1,258	1,182
India	1951	1,056	1,171	1,109
	1961	1,063	1,184	1,114
	1971	1,075	1,166	1,084
Nepal	1961	973	1,129	1,160
	1971	1,014	1,173	1,158
	1981	1,050	1,152	1,097
Pakistan	1961	1,111	1,301	1,171
	1968	1,138	1,227	1,079
	1981	1,105	1,153	1,043
Sri Lanka	1953	1,115	1,310	1,175
	1963	1,082	1,179	1,090
	1971	1,061	1,133	1,069
	1981	1,040	1,096	1,054
MIDDLE EAST AND NORTH AFRICA				
Afghanistan	1979	1,059	1,093	1,032
Algeria	1966	1,008	999	991
Egypt	1960	1,012	1,037	1,025
	1976	1,037	1,054	1,016
	1986	1,047	1,057	1,009

Table 3.3 (*cont.*)

Region/country	Year	Sex ratio: males per 1,000 females		
		National	Urban	Adjusted Urban[a]
Iran	1956	1,036	1,065	1,028
	1966	1,073	1,085	1,011
	1986	1,046	1,057	1,010
Iraq	1957	1,010	1,069	1,059
	1965	1,041	1,067	1,025
	1972	1,015	1,029	1,014
	1977	1,063	1,085	1,021
Morocco	1951	983[b]	935[b]	951[b]
	1960	998	984	986
	1971	1,002	958	955
	1982	1,002	1,006	1,003
Syria	1960	1,056	1,064	1,008
	1970	1,053	1,071	1,018
	1981	1,045	1,069	1,024
Turkey	1955	1,034	1,211	1,171
	1960	1,042	1,195	1,147
	1966	965	975	1,010
	1970	1,023	1,148	1,122
	1980	1,030	1,096	1,064
	1985	1,027	1,090	1,061
SUB-SAHARAN AFRICA				
Cameroon	1976	960	1,077	1,122
Côte d'Ivoire	1975	1,074	1,177	1,096
Ethiopia	1968	1,025	903	881
	1984	994	867	872
Ghana	1960	1,022	1,062	1,039
	1970	985	996	1,011
	1984	973	949	976
Kenya	1969	1,004	1,386	1,380
	1979	985	1,216	1,234
Mozambique	1980	945	1,097	1,161
Nigeria	1963	1,020	1,149	1,127
South Africa	1951	1,031	1,192	1,156
	1960	1,010	1,150	1,138
	1970	973	1,119	1,151
	1980	1,035	1,068	1,032
	1985	975[c]	1,007[c]	1,032[c]
Sudan	1973	1,023	1,131	1,105
	1983	1,031	1,133	1,098
Tanzania	1967	955[d]	1,180[d]	1,236[d]
	1973	969	1,078	1,112
	1978	962	1,075	1,117
Uganda	1969	1,019	1,191	1,169
Zaïre	1984	988	992	1,004

Table 3.3 (*cont.*)

Region/country	Year	Sex ratio: males per 1,000 females		
		National	Urban	Adjusted Urban[a]
Zimbabwe	1969	1,012	1,412	1,395
	1982	960	1,140	1,188
LATIN AMERICA				
Argentina	1947	1,053	974	925
	1960	1,001	953	952
	1980	969	936	966
Brazil	1950	994	915	920
	1960	997	927	930
	1970	990	939	949
	1980	991	952	961
Chile	1952	963	873	907
	1960	961	889	926
	1970	956	906	947
	1982	961	920	957
Colombia	1951	988	856	866
	1964	971	884	910
	1973	945	868	918
	1985	980	912	931
Cuba	1953	1,050	963	917
	1970	1,052	983	935
	1981	1,022	975	954
Ecuador	1950	992	906	914
	1962	999	917	919
	1974	1,004	930	926
	1982	995	945	950
Mexico	1960	994	944	950
	1970	996	962	966
	1980	977	950	972
Peru	1961	989	992	1,003
	1972	1,005	1,000	995
	1981	997	997	1,000
Venezuela	1950	1,028	993	966
	1961	1,032	1,001	970
	1971	996	961	965
	1981	1,000	977	977

[a] The adjusted urban sex ratio is the urban over the national figure, multiplied by 1,000; slight divergences are due to rounding.
[b] Excluding foreigners.
[c] Excluding Bophuthatswana, Ciskei, Transkei, and Venda.
[d] Tanganyika.

Sources: Ratios calculated from national censuses and estimates for the 1950s, 1960s, and 1970s compiled by the United Nations and provided to the author on tape; and from census data in United Nations (1984, 1986, 1990).

Men outnumber women in the cities by a large margin in every country in South Asia. The imbalance is somewhat reduced, but remains substantial, when urban sex ratios are adjusted for the fact that men outnumber women also in the total population of each of these countries. A study of regional differences within India found that where women enjoyed higher status, as measured by such variables as infant mortality, literacy, age at marriage, and fertility, they were more likely to be found in the urban labour-force and among rural–urban migrants (Ferree and Gugler, 1983).

In most countries in the Middle East and North Africa, men also outnumber women in the urban population, but by a smaller margin than in South Asia. Again, this margin is reduced when urban sex ratios are adjusted for national sex ratios.

In most East and South-East Asian countries, urban sex ratios are quite balanced. But there are notable exceptions. In China, considerably more men than women are found in urban areas, but much of the difference can be accounted for by the imbalance in the country's sex ratio. In Vietnam, on the other hand, the high urban sex ratio becomes even more unbalanced when we taken into account the low national sex ratio, a legacy of the many years of murderous war visited upon the people of Vietnam. The Philippines, finally, were unique in the region until recently: women outnumber men in the urban population by a substantial margin.

In sub-Saharan Africa, men outnumber women in the urban population of many countries, sometimes by a large margin. Such male preponderance is, however, far from universal in the region. Unmarried, separated, divorced, and widowed women move to the cities on their own, and this movement appears to have accelerated in recent decades (Gugler and Ludwar-Ene, 1990).

Latin America stands in sharp contrast to other Third World regions. It resembles developed countries in that women outnumber men in the urban areas. The pattern holds for every single country in the region. In some countries women predominate by a substantial margin. It appears, then, that large numbers of women move on their own. The uniformity of the pattern, and its repetition in the Philippines, the only Latin Third World country outside the Americas, suggest a cultural interpretation. While most Third World cultures encourage early marriage and childbearing, the common heritage of Latin countries includes a religious ethos that exalts the status of the single woman.[17] In terms of our discussion of age selectivity in migration, the interlude between puberty and commitment to the role of wife and mother is relatively extended in Latin cultures. Young Latin women thus are potentially mobile, independent of a spouse. Faced with limited rural opportunities, they turn to the cities where many households can afford to offer them the low pay and limited benefits that go with domestic service.[18]

Urban sex ratios are becoming more balanced in most Third World countries. In part, this reflects the growing share of urban-born with their quite balanced sex ratio in the urban population. But the preponderance of men in the urban population has declined so abruptly in a number of countries as to indicate a change in the sex selectivity in rural–urban migration. Two factors are at work. On one hand, it has become more common for single women to migrate. The interplay between a weakening in pressures for early marriage and childbearing and expanding job opportunities for women—most strikingly in factory jobs in newly industrializing countries—remains to be elucidated. On the other hand, it has become less common for men who work in the city to leave their wives and children in the village.

Migration Strategies

The movement of individuals is the focus of much migration analysis. Such a focus is fostered by the fact that migration frequently involves young single persons. However, as we have seen, migration is embedded in social relations: the option to migrate is informed by the experience of others, the decision to migrate is taken in a family context, and the migratory move is assisted by relatives, co-ethnics, friends.

In many cases migration is not just a 'once for all' move. Rather there are a series of moves over a lifetime. Such a migratory career has to be understood with reference to family and community. Four principal strategies of rural–urban migration in the Third World stand out:

(*a*) circular migration of men;
(*b*) long-term migration of men separated from their families;
(*c*) family migration to urban areas followed by return migration to the community of origin;
(*d*) permanent urban settlement.

These statuses are not fixed. The man who leaves his family behind, subsequently decides to have them join him. The family that expects to return to its community of origin, eventually settles down in the city. Changes in migratory status typically strengthen the commitment to the place of destination. They are affected, however, by changing circumstances in both the urban environment and the community of origin, for example deteriorating urban conditions may force men to send their families to the village.[19]

The preponderance of men over women in the cities of South Asia, the Middle East and North Africa, and in much of sub-Saharan Africa (table 3.3) reflects the tendency of male migrants to leave wives and children in their rural area of origin. If the Industrial Revolution engendered the distinction of workplace and home, the separation of men from their wives and their children has been drastically magnified in the Third World.[20] Support by kin typically facilitates such dual involvement in the urban and the rural economy. A wife manages the farm, holding her own in a male-dominated environment, with the support of male relatives who assist in certain tasks and provide protection.

Throughout Latin America the predominance of women in the urban population indicates that the temporary migration of men to cities is unusual. The lack of kin support in the rural setting in much of Latin America may be part of the explanation. The crucial role rural social structure plays in establishing and maintaining specific migration strategies is suggested by the fact that some Indian communities present an exception to the Latin American pattern. Laite (1981, 1988), in his study of labour migration in highland Peru, describes and analyses a widespread pattern of temporary migration of men. The majority of miners and refinery workers in the copper industry are migrants, and nearly all these migrants maintain village interests. Their most important interest is village land, even though since the nineteenth century land has increasingly become a commodity to be bought and sold. Three-generation families continue to control property and organize production. The senior generation owns the resources, other members of the family work them. So, whilst junior members of the household are 'landless', they have access to land. The migration of one, or several, men provides outside resources to meet household needs. During their absence, the women do the work or, at planting and harvest time, recruit labour. Their task is facilitated by co-operative practices well established in Andean peasant culture.

It is tempting to speculate that the temporary migration of men is a 'traditional' society's response to new opportunities to earn wages and acquire manufactured goods—to visualize 'tribesmen' making forays into an alien environment. The facts indicate otherwise. Migration has been an enduring feature of highland Peru since peasants went to work in the mines during the Inca period. Half a millennium later a pattern of temporary migration persists. In Indonesia, temporary migration was well established during the colonial period, and has since greatly increased (Hugo, 1983). And a review of migration research in the Pacific islands concludes by emphasizing the persistence of temporary migration during 150 years of European contact (Bedford, 1973: 126).[21]

The migration of individuals—whether single or separated from their families—has distinct economic advantages: it optimizes labour allocation, and it minimizes the cost of subsistence, at least in rural–urban migration. Employers save on wages and retirement benefits. Public authorities face less demand for housing and infrastructure. But migrants also have reason to accept separation from their family. Living costs in the city are high, urban earning opportunities for women frequently narrowly circumscribed.[22] Typically wife and children remain on a family farm growing their own food, perhaps raising cash crops as well. Where land is communally controlled, such as is the case in much of Tropical Africa, there is no compensation for those who give up farming. A wife who comes to town abandons a source of food and cash to join a husband on low wages.[23]

Family separation frequently takes the form of circular migration. After six months of employment, a couple of years at most, the migrant returns for an extended stay with his family. In the ideal case, his return coincides with peak labour demands on the farm. In some areas, such migrants go as contract labour. They are recruited for a fixed period of time at, or close by, their home place, and provided with return transportation. Repetition of the circular movement is common, and many migrants build up extended urban experience.

Circular migration is a function of the recruitment of men at low wages. Where employees have only bachelor accommodations, both aspects—the cheap labour policy and the exclusive recruitment of men—are brought into sharp relief. Most strikingly, in environments thus characterized by circular migration, some employers did establish a more stable labour-force by providing conditions that encourage workers to bring their wives and children.

Circular migration was common in Tropical Africa in colonial days. However, the Union Minière du Haut Katanga changed its labour policy in the copper mines in what is now Shaba Province, Zaïre, as early as 1927. A measure of compulsion was involved: workers had to bring their wives, but the region soon boasted a stable labour force. By 1957 the average length of service of African employees in the Katanga mines was eleven years. On the Copperbelt, in what is now Zambia, a policy to establish a permanent labour-force was initiated in 1940. Employers and colonial government provided permanent accommodations for married employees, adequate schooling in the urban areas, and a pension scheme. The average length of employment of African workers increased from four and a half years in 1956 to seven in 1964.[24] In sharp contrast, the South African gold mines continued to recruit labour on short-term contracts from adjacent areas throughout these years (Wilson, 1972*a*: 123–7).

In parts of British colonial Africa, as well as on some Pacific islands, circular migration was not just the outcome of cheap labour policies pursued by employers and colonial authorities: temporary sojourn of indigenous workers in urban areas was politically intended, and policies were adopted to that purpose. South Africa presents the extreme case of temporary migration imposed by law on a large part of the urban labour-force for over a century. The origins of pass controls can be traced back to the period of slavery in the Cape in the eighteenth century. After the abolition of slavery in the early nineteenth century, pass controls were adapted to support various forms of indentured and child labour. When the diamond and gold mines were established in the late nineteenth century, pass controls were refashioned to institutionalize a system of temporary migrant labour. They were used to promote organized recruitment, enforce contracts of limited duration, and prevent permanent settlement of workers in the mining districts. This 'cheap labour power system' benefited the mines and other sectors employing migrant labour: the costs of reproduction of labour power were borne by the rural sector (Hindson, 1987: 15–27).

Agricultural production in the South African 'reserves' declined sharply, not only in per capita, but even in absolute terms, in the late 1950s. By 1970 agricultural production as a proportion of subsistence requirements was negligible (Hindson, 1987: 55, 72). The 'cheap labour power system' had collapsed: the cost of reproduction of the urban labour force could no longer be displaced on to the rural sector. Pass controls continued, however, to serve the interests of *apartheid*: not only was urban surplus labour resettled in the 'bantustans', but a large proportion of the urban labour force continued to be recruited on temporary contracts while dependants were forced to live in the 'bantustans'. Thus the gold mines were prohibited by law from providing family accommodations for more than 3 per cent of their African work-force until 1986 (Crush and James, 1991). Many men were recruited on short-term contracts not only for the mines, but in various other sectors of the economy. Turok and Maxey (1985: 252–3) estimate that substantially more than half the Africans in registered employment were migrant workers in the 1970s. Racial oppression had given rise to a paradox: the continent's most industrialized country—where large numbers of Africans have worked in mines, factories, and urban services for several generations—had the highest proportion of short-term recruits in its labour-force.

In China, stringent controls on rural–urban migration have kept permanent migration extremely low while meeting the labour requirements of the urban economy through contract labour. (See Chapter 8 for further discussion of migration control.) Shanghai was reported to have 1.8 million temporary migrants, Beijing and Guangzhou more than 1 million each in recent years (Goldstein, 1990). Blecher (1988) provides an account of the temporary recruitment of workers from rural areas to work in urban jobs in Shulu, a county in Hebei Province. Between 1964 and 1978 the industrial labour-force in Shulu trebled. Three-quarters of the new workers came from rural areas on temporary contracts. By 1978, contract workers constituted over half of the work-force in county-level industry. They received somewhat lower wages than regular workers. And they had no claim to the fringe benefits enjoyed by regular workers, such as free medical insurance, compensation for accidents at work, pensions, and sick-leave. Most important, contract workers were housed in dormitories, and had to leave their families in the rural areas. In a period of rapid industrialization, urban growth was thus reined in. Only workers needed in production were authorized to come to urban areas. When the system was effectively implemented, unemployment was avoided. The proportion of dependants was kept extremely low in urban areas. At the same time, inequities were created between regular workers and contract workers, as well as between the latter and rural workers. Rural production teams, however, received a share of contract workers' wages. Inasmuch as contract workers and their families remained part of the rural population, the average income of that population was higher than it would have been if they had left altogether. The system can be argued to have been more beneficial to the rural masses than if some in their midst had left permanently and cut all ties.

Circular migration has become the exception throughout the Third World. Because of the appearance of substantial urban unemployment, circular migration is no longer a viable option. The search for a job may take months, and the outcome is aleatory. The migrant who has secured regular employment now has good reason to hold on to it.[25] Long-term migration replaces circular migration. For more than a century men from the Maharashtra Deccan have come to work in Bombay. Initially they pursued a circular migration strategy, coming to the city for the six months of the year following the wet season when agricultural tasks were few. But by 1942 temporary jobs had become much more difficult to obtain, and men began to remain on the job in Bombay for the whole year. The migration strategy that now emerged encompasses the entire life-cycle of a man. He leaves the village at age eighteen or twenty, spends all his working life in the city, and returns to the village upon retirement, bringing his savings to invest in the village. Very few leave the village permanently (Dandekar, 1986: 219–31).

Many such long-term migrants leave their wives and children behind in the village. Short visits to the family replace the extended stays that characterize circular migration.[26] Instead of an economic cost to employers—a labour-force characterized by high turnover and absenteeism—there is now a social liability for workers: the separation strains their relationship with wife, children, extended family, and village community. The frequency of visits varies a great deal with employment conditions and distance. Faster and cheaper transport may allow monthly or even weekly commuting. But in a country such as India, many migrants cover considerable distances, and can visit their families only during their annual leave. Dandekar (1986: 225) speaks of the emotional stress and hardship in the personal lives of individuals, particularly in the relationship between husband and wife. And a folk-song in rural Uttar Pradesh laments:

> The field is turning a jaded wasteland.
> In their home the flowers are withering.
> She is fading in her father's place,
> Her husband is wasting away in Calcutta.[27]

The preponderance of men in the urban population has declined abruptly in a number of countries over the last two decades (table 3.3). This reflects in part that family separation is becoming less common. As the period of urban employment has lengthened, family separation has become a less satisfactory pattern. For some, increases in urban wages have diminished the significance of the rural income forgone. Also, in many areas, the opportunity cost of abandoning farming has diminished: land shortages entail a decline in output; the, often related, breakdown of communal control over land transforms it into an asset that can be realized. Last, but certainly very important, urban earning opportunities for women have become less restricted.

Settling down in town with a family is usually for the long term, perhaps for a working life. However, it does not necessarily signify a permanent move. Strong ties to members of the extended family, and to the village community, can make an eventual return attractive. At the same time most migrants, even when they manage to support a family in the urban setting, enjoy little economic security. Unemployment and underemployment are widespread, but few qualify for unemployment compensation. And social security systems covering disablement and old age are still in their infancy. For many urban dwellers, the solidarity of the village provides an alternative social security, meagre but reliable. Plentiful land under communal control is still a common pattern in Tropical Africa and the Pacific.[28] In such a situation the migrant can maintain his position in the rural community, and even during an extended urban career remain assured of access to land on his return.

I encountered such a pattern in what was then Eastern Nigeria in 1961–2 (Gugler, 1971). Urban dwellers invariably stressed that they were strangers to the city. Regardless of his birthplace, every man could point without hesitation to a community he considered his 'home place', i.e. the community in which his forefathers lived, even if he himself was not born there. Women, upon marriage, would adopt the husband's 'home place'. Most urban dwellers identified with their rural home, felt that they belonged there, affirmed their allegiance. The home community conversely referred to the men in the city as 'our sons abroad'. They were expected to maintain contact and to return eventually. Only a few had broken contact altogether, and the hope that they would return one day was not abandoned until word of their death was received.

These same people were fully committed to urban life. Most of the married men had a wife and children living with them in town. Nearly all expected to spend their entire working life away from their home places. Losing their urban employment or trade was the worst calamity that could befall them. I concluded that Eastern Nigerians in the city lived in a 'dual system': they belonged not only to the city in which they lived, but also to the village from which they or their husband had come. To put it in terms of social network analysis: the social networks of Eastern Nigerians resident in urban areas were focused on two poles, and these poles were well connected.

In 1987 I returned to Enugu, the focus of much of my earlier research, to replicate a survey I had carried out in 1961 (Gugler, 1991). I found that Enugu dwellers of the next generation maintain strong ties with the community of origin as well. They also visit, contribute to development efforts, build houses, intend to retire, and want to be buried 'at home'. The principal factor explaining the endurance of the 'dual system' is that the city offers even less economic security now than in the early days after independence. Public-sector workers are paid irregularly, pension payments are in arrears, senior civil servants have been dismissed overnight by a military government. In these circumstances the village community continues to provide the only reliable security.

'Life in a dual system' is founded on economic necessity, but it is also an established cultural norm. Enugu residents articulate their home ties invariably in terms of loyalty. The great majority, dependent as they are on the security the home community provides, have no reason to question the norm. But even the urban élite embrace the norm to varying degrees. They are accustomed to a life-style well beyond what a return to the village economy could offer. They own real estate and businesses in the city. Nevertheless, most of them remain committed to the village, practise loyalty to the home community, continue to set store by what home people think of them.

Even where such a 'dual system' is firmly established, there will be exceptions. Migrants securely established in the urban economy may want to reduce their commitment to the village—there was some evidence of that in Enugu in 1961, but not in 1987. For women in patrilocal societies village ties are mediated by men. If they are never-married, separated, divorced, or widowed, these ties become problematic. And the relationship of women with the village of origin is often tenuous. Thus, where temporary migration is the rule, those women who come to the city are more likely than men to stay permanently (Gugler and Ludwar-Ene, 1990).

In parts of Africa, in much of Asia, and virtually everywhere in Latin America, most migrants have little prospect of maintaining access to agricultural land because of population pressure and/or institutional constraints. Wholly dependent on their urban earnings, they have become proletarians. Instead of planning for a return to the village, they press for the provision of social security. And they search for sources of earnings other than paid employment. Escaping the vagaries of employment is a major motive in establishing one's own business. Ownership of a home, however rudimentary, assures accommodations and the possibility of income from rent. The strength of squatter movements in Latin America—which we discuss in Chapters 5 and 7—can be understood, in part, in terms of migrants who have severed their rural connection and want to establish an urban base.

The demands made by migrants on the urban system thus vary according to their plans for the future. Single migrants expect little. Indeed, the limited prospects they face are the key reason the married left their families behind. They will tend to opt for a minimum of expenditure for housing. Such frugality allows migrants either to limit their urban sojourn or to increase their remittances and savings. Migrants who anticipate eventually returning to the area of origin, will remain concerned with conditions there. For permanent migrants the provision of security in the urban setting, especially in old age, becomes crucial.

In delineating these four migration strategies I have given considerable emphasis to intended, as against actual, migratory behaviour. There is some question, however, whether a large proportion of migrants realize their plans to return to the area of origin. Indeed, it is frequently assumed that the majority will end their lives in the cities (Lloyd, 1979: 136). I am not so sure. What little data there are suggest that many migrants do retire to the rural community. In any case, many of the implications of return as distinct from permanent migration hold, whether the intention to return is realized eventually or not; they hold as long as migrants act on the assumption that one day they will settle down 'back home'.

4

The Urban Labour-Market

Urbanization is frequently assumed to be intimately connected with industrialization. In developed countries, cities may be seen to shelter the multitudes required for the operation of industry. And, as table 1.2 showed, world-wide there is a close correlation between the level of urbanization and the level of development, i.e. the richer countries tend to have a greater proportion of the population living in urban areas. However, the relationship between urbanization and industrialization is not as clear-cut as it may appear at first sight. Before the Industrial Revolution, various regions around the globe boasted cities which were the locus of religious activities, the seats of governments, the centres of trade.[1] While an urban population depends on the acquisition of an agricultural surplus, such a surplus need not be traded for industrial products, but may be obtained through offerings to priests, as tribute to rulers, or in exchange for surplus trade goods.

Colonial cities similarly had little industry, but primarily performed functions of control and commerce. In India and the larger Latin American countries some industries were established early, and a substantial range of industries had developed by the 1940s. Most Third World industrialization, however, came only following the Second World War or even later—for instance around 1960, the time of independence, in Tropical Africa. At present, and into the foreseeable future, industrial expansion continues to be severely limited by the structure of the market in most Third World countries. Much of the population is so poor that it can afford few industrial products: textiles and footwear, food and drink, tobacco, enamelware, paint, plastic utensils, and bicycles are the more common industries for the mass market.[2] Producers of most consumer durables, in particular the automobile industry, face the fact that only the small middle- and upper-income groups have the means to acquire their output. In very large countries, of course, even a relatively small middle class constitutes a sizeable market, for example in India, Indonesia, Brazil, Mexico, and perhaps Nigeria. A few small countries have established export markets for a large part of their industrial production—Hong Kong, Singapore, South Korea, and Taiwan are the famous success stories. These are the exceptions; few Third World cities are hubs of industry. Some are mining centres, but most are primarily nodes in networks of trade and transport and/or seats of government.[3]

The industrialization of most Third World countries is quite limited. Furthermore their factories provide little employment compared to the factories established during the industrialization of the West. Throughout the Third World, industry does not use the technology characteristic of that earlier industrialization, but imports technologies recently developed in the West. There are advantages to having access to such advanced technology, but this technology is usually inappropriate for poor countries: it is geared to the availability of the various factors of production in highly industrialized countries. As capital stock expanded during the industrialization of those countries, the productivity of labour rose dramatically, and labour became relatively expensive. Hence modern technology increasingly sought to minimize the role of labour. Automation and data-processing are only the most recent responses to the increasing cost of labour in industrialized countries. Transposing this capital-intensive technology wholesale to poor countries entails a heavy drain on scarce capital and foreign-exchange resources, and provides employment for only a few from a large and rapidly growing, low-productivity labour-force.[4] Nevertheless most Third World countries foster capital-intensive industrialization. They pursue policies that explicitly encourage such industrialization, and they implicitly subsidize the importation of capital goods through overvalued exchange rates.

Third World governments and private investors alike might be thought to have a strong interest in technologies that take full advantage of the cheap labour abundantly available, but the obstacles are several. For one thing, in many cases labour-intensive technology is simply not available. Research and development, as well as the machine-tool industry, are concentrated in rich countries. They are directed by the large demands from these countries, rather than the relatively small investments in poor countries. The efforts of international agencies to develop what has come to be called 'intermediate technology' have had limited results so far. However, an increasing number of multinational corporations have emerged from firms indigenous to Third World countries. Athukorala and Jayasuriya (1988) show for Sri Lanka that in some industries such 'new' multinationals are less capital-intensive than the still predominant 'traditional' multinationals based in developed countries. Second, national élites are typically concerned that their country should have the very best equipment, and that it should be up-to-date; this particularly affects the choice of technique by the public sector.[5] Third, most industrial production takes place in large production units operated by major companies. They tend to provide higher wages, more generous benefits, and considerable job security for political reasons I shall discuss in Chapter 7. Thus the cost of their labour is increased. Fourth, foreigners are responsible for much of the investment in most Third World countries, and they tend to opt for the more capital-intensive techniques. In a study of the large-scale manufacturing sector in India, Agarwal (1976) found foreign firms to be somewhat more capital-intensive than domestic firms. He suggested several reasons: foreign firms paid substantially higher wages, had better access to capital and financial markets, were less constrained by foreign-exchange and import regulations, and had greater experience in capital-intensive technologies and operations; domestic firms were less hesitant to hire large numbers of workers because they were more 'at home' in the indigenous labour-market.[6]

Unemployment, Underemployment, and Misemployment

A generation or two ago employers in many parts of the Third World clamoured for workers, complaining about high rates of turnover and widespread absenteeism. Even where rural areas were effectively incorporated into the world economy, labour shortages were common, and the urban work-force was frequently described as uncommitted. The level of urban wages, and working and living conditions, were at stake. Thus, a cheap-labour policy characterized much of colonial Africa, encouraging the circular pattern of labour migration discussed in Chapter 3. The policy was buttressed by the proposition that migrants would work less at higher wage rates—they would stay in town only as long as necessary to meet a fixed objective: they were 'target workers'. In most settings, the contention that therefore the labour-supply function was backward-sloping bore little relationship to reality; rather, it was a myth that provided the ideological underpinning for a cheap-labour policy (Berg, 1961).

Labour shortages are a thing of the past. Turnover and absenteeism are no longer of concern; in the major firms they are frequently at levels below those prevailing in industrialized countries. Sabot (1979) describes and analyses the transformation in Tanzania. The colonial labour-market was characterized by a persistent shortage of labour. Government and private employers paid low wages, arguing that at higher wages migrants would work for shorter periods. Sabot could find no evidence in the historical record to support this proposition. Moreover, even if higher wages were to induce the individual migrant to work for a shorter period, they could also be expected to attract a greater number of migrants. It appears then that a small number of large estate owners and the government took advantage of their oligopolistic position to administer wages at a level below that which would attract a sufficient number of workers. The shortfall was made up in part through forced labour until the 1920s and again during the Second World War, and through unscrupulous methods of recruiting labour in distant regions. After the Second World War urban wages increased steadily while rural incomes stagnated. When independence was granted in 1961, the rise in wages accelerated. Growth of the rural labour-force, and the widening gap between income from farming and from urban wage employment, increased the size of the migrant stream, and labour surplus replaced labour scarcity. At the same time a labour-force that had been characterized as 'uncommitted' was stabilized: turnover and mobility rates became low by international standards.

Sabot explains the dramatic rise in wages, and the transformation of the problem of labour scarcity into a problem of urban unemployment, in terms of the convergent interests of employers, trade unions, and government. As Tanzania became independent, a burst of import-substituting industrialization occurred. These relatively capital-intensive factories had to invest in training the industrially disciplined, semi-skilled workers they required. In order to secure this investment, the labour-force had to be stabilized, and employers raised wages to that end. The trade unions were identified with the nationalist movement, gaining in strength with the emergence of educated leaders and the stabilization of the labour-force. They had begun to develop only in the 1950s, but more than half the labour-force was unionized by 1965. The unions brought pressure to bear on wages through both political action and collective bargaining. The government legislated a minimum wage in 1957, and increased the wages of low-level government employees.

For several decades now, urban labour-markets throughout the Third World have experienced an excess of labour with limited skills. Open unemployment is only one facet of urban surplus labour. A second element is underemployment, i.e. labour is underutilized. Finally, substantial numbers, while perhaps fully employed, produce goods or provide services that can be judged to contribute little to social welfare; such labour may be labelled 'misemployed'.

Information on open urban unemployment in developing countries is notoriously problematic. First of all, there are few data. A recent unpublished compilation by the International Labour Office provides estimates of open urban unemployment for seventeen Latin American and Caribbean countries throughout the 1980s. According to these estimates, 6.5 per cent of the urban labour-force were unemployed in 1988. For Africa, estimates are available for only eight countries in the early 1980s. For Asia, the only data on urban unemployment are for China where offical sources indicate a decline from 5.4 per cent in 1979 to 1.9 per cent in 1984. Urban unemployment in China has been limited by controls on rural–urban migration and by 'sending-down' campaigns, most notably the rustication of middle school leavers that took around

17 million youths to the countryside during the Cultural Revolution (Kirkby, 1985: 21–53). However, when these policies were relaxed or reversed altogether, substantial urban unemployment appeared; a low estimate put unemployment at 10 per cent of the labour-force in the non-agricultural sector in 1979 (Emerson, 1983).

Second, there is good reason to doubt how completely urban populations are covered by censuses, how accurately they are represented in surveys. There is probably a systematic bias in that low-income groups tend to go underreported; in so far as their unemployment rates diverge from the average, the unemployment rates reported overall are affected.

Third, the extent of reported unemployment is very much a matter of definition. Is it restricted to those actively seeking work? Or does it cover all who are available for work, including those who have become discouraged about finding work? The distinction is likely to have a particularly strong effect on the unemployment rate reported for women. This is even more the case for a further issue of definition: are those searching/available for part-time work to be included? Finally, does part-time work disqualify a person from being considered unemployed?

The unemployed are obviously unproductive, but they are usually not representative of the most desperate urban living conditions. In countries where very few qualify for unemployment benefits, it is only the not-so-poor family that can support an unemployed member.[7] If unemployment is frequently reported higher among the urban-born than among immigrants, it is because the families of the urban-born are more likely to be already well established in the urban economy. Given family support, an extended search for a satisfactory job can be a rewarding strategy, especially for those with better qualifications. Higher levels of unemployment among the more educated, a common pattern,[8] thus appear as a function of both the potential rewards for the better educated of an extended job search and the fact that they tend to come from families which are able to support them through a lengthy period of unemployment. In contrast, the poorest, whose relatives and friends cannot help them, and those recent immigrants who have nobody to turn to, are forced to find some livelihood in a hurry; unemployment is a luxury they cannot afford.[9] But productivity is low for the many who are underemployed.

We define 'underemployment' as the underutilization of labour.[10] Such underutilization is most conspicuous where labour is idle part of the time. While this is a common pattern in agriculture, it is not unknown in the urban sector where seasonal fluctuations are marked in industries related to the agricultural production-cycle, in construction, and in the tourist trade. Underemployment is not limited to these sectors, however, but is much more pervasive.

Underemployment takes three distinct forms. In one guise it is related to fluctuations in economic activity during the day, for example at markets; over the week or month, for example in recreational services; or seasonally, for example in tourism. As activity ebbs, casual labour is laid off and many self-employed are without work. Underemployment takes a second form where workers are so numerous that at all times a substantial proportion are less than fully employed, i.e. a reduction in the number of workers will not decrease aggregate output. In terms of numbers affected, street vendors constitute the most important category in many countries.[11] A third type of underemployment is what may appropriately be called 'hidden unemployment': solidary groups continue to employ all their members rather than discharging them when there is insufficient work to keep them fully occupied. Such guaranteed employment is typical of family enterprise, but social ties other than kinship proper, such as common origin or shared religion, can also provide a commitment to maintain every member of the community.[12]

Finally, there is what I have called 'misemployment' (Gugler, 1988*b*). Labour may be employed full-time, but the tasks performed contribute little to social welfare. Begging is a clear-cut example.[13] More respectable, but hardly more productive, are the hangers-on to be found in the entourages of the more powerful and affluent. There is scope for debate: the thief who redistributes resources from the wealthy to his poor family can be argued to perform a service not dissimilar to that of many bureaucrats in a welfare state.[14] And indeed, ultimately the productivity of an activity is socially defined.

The notion of unproductive labour dates back at least as far as Adam Smith's *The Wealth of Nations* (Katzman, 1977: 168–72). It was part of his polemic against the mercantilist state, whose purpose was to redistribute income from its more productive subjects to the sovereign. The political élite, the religious estate, and the cultural and intellectual superstructure were perceived to have a basically parasitic relationship to the productive classes. Presumably, the larger the surplus income generated by the productive sector, the larger the number of retainers and other parasites that could be supported by the ruling class. Substantial numbers of public administrators in many contemporary societies appear similarly misemployed.[15] As in the courts of yore, their role of hangers-on has become institutionalized.

Much misemployment is based on getting crumbs from the table of the rich. The member of the local élite or middle class, the foreign technical adviser, or the tourist is beseeched for a morsel, or made to maintain a company of sycophants, or has his wallet snatched away. The relationship is vividly portrayed by three activities: the army of domestics who clean and beautify the environment of the privileged;[16] the prostitutes who submit to the demands of those able to pay and expose themselves to AIDS;[17] and the scavengers who subsist on what the more affluent have discarded, who literally live on crumbs from the rich man's table. As Lomnitz (1977: 208) put it in her study of a shanty town in Mexico City:

The settlers of Cerrada del Cóndor may be compared to the primitive hunters and gatherers of preagricultural societies. They go out every day to hunt for jobs and gather the uncertain elements for survival. The city is their jungle; it is just as alien and challenging. But their livelihood is based on leftovers: leftover jobs, leftover trades, leftover living space, homes built of leftovers.

Admittedly, most forms of misemployment make some contribution to social welfare. Waste paper provides a third of raw material requirements for the paper industry in Cali, Colombia, and some 60 per cent of that waste paper is collected by garbage pickers working on their own (Birkbeck, 1979). In domestic service, the contribution can be substantial where women with qualifications that are in short supply are released from household work. What is at issue here is that large numbers of people are employed in a wasteful manner because their labour is so cheap—relative to the incomes of the élite, the middle-class, and foreigners.[18] The point is well demonstrated by the fact that the requirements of middle-class households for domestic help rapidly decrease as domestic wages rise.[19] Kate Young (1982: 173–4) writes of domestics in Mexico:

Domestic service for these girls is . . . a waste of human resources. It permits little awakening of their latent capacities, and provides no possibility for them to learn a wide range of skills, to value their own work, or to develop an independent and enquiring personality . . . Domestic service, in reinforcing dependence, does not encourage the questioning of a system in which wealth permits certain categories of people to condemn others to a life of servicing them. Nor does it encourage them to question a system of relations between the genders in which women are essentially seen either as playthings to entertain men or as drudges to service them.

Unemployment, underemployment, and misemployment, adding up to a massive waste of labour in Third World cities, relate to three dimensions of inequality. The sharply differentiated incomes of the affluent and the masses, allowing the few to pay so little for the services of the many, foster misemployment. The privileged position of the protected labour-force compared with that of the bulk of urban workers causes unemployment where it encourages workers to hold out in the hope of joining the protected labour-force; and it leads to underemployment as the mass of urban workers, with little capital equipment, compete for limited markets.

Finally, rural–urban inequality brings large numbers of migrants from rural areas to Third World cities. Most of them are ready to enter the labour-force. If they contribute about two-fifths of urban growth in the Third World (Preston, 1988), they constitute a considerably higher proportion of the new entrants into the labour-force.[20] They come to cities where unemployment is widespread and underemployment common; their migration entails the loss of potential rural output—in agricultural production as well as non-farm activities; and they require more resources for their survival in the city than they would in the countryside. These three consequences of the large-scale rural–urban migration prompted by rural–urban inequality may be taken to define over-urbanization (Gugler, 1988*b*). This is not to deny that migrants rationally maximize their benefits. The seeming paradox is resolved when the redistributive effect of rural–urban migration is taken into account. Rural–urban migrants lay claim to a share in urban income opportunities, they gain some access to urban amenities. Rural families send their sons and daughters to the city so that they will be able to partake, however little, of its riches.

Any attempt to alleviate the problem of urban surplus labour has to confront the prospect that its very success will attract additional migrants from rural areas. The urban employment problem cannot be solved within the urban arena, unless it is sealed off by a break in the rural–urban connection. The success stories of the city state of Singapore and the city colony of Hong Kong are in part based on the fact that they were cut off by an international boundary from their hinterland, Malaysia and China respectively: the rapid increase in the productivity of their labour-fource was not diluted by rural–urban migration. The equivalent policy elsewhere would be to halt rural–urban migration. Such a solution has little to recommend to it. Not only do even quite authoritarian regimes have great difficulty in enforcing controls on internal migration, but the improvements in the urban labour-market that might be achieved translate into worsened conditions in rural areas left to absorb the entire natural population growth.

Where the rural–urban flow proceeds unimpeded, no solutions to the employment problem can be effected within the urban labour-market. Rather, the issue of rural–urban inequality has to be addressed. This is of necessity a long-range proposition, and it faces formidable obstacles everywhere. Yet if there is an urban employment problem, it can be solved—in all but the largely urbanized countries—only in the rural areas.

The Informal Sector

A common approach to the unemployment problem contrasts two sectors that, it is argued, differ sharply in their labour requirements.

Dualistic conceptions of Third World economies have long been current. An early distinction was between 'modern' industry and 'traditional' artisanship. The unsatisfactory nature of these labels soon became obvious. 'Modern' activities, such as servicing and repairing imported automobiles or television sets, are frequently carried out in a quite 'traditional' manner, i.e. in small, poorly equipped workshops.

In the 1960s renewed attention was drawn to the 'murky' sector. In country after country substantial additions to the urban labour-force failed to show up in employment statistics. The concern about unemployment was accompanied by an increasing recognition that a large and growing number of people were engaged in non-enumerated activities. They were thought to be working in the service sector. However, statistical enumeration is primarily a function of the size of a firm's work-force, and small enterprises in the primary sector, such as peri-urban gardening, and in the secondary sector, shoe-making for example, are just as likely to go unenumerated as are street vendors.

Hart's (1973) classic paper, first presented in 1971, introduced a new terminology, distinguishing an 'informal' from a 'formal' sector. On the basis of research in a low-income neighbourhood in Accra, Ghana, Hart emphasized the great variety of both legitimate and illegitimate income opportunities available to the urban poor. Subsequently McGee (1976) explored various approaches towards what he called the 'proto-proletariat'. The response to Hart's plea that a historical, cross-cultural comparison of urban economies in the development process must grant a place to the analysis of 'informal' as well as 'formal' structures was nothing less than overwhelming.[21] A great deal of research was carried out on the 'informal sector' in the 1970s. It served to direct attention to a work-force that is typically under-enumerated, commonly characterized as unproductive, and all too often dismissed altogether as making little, if any, contribution to the urban economy.

Thus Quijano's more sophisticated three-sector model for Latin America maintained the notion of a marginalized labour-force:

> A growing sector of the labour-force is produced which with regard to the employment needs of the monopolistically organised hegemonic levels of activity is *superfluous*; and with respect to the intermediate levels organised under the competitive form and consequently marked by permanent instability of its weakest enterprises and its peripheral occupations, this labour-force is *floating*, since it tends to be intermittently employed, unemployed or under-employed according to the contingencies affecting this economic level. As a result, it inevitably tends to be forced to take refuge in the roles characteristic of the 'marginal pole', where it fluctuates among a numerous range of occupations and labour relations. In this sense, the principal tendency of this labour-force is to turn 'marginal' and to differentiate and establish itself as such within the economy (Quijano, 1974: 414–15)

Hart's terminology was adopted by a mission to Kenya organized by the International Labour Office. It argued that the informal sector provided a wide range of low-cost, labour-intensive, competitive goods and services, and recommended that the Kenya government should promote the informal sector (ILO, 1972: 223–32). The report characterized the two sectors in the following terms (ILO, 1972: 6):

> Informal activities are not confined to employment on the periphery of the main towns, to particular occupations or even to economic activities. Rather, informal activities are the way of doing things, characterized by—
>
> (*a*) ease of entry;
> (*b*) reliance on indigenous resources;
> (*c*) family ownership of enterprises;
> (*d*) small scale of operation;
> (*e*) labour-intensive and adapted technology;
> (*f*) skills acquired outside the formal school system; and
> (*g*) unregulated and competitive markets.
>
> Informal-sector activities are largely ignored, rarely supported, often regulated and sometimes actively discouraged by the government.
>
> The characteristics of formal-sector activities are the obverse of these, namely—
>
> (*a*) difficult entry;
> (*b*) frequent reliance on overseas resources;
> (*c*) corporate ownership;
> (*d*) large scale of operation;
> (*e*) capital-intensive and often imported technology;
> (*f*) formally acquired skills, often expatriate; and
> (*g*) protected markets (through tariffs, quotas and trade licenses).

This characterization of the formal sector is quite persuasive. In every Third World country large-scale enterprises play a major role in various sectors of the economy. Their status ranges from public-sector companies to multinational corporations to locally owned firms, but they are invariably closely related to the state. Their power *vis-à-vis* the state and other participants in the economy raises questions about their impact on the autonomy of the national polity and the threat of monopolistic control over sectors of the economy. Many formal-sector workers enjoy a measure of protection through legislation and/or collective bargaining. Their wages and benefits, working conditions, job security, and social security coverage as a rule compare favourably with those of other workers. The establishment of such a 'protected' labour-force has political implications that I will address in Chapter 7. Here our concern focuses on the employment implications of the tendency of large-scale enterprises to rely on imported, and hence capital-intensive, technology. Their recruitment of expatriate personnel for the highly skilled positions adds a further irritant.

In contrast, the informal sector appears to offer a panacea for the urban employment problem. At the same time it may provide scope for the emergence of local entrepreneurial talent. The size of the informal sector is impressive enough. Estimates for cities in six Latin American and two Asian countries suggest that between two-fifths and two-thirds of the urban labour-force work in the informal sector (Souza and Tokman, 1976; Mazumdar, 1976). Admittedly, such estimates are problematic. Not only, as we shall see, is the informal sector quite impossible to delineate, but many of its workers have reason to evade attempts to record them. Nevertheless, there can be no doubt that in most Third World countries a large proportion of the urban work-force is found in this sector.

The view that the informal sector provides the answer to the urban employment problem is reinforced by the notion that it receives little support from government but, to the contrary, is subject to inaction, restriction, and harassment. This is what the International Labour Office (1972: 226) mission argued in the case of Kenya. It should be noted though that Kenya gained independence only in 1963, and that its European settlers had established regulations expressly keeping African enterprise out of the cities.[22] More generally, while formal-sector firms typically enjoy privileged access to credits, foreign exchange, and tax concessions, entrepreneurs in the informal sector can be seen to enjoy competitive advantages *vis-à-vis* large-scale industry in so far as they escape taxation, social-security levies, and government regulation of wages, working conditions, and job security. Eckstein (1988*a*: 141–7) describes such a pattern in Mexico, suggesting that the implied support for the informal sector is intended by the government.

The basic difficulty of the two-sector model is that any multidimensional definition can be applied to only one of the sectors, leaving the other sector as a residual category. A multidimensional definition of the formal sector, such as that advanced by the Kenya report, fits large-scale industry reasonably well. It characterizes also, by and large, major commercial, financial, and service organizations, and government administration. If we accept it, we are left with an informal sector which does not conform to a similarly distinctive ideal type, but rather covers a varied range of activities.

The postulate that the informal sector has the obverse characteristics of the formal sector is clearly untenable. To take the characteristics used in the Kenya report, entry into much of the informal sector is far from easy,[23] the self-employed repairers of audio and video equipment are dependent on imported supplies; it is quite common for small-scale enterprises to employ non-family labour; illegal activities may be organized on a large scale; the owner-operators of taxis use capital-intensive technology which has not been adapted to the resource constraints of a poor country; skills are acquired in the formal school system and in formal-sector employment; and there is sometimes monopolistic control of markets.[24]

The death knell of the informal-sector concept was sounded when Castells and Portes (1989) introduced a collection devoted to the 'informal economy'. They entirely abandoned the attempt to define such economy or sector, focusing instead on unprotected labour, i.e. labour protected neither by unions nor the state. Using the formal–informal distinction in this fashion to contrast conditions of employment, Roberts (1989) was led to emphasize that informal employment can be found in the largest and most modern firms in Mexico. Thus in a labour-market survey in Guadalajara, 20 per cent of respondents working in firms employing more than 500 workers, reported not being covered by social security and having, at best, a purely temporary contract.

The shortcomings of the informal-sector concept are not just of analytical concern; rather, they invalidate any attempt at policy prescription. As Bromley (1978: 1034) puts it:

> It is often mistakenly believed that a single policy prescription can be applied to the whole informal sector, so that governments should adopt similar programmes towards artisans making furniture, towards artisans illegally manufacturing fireworks, towards sellers of basic foodstuffs, and towards prostitutes or drug-peddlers. The informal sector is large enough to permit and diverse enough to necessitate a wide range of different policy measures, allowing governments to mix incentives, assistance, neglect, rehabilitation and persecution within the total range of policies.

As for informal-sector incomes, any generalizations are bound to be misleading in several respects. Earnings vary dramatically across the wide range of activities commonly subsumed under the concept, for example between street-vendors and prostitutes. Even for the same activity, earnings vary substantially according to the clientele, for example between street-vendors peddling their wares to the poor and those catering to tourists. And earnings also vary significantly between employers and employees: Portes *et al.* (1986) found in a survey of the low-income neighbourhoods of Montevideo, Uruguay, that average incomes were virtually identical in the formal and the informal sector, but that informal-sector employers reported incomes more than four times as high as those of informal-sector workers.

The widespread assumption that the informal sector is the province of the poor is thus subject to important qualifications. On the one hand, earnings vary a great deal across the informal sector. On the other, low incomes are common in the formal sector: many large firms employ substantial numbers of casual workers at low wages, without fringe benefits, and unprotected by social security. Any comparison of incomes in the formal and the informal sector has further to allow for the fact that the informal sector disproportionately recruits the very young and the old, women—whose earnings are lower than those of men for reasons we will discuss shortly, and the less educated.

The Kenya report dichotomized the urban economy into the formal and the informal sectors and recommended strengthening the linkages between the two sectors as a strategy of promoting the informal sector (ILO, 1972: 228–31). Since then studies in other countries have shown significant existing linkages.[25] Such linkages provide the basis for a good deal of informal-sector activity. However, the informal-sector participants in such relationships tend to be in a subordinate position. In particular, the informal sector can be argued to subsidize the formal sector: its low-wage labour produces low-cost inputs for the formal sector, and provides cheap goods and services for formal-sector workers (Portes and Walton, 1981: 67–106).

Benería (1989) describes such linkages for a multinational corporation producing electrical appliances in Mexico City. It employs 3,000 workers, but sends out 70 per cent of its production to 300 regular and 1,500 occasional subcontractors. One of these subcontractors, owned by Mexican capital, employs 350 workers and sends out 5 per cent of its production. Typical for this subcontractor's subcontractor is a sweatshop that operates illegally in the owner's residence. This sweatshop employs 6 workers between the ages of 15 and 17 on a temporary basis, and sends out work to a fluctuating number of homeworkers. Average monthly wages ranged from 12,000 pesos for manual workers at the multinational corporation to 8,500 pesos at the Mexican firm, to 6,000 pesos at the sweatshop, to 1,800 pesos for the homeworkers. There is such a clear hierarchy of wages, quite apart from benefits, and the drop is sharpest for the—women—homeworkers. In India, according to a review of case studies in four industries, women's incomes decline from Rs. 300–500 a month in factories to Rs. 200–300 in workshops to less than Rs. 100 in home production (Baud, 1987).

Exploration of the linkages between the informal and the formal sector has focused attention on employment relationships. Bromley (1988) has conceptualized a continuum of employment relationships that range from career wage-work to career self-employment (figure 4.1). Between these two types of career work, he distinguishes four types of casual work that range from short-term wage-work, through disguised wage-work and dependent work, to precarious self-employment. These distinctions serve to make the point that most of those engaged in the least stable and least secure work, while seemingly self-employed, in fact enjoy little autonomy and have rather inflexible working regimes and conditions. 'Disguised wage-work' is paid according to output, like much wage-work—the difference is that it is conducted off-premises.

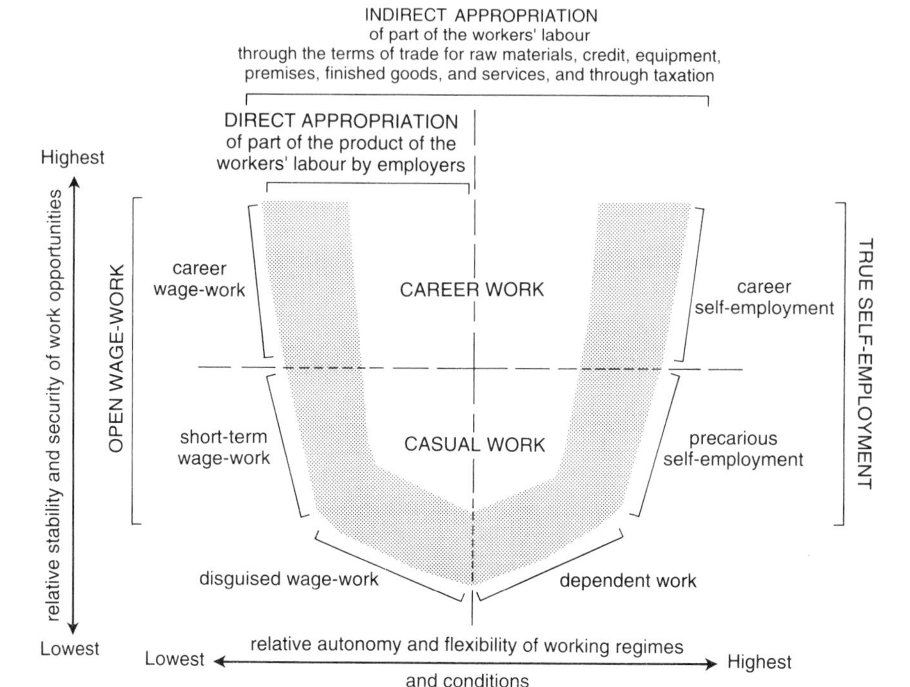

FIG. 4.1 The continuum of employment relationships
Source: Bromley (1988: 167)

And 'dependent workers' have contractual obligations that substantially reduce their freedom of action: they have to pay rent for premises and/or equipment, to repay credit, and to purchase or sell at disadvantageous prices.

Clearly, any assessment of the prospects for the 'informal sector', and of policy options, has to be both specific and comprehensive, i.e. it has to focus on particular activities and those engaged in them to take full account of linkages, with the formal sector in particular.

The Fragmentation of the Labour-Market

Different categories of people enjoy differential access to earning opportunities. This is obvious in geographical terms: the better opportunities tend to be concentrated in urban areas, inaccessible to those who are beyond commuting distance. But any one local labour-market is fragmented in a variety of ways as well. Three major sources of fragmentation stand out: labour-markets are stratified by education, segmented by patronage, and segregated by gender.

Wages and benefits, job security, social security, and working conditions stratify the urban labour-force. That stratification is highly visible as it translates into differences in dress, mode of transport, and housing. The labour-market appears composed of distinct layers of earning opportunities. Formal education qualifications are usually a prerequisite for entrance at various levels. They may be thought to be related to the functional requirements of a given job; however, beyond certain minimum requirements, it is probably more accurate to see such entrance prerequisities as a function of the educational characteristics of the labour pool. Across countries the educational prerequisites for a given job tend to be higher in those countries where the labour-force is more educated, and in any given country the demands put on applicants typically increase along with the expansion of the educational system. 'Credentials' thus appear as a screening device, and their relationship to functional job requirements becomes increasingly tenuous as first primary, then secondary, and finally tertiary education reach an ever larger proportion of the population.

The labour-market may be compared to a geological formation composed of a sequence of distinct horizontal layers. This image corresponds closely to the patterns established by formal organizations. And public discourse tends to focus on the access various population groups have to formal education as the prerequisite for entry at various levels. However, labour-markets are not only thus stratified, they are also segmented vertically. Rather than one, there are several geological formations, separated to the extent that entrance to each is controlled by a network of patronage, granting privileged access to a certain category of people and denying access to others.[26]

Most migrants obtain assistance from urban contacts, as we have seen in Chapter 3. The urban host, to help the new arrival, to relieve the burden of housing and perhaps even of feeding him, has good reason to find him work. Thus migrants who have secured employment introduce their relatives and other people from 'home' to their firm. Many employers find such 'family brokerage' convenient and even advantageous. They know that skills and knowledge are not as important for many positions as other qualities: dependability, potential for training, persistence, and initiative. Further, in many cases, job advertisements will generate all too many applications from people with similar qualifications. In such circumstances the employer prefers to use a 'broker'. He selects among his employees one or two persons he trusts, and asks them for suitable candidates, whom they will probably have to train. The broker will look to his extended family for suitable candidates and draw up a short list. He may coach a candidate on how to fill in the application forms, and on how to react at the interview. A close and complex relationship thus arises between the employer, the broker, and the new employee. The broker has increased the socio-economic position of his kin group and his own standing within it, the unemployed has obtained a job, and the employer can exert leverage over his employee through the broker (ILO, 1972: 509–10). In small-scale enterprises the owners themselves may initiate their relatives and other people from 'home' into their trade.[27]

Because of such particularistic recruitment patterns, migrants of common origin tend to cluster in certain jobs and trades. In a survey of fourteen villages in West Java, Indonesia, Hugo (1985) found that the two narrowly defined occupational categories most common among migrants and commuters from any one village accounted for over two-fifths of them (table 4.1). In Accra the timber market was controlled by immigrants from one village in faraway Niger at one time (Rouch, 1956). And migrants from the community of Mexticacan are involved in the manufacture and sale of ice-cream throughout Mexico (Rollwagen, 1971).

The stratification and segmentation of urban labour-markets is mirrored in the stream of migrants. Their prospects vary widely but, at least, they are reasonably well defined for many. The integration of new arrivals into the urban labour-market is thereby eased. Discontent over discrimination does not crystallize as long as the criteria for privileged access vary from one little niche in the urban economy to the next. In many countries, however, large segments of the labour-market appear as the exclusive preserve of an ethnic group. Such a situation will shape the contours of political conflict, an issue I will address in Chapter 7.

So far we have focused on the stratification and segmentation of labour-markets, i.e. the differential access to earning opportunities conferred by education and patronage. Gender stands out as a third criterion that affects access to labour-markets throughout the world.

Women in the Labour-Force

The distinction between domestic tasks and outside work is solidly entrenched in everyday language, in official statistics, and in the social sciences (Jelin, 1982). Usually 'work' refers to what adult persons do for a living, but does not include unpaid domestic tasks. Yet, both the daily maintenance of the labour-force and its reproduction in the next generation depend on the performance of domestic tasks within the household.[28] While there is considerable variation across cultures, domestic work is universally defined as incumbent primarily on women.

The domestic responsibilities of women affect their labour-force participation. Many are released from these responsibilities at some stage in their life-cycle, when other women in their households are in

Table 4.1

Occupational clustering of migrants and commuters from fourteen West Java villages working in urban areas, 1973

Village	Number of working migrants and commuters	Proportion of migrants and commuters				
		(*a*) in most common occupation	(%)	(*b*) in second most common occupation	(%)	(*a*) + (*b*) (%)
1	74	Groundnut hawker	65	Government/Army	15	80
2	55	Cooked food/cigarette hawker	35	Day labourer	22	57
3	91	Cooked food hawker	43	Jewellery hawker	21	64
4	70	Pedicab driver	57	Day labourer	16	73
5	82	Pedicab driver	41	Factory worker	34	75
6	100	Labourer	35	Hospital worker	13	48
7	87	Kerosene hawker	32	Household domestic	15	47
8	77	Airline/hotel worker	32	Household domestic	10	42
9	87	Kitchen utensils hawker	60	Government/Army	12	72
10	88	Driver	27	Government/Army	26	53
11	87	Pedicab driver	38	Construction worker	20	58
12	92	Carpenter	49	Government/Army	28	77
13	99	Barber	31	Bamboo worker	20	51
14	104	Bread hawker	42	Driver	32	74

Source: Hugo (1985: 94).

charge of domestic tasks. Thus, unmarried daughters in Hong Kong take jobs to contribute to household income (Salaff, 1981). Also, some women can afford to hire domestic help, while they work outside. But many other women are burdened by a 'double' working day: in spite of heavy domestic responsibilities, they are gainfully employed, because the earnings of other members are too low to support the household, or indeed because they have been abandoned, divorced, or widowed and are the only ones to support themselves and their young children in the absence of any, let alone adequate, welfare provisions.

Variations in household composition across the Third World affect the extent and allocation of domestic tasks. These tasks decrease where fertility declines. They can be allocated so as to free some women where daughters marry late, where grandparents live in, or the families of married brothers live together. Cultural norms affect the role women can play outside the house: the Islamic rule of seclusion is the most striking example, but in fact there is considerable variation in its application across Muslim countries.

International comparisons of the labour-force participation of women are problematic because it is not uniformly defined. Housework goes unreported everywhere; and the reporting not only of agricultural work but also of much work in the urban 'informal' sector—in which women are disproportionately represented—varies a great deal. These problems of definition are minimized when comparisons are made within a country. Thus, differences in the labour-force participation of women in Indian cities correspond to a long-established North–South divide in the position of women: the stronger position of women in the South goes hand in hand with a higher labour-force participation (Ferree and Gugler, 1983).

Reported unemployment is usually considerably higher for women than for men. This is the case, even though many women do not actively look for work, because they are aware that the opportunities open to them are severely limited. As for underemployment, disproportionate numbers of women are affected by it in many countries. Indeed, the very activities in which they predominate tend to be characterized by underemployment: the women traders of West Africa are legendary. And many women are misemployed. Substantial numbers make their living from prostitution.[29] Domestic workers constitute the largest category of employed women in many countries.[30] Homekeepers are similarly misemployed to the extent that part of their labour is lavished on superfluous chores. For many, however, the very essentials of maintaining their household constitute a full-time task: women often service a large household that includes several children as well as various other dependants, and makes high demands on labour inputs: fetching water at a distant tap, gathering and transporting fuel, buying food from a variety of traders, grinding grains, cooking on a charcoal fire, laundering at the public water-tap.

Domestic demands affect the kind of additional work women can take on.[31] Those who need to look after young children are frequently reduced to accepting industrial homework put out at extremely low rates (Benería and Roldán, 1987). In China, women disproportionately work in neighbourhood workshops rather than state enterprises (G. White, 1988). They usually enjoy neither the wages nor the fringe benefits that come with comparable work in state enterprises, but they can stay close to home. The earning opportunities of women are also restricted by cultural norms of their role in the public domain. Muslim women in Northern Nigeria trade from their houses, employing their children as intermediaries (Schildkrout, 1982; Callaway, 1987: 68–82).

Women are disproportionately found in the least remunerative and/or lowest status occupations.[32] Nici Nelson's (1988) detailed study of the economic activities of women in a squatter settlement in Nairobi shows women to be much more restricted than men in their choice of economic activity. Few of the local business establishments are run by women. Illegal beer-brewing and prostitution are the principal sources of income for women. They are at a disadvantage because they are less well educated than men, have fewer skills of commercial value, and support and care for children. The last point is demonstrated by the fact that a disproportionate number of successful women entrepreneurs are childless, that other women begin to expand and consolidate their business only in their late forties, when most or all of their children have grown up and perhaps contribute to household income. The handicaps experienced by women may be seen as structural constraints, but they are a function of the cultural context: the education and training thought appropriate for girls; the occupations considered suited for women; the emphasis on women as mothers who bear children, many children, and have the primary responsibility for raising them.

Jellinek's (1988) account of one woman's struggle to support herself and her dependants in Jakarta is unique, because it is based on regular contact over a period of fifteen years. Jellinek thus followed an unfolding chain of events, rather than reconstructing a distant past from what informants remember and are prepared to tell. Her account of the dramatic ups and downs of this woman's street sales of prepared food illustrates the precarious nature of even a flourishing trade, and the dramatic impact of economic change and administrative fiat—in the case of Jakarta the repressive measures taken against street-traders. The changing fortunes of commerce are related to changes in the composition of the household as well, affecting even the very membership of the family.

Women have come to constitute a large share of the greatly expanded

Table 4.2

Proportion of women among paid employees in manufacturing, 1980s[a]

Asia	(%)	Africa	(%)	Latin America and Caribbean	(%)
China (1989)	41[c]	Egypt (1984)	8	Bolivia (1989)	39
Hong Kong (1989)	49	Kenya (1988)	11	Brazil (1985)	26[c]
India (1988)	9	Mauritius (1989)	58	Chile (1989)	27
Malaysia[b] (1979)	45	Tanzania (1984)	10	Colombia (1989)	39
Philippines (1989)	41	Zimbabwe (1988)	7[d]	Costa Rica (1989)	35
Singapore (1989)	47			Cuba (1988)	34
South Korea (1989)	43			El Salvador (1983)	36
Sri Lanka (1986)	45			Haiti (1988)	43
Thailand (1986)	45			Jamaica (1987)	33
Turkey (1985)	13			Mexico (1986)	26
				Uruguay (1988)	37
				Venezuela (1984)	26

[a] Third World countries with more than 50,000 paid employees in manufacturing for which a breakdown by gender is published by the International Labour Office.
[b] Peninsular Malaysia.
[c] Including mining, quarrying, electricity, gas, and water.
[d] Including armed forces.

Source: Proportions calculated from data in International Labour Office (1988: 507–19; 1990: 467–78).

industrial labour-force in several rapidly industrializing countries (table 4.2). In East and Southeast Asia they make up close to half, in Mauritius more than half, the employees in manufacturing. In sharp contrast, their employment in manufacturing in Africa, India, and Turkey appears to be extremely low. In Latin America and the Caribbean, women hold an intermediate to high share. The rate of mobilization of female labour into industry has been fastest where the rate of growth of industrial employment has been the most rapid, i.e. in the countries that have dramatically increased their exports of manufactures: light industrial consumer goods, produced in factories using labour-intensive techniques and employing large numbers of women. As Joekes (1987: 81) puts it: Third World industrialization in the post-war period has been as much female-led as export-led.

Over the last decade a great deal of scholarship has been devoted to women working in export industries in the Third World.[33] The early generalizations about the characteristics of the female labour-force are giving way to differentiated accounts, and simplistic interpretations are replaced by sophisticated analyses. In Asia, where the bulk of the export industries are located, most of their women workers are young and unmarried. In Thailand and the Philippines, however, many more older married women enter or stay in the labour-force. A survey of semiconductor factories reports an average age of 24 years for women workers in Penang, Malaysia, 27.5 years in Singapore. In Penang, 24.5 per cent of the workers were married or had become separated, widowed, or divorced. In Singapore, such was the case of 46.5 per cent. In both locations about four-fifths of the married workers had children. In Latin America and the Caribbean, export industries employ more older women than in Asia, and more who are married or have been married. Explanations for such regional variations focus on the level of family income, cultural norms, and the availability of child care, as well as the market conditions faced by employers (Lim, 1990; Lin, 1987).

The characteristics of women workers vary by industry as well: the textile and garment industry in Asia has more older and married women than the electrical and electronics industry. And over time, the proportion of older married women workers has increased in all countries and industries in the region (Lim, 1990).

Women are less likely than men to take up membership in a trade union. The reasons are cultural as well as structural. That women should be modest, self-effacing, and deferential to their superiors to a greater extent than men is part of the definition of the female role in most if not all cultures. And many women are less concerned with their jobs than men because they see them as temporary. Lin (1987), however, suggests that women workers in Asia, rather than constituting a temporary work-force whose lives are manipulated at the whim of management, emerge as a new working class. Lim (1990) notes considerable variation in unionization across countries. In many countries female-intensive export industries are more heavily unionized than sections of the male labour-force. This is the case, for example, of the electronics industry in Singapore and the textile industry in South Korea. Indeed, women textile workers are considered among the most militant workers in South Korea's trade unions.

Multinational employers tend to match if not better local wages and working conditions. Wages, working conditions, and job security in the export factories are comparable if not superior to those found in women's, and even men's, jobs in most other sectors.[34] Wages earned by women in export factories are usually higher than what they could earn as wage labourers in alternative low-skilled female occupations, such as farm labour, domestic service, most informal-sector and other service-sector activities, small-scale local industry, and in some countries, even white-collar and 'pink-collar' jobs such as hairdresser, beautician, or sales clerk (Lim, 1990).

Ong (1990) found significant differences between American and Japanese firms operating in Malaysia in corporate ideology and in the impact of their policies on Malay notions of gender relations and sexuality. American companies encourage individualistic practices.

Their cosmetic shows and beauty contests promote Western notions of femininity. Japanese factories emphasize group co-operation and subordination to male authority:

> Japanese ideals of male domination and female obedience are produced and reproduced in the daily interactions between foremen and operators on the shop floor. The foreman-operator relationship, based on the male–female authority system in Japanese culture, is the mechanism by which women workers become infected with ideas of female inferiority and servility to men, and the process by which high production levels are attained. Because of low labor costs and consistently high production rates, the Malaysian subsidiaries of Japanese corporations are more profitable than parent companies. Nevertheless, Japanese managers feel that in order to compete successfully with American firms, they have to push continually for higher production targets for Malaysian workers. Again, the image of family claims is invoked: 'Parents do not say that they are satisfied with their children; every time parents hope for more from their children.' (Ong, 1990: 403)

Not all labour-intensive industries employ women, but all female-employing industries are labour-intensive. Clothing and electronics are the two industries that almost universally employ disproportionately large numbers of women. Even in China, where women are more widely employed throughout industry than in most other countries, women are nevertheless concentrated in food-processing, textiles, and, especially, the clothing and leather industries, where 59 per cent of the total labour-force in state-owned units was female in 1980, compared with 32 per cent of the industrial labour-force as a whole (Joekes, 1987: 83–4, 95).

Joekes (1987: 90–1) proposes a social theory of male cartelization. Men work in occupations where employers can pay higher wages for one or both of two reasons: in highly capital-intensive industries the wage bill is less important, and/or final product prices are not under pressure in uncompetitive markets. Third World countries exporting manufactures confront the very opposite conditions: they employ highly labour-intensive technologies—this is their advantage *vis-à-vis* the high-wage producers in developed countries, and they sell in highly competitive international markets. The availability of cheap female labour has thus been crucial to the rapid growth of manufactured exports from the newly industrializing countries.

Male cartelization is based on ideology and political power. The common notion that men are responsible for the material support of family dependants, the 'breadwinner' ethic, casts women in the role of 'secondary' workers, uninterested in promotional prospects in employment and satisfied to bring home a mere supplementary income. This ideology refuses to acknowledge two well-established facts. In many countries substantial numbers of households are headed by women, perhaps as many as half of urban households in Latin America and parts of Africa (Moser, 1989). And even where men as well as women contribute to household income, women typically contribute a larger part of their earnings to household expenses: Roldán provides a remarkable account of money allocation in the households of home-workers in Mexico City (Benería and Roldán, 1987: 113–22).[35]

Women are subject to serious discrimination in the labour-market. At its most obvious, women's earnings are less than men's in the same occupation and for the same job. In the Moroccan clothing industry, where women work side by side with men on the assembly line doing identical jobs, they are paid only about 70 per cent of the male wage, even though they are better qualified in terms of general education, are said to do better quality work than men, and—contrary to the stereotype—tend to stay longer in a given job (Joekes, 1982: 82).

Gender discrimination affects women in a less obvious fashion through job placement: women tend to be placed in jobs that involve repetitive, short-cycle, and relatively quickly learned tasks: the assembly line is becoming a female institution. Typically these jobs have no promotion lines leading on to more varied and rewarding work within the enterprise—work that would entail training to higher levels of complexity and that would be seen as more responsible and accordingly better paid. Furthermore, such promotional opportunities as do exist, for example to supervisory positions, tend to be monopolized by men even where women predominate in the labour-force. Joekes (1987: 87) concludes:

> All such apparently anomalous cases [where the real skill levels of 'female' jobs are clearly higher than those of more highly paid male jobs] can be explained by reference to the prior existence of different wage rates by sex. Differing male and female wage rates are the product of market forces in segmented labor-market conditions resting on discriminatory social gender relations, such as are reflected in the 'breadwinner' ethic. Discrepancies between skill levels and wages in women's jobs compared with men's can be explained as a consequence of the employers' being able to pay women the going female wage rate and *then* determining the nominal (as opposed to real) skill level of the job in conformity with that lower wage.

Gender discrimination affects the very organization of work and classification of jobs. Humphrey (1987: 157–9) describes the situation in an electrical plant in Brazil where 90 per cent of the women in production and quality control jobs were classified as 'production assistants', the lowest-paid job in the plant. They earned even less than unskilled workers in the canteens and kitchens, and their wages were close to the minimum payable in the industry. Some of the women did simple but exacting jobs requiring a degree of manual dexterity, good eyesight, and considerable concentration. Other women used sophisticated machinery in various stages of production. They were experienced, carried responsibility, and had to work with extreme care and attention. Inspection work was even more demanding. The best young female workers, chosen from the assembly lines, required between 4 and 6 months of intensive training and constant supervision in order to be able to spot the main defects that arose. The women working in inspection represented a considerable expenditure in terms of training for the company, and also an asset in terms of their experience and knowledge. However, in spite of this, and in spite of the need for high standards of care, cleanliness, attention to detail, and discipline, all the female workers in inspection were classified as unskilled 'production assistants'. Without exception, they were being paid the lowest wage rate in the plant.

The occupational structure would look very different, Humphrey suggests, were the plant to employ men. Indeed, prior to the construction of the new factory and the introduction of more advanced technology, both men and women had been employed on the basic production processes—and the relatively high turnover rate of men dissatisfied with low wages and poor promotion prospects had been a problem. Employing men, then, would entail recognizing that certain jobs require more training, responsibility, experience, and dexterity, and rewarding them more highly. Mobility chains would have to be constructed linking the less-skilled occupations to the more-skilled, and thus providing an incentive for the lower-paid and less-skilled workers to stay in the plant and try for promotion.

Humphrey stands dual labour-market theory on its head: women are not left at the bottom of job hierarchies because they lack stability in the job and are not, therefore, selected for training that would give them access to better jobs; rather, women's position at the bottom of job hierarchies is due to their greater stability in the absence of promotion, which encourages management not to formally recognize and reward their training, experience, and skills. He concludes:

> Women's disadvantaged position arises from two processes. On the one hand, women are marginalized from more productive and prestigious work and the training opportunities which lead to it. On the other, employment policies are constructed differently by employers for the male and female labour forces. Even when women have skills or require training, their occupations are often not classed as skilled, and their stability in low-paid employment works against them, since employers do not need to adopt the kinds of stability policies applied to men. Both these processes which work to women's disadvantage arise from the construction of gender identities in factories which devalue women's work. In this way the male and female labour forces are differentially rewarded. The commensuration of female and male labour, which would in

principle undermine this differential reward is prevented by job segregation by sex, rules which regulate and maintain sexual hierarchies, and the prevalence of notions of work suitable for men and for women. (Humphrey, 1987: 175)

In most Third World countries the industrial labour-force is small, and women are underrepresented in it. Most women work in services and their earnings are usually very low. One consequence, as we have seen in Chapter 3, is the separation of families as men migrate alone, leaving their wives in the rural economy. A second consequence arises to the extent that women are relegated to low productivity jobs, for example in many countries large numbers of women traders effect only very small sales, or are discouraged altogether from joining the labour-force.

If women were more effectively integrated into the urban economy, a smaller population would have to be accommodated in urban centres to perform the same economic tasks. Accordingly a lower investment would be required in key elements of infrastructure, such as housing and sewage disposal, which are considerably more expensive than their rural equivalents. Furthermore, there would be savings in the requirements for services such as the provision of fuel, the distribution of staple foods, and garbage disposal, which are also more costly in the urban setting. The wives and daughters of the predominantly male work-force require infrastructure and services, but they remain largely unemployed, underemployed, or misemployed (Boserup, 1970: 206–8). Gender discrimination thus imposes high costs on the collectivity. That substantial numbers of men are unemployed or underemployed as well does not invalidate the argument: continued preferential treatment for men in the urban labour-market boosts the urban growth rate by fostering the rural–urban migration of men.

The high costs of discouraging part of the urban population from realizing its full potential contribution to the economy are readily apparent to policy-makers in centrally planned economies. In China, urban female employment began to increase in the late 1950s, becoming almost universal by the late 1960s. By the 1980s the vast majority of women in their twenties and thirties worked outside the home: not in part-time jobs, as in many other societies, but in full-time jobs. According to the 1982 census, 84 per cent of the women in cities aged 16 to 54 were employed (Parish, 1987).

Child Labour

Not only are many women overburdened by a 'double' working day, many families have no choice but to put a heavy work-load even on quite young children (Leiserson, 1979). Child labour is common in the

Third World, but it continues to receive little scholarly attention.[36] Of course, most children work. Throughout the world, children are expected to make a contribution to domestic tasks. Many children, working alongside a parent, train for a future occupation. Such is the rule in agriculture. And it is a common pattern among artisans. In Malaysia, not only do many children contribute housework and home products as well as market work to their family's well-being, but—contrary to a common assumption—children from poor families neither participate more in productive activities, nor work longer hours when they do participate, than children from more well-to-do families (De Tray, 1983).

Child labour is a matter for serious concern in three respects. First, large numbers of children are trapped in highly exploitative and abusive employment relations such as domestic work and bonded labour. Second, many work in dangerous and hazardous activities such as brickmaking, construction, and mining. Third, many more work for excessively long hours and do not receive adequate nutrition, health-care, and education (Bequele and Boyden, 1988).

Kothari (1983: 1191–2) reports from the match and fireworks industry in India:

> 45,000 children work in these factories, most situated in and around Sivakasi . . . The ages of the children range from 3½ years to 15 years . . . The working conditions in the industries are unsafe and detrimental to the mental and physical health of the child[ren]. Staying a total of fifteen hours away from home, twelve at work, they work in cramped environments with hazardous chemicals and inadequate ventilation. Dust from the chemical powders and strong vapours in both the store room and the boiler room were obvious in practically every site we visited . . . As the piece-rate system is prevalent in Sivakasi, the children work feverishly to ensure maximum output.

Children working in underground mines have been reported in India and Colombia. In Morocco, children from the age of 8 work up to 72 hours a week in the carpet industry (Anti-Slavery Society, 1978). In Seoul, shoeshine boys are integrated into crime syndicates. They work twelve hours a day, seven days a week, every week of the year, stopping only if they are sick or if it is raining. They are part of a team that works together, eats together, and lives together. They associate only with one another. The boys' lives are totally controlled by the syndicate that houses and feeds them, provides them with the implements of their trade—and takes about 85 per cent of their earnings (Kang and Kang, 1978). Even from China there have been reports of children, predominantly girls, some only 11 years old, working in factories (All-China Federation of Women, 1987).

If women provide cheap labour, the earnings of children are extremely low as a rule.[37] This very fact makes them attractive to employers who otherwise might hire adults from the large pool of unemployed and underemployed usually available. Also disturbing is the fact that a large proportion of working children are misemployed: they beg, scavenge, steal, deal in drugs, and become prostitutes.

The pitiful earnings of children make a significant difference for poor families. For many a woman, left alone to raise her children, the contribution of the first-born is crucial to the very survival of her family. Other families, unable to support all their children, send some away to be more or less well fed and housed by relatives or strangers. Such 'child fostering' at times resembles slavery, and many an 'apprentice' is exploited and ill-treated.

Some families are so hard pressed that they abandon children altogether.[38] In Bogotá thousands of *gamines* survive by their wits: eight- to fourteen-year-old boys beg, steal, and rob, band together, and sleep in the streets, covered with cardboard and plastic sheets. If the phenomenon has been known in parts of Latin America for many years, it is now also appearing on a significant scale in Asia and Africa. Some estimates put the number of street children in the Third World at 30 million. Children working and children abandoned—in both cases the consequences for their health and education are often devastating.

5

The Housing of the Urban Poor

Housing is a highly visible dimension of poverty. Perhaps that is why it constitutes such an emotive issue in so many Third World cities. The sight of thousands, or millions, of people huddled in shabby accommodation with a minimum of servicing is certain to evoke some reaction from politicians and public alike. Why do such conditions persist and how can we explain the policies of governments towards low-income housing? These are the principal themes of this chapter. In order to understand these themes, however, we need to find answers to a number of additional questions. How successful are the poor in building their own homes, what are their main priorities, and how do they order those priorities? To what extent is the success of so-called self-help housing limited by the structural conditions of Third World cities? To what extent are there signs of improvement or deterioration in urban housing conditions? How have governments reacted to the housing issue and why have these policies been adopted? To what extent have governments acted in the interests of the poor and how far have their actions worsened urban conditions? By asking such questions, I hope to provide not merely a description of Third World urban housing conditions but also an explanation of why such conditions persist.

The Dimensions of Poverty

It is only too easy to demonstrate the effects of poverty on housing in most Third World cities. On any index of service provision, room density, or physical quality, a majority of the urban population is living at standards that are clearly unacceptable when compared to the way most Europeans or North Americans live. In urban China homes are so crowded that every person occupies an average area of only 4.8 square metres (Fujima, 1987). In Greater Bombay 77 per cent of households, with an average of 5.3 persons, live in one room (Misra, 1978: 375–6) and many others are forced to sleep on the pavements at night (Ramachandran, 1974). In Ghana room densities range from 2.5 to 3.2 in the cities of Takoradi, Kumasi, and Accra (Hinderink and Sterkenburg, 1975). In Nigeria the average density in Lagos dwellings is 4.1 persons per room (Ayeni, 1981).

In terms of service provision, the situation is equally alarming. In Jakarta only 32 per cent of homes had running water in 1985 (ISHOC, 1987). In Cape Coast, Ghana, 73 per cent of houses lacked water and 25 per cent electricity (Hinderink and Sterkenburg, 1975: 293).[1] In Brazil, in 1970, 47 per cent of Recife's houses lacked running water and in Greater São Paulo, 41 per cent (IBGE, 1970). In Calcutta 77 per cent of all families share lavatories with other families and more than 10 per cent have no facility at all (Lahiri, 1978). In the Mexican tourist resort of Acapulco, 45 per cent of homes lacked piped water in 1980 (Mexico, 1984).

I could continue to list figures of this kind but the level of neglect is already more than clear. Instead, I shall use the space to offer the warning that three points need to be borne in mind in evaluating data of this kind. The first is that services in Third World cities compare very favourably with those of the surrounding rural areas. In Malaysia, for example, 61 per cent of urban dwellings lack an inside flush toilet but the proportion in rural areas is 91 per cent; while 17 per cent of urban dwellings lack an electricity supply, in the rural areas it is 69 per cent (Wegelin, 1977: 61). In 1980, 80 per cent of urban Brazilians used piped water compared to only 5 per cent of their rural counterparts; in Chile the respective figures were 92 and 16 per cent, and in Colombia 78 and 19 per cent (Wilkie and Perkal, 1984: 283). While this comparison in no sense condones urban conditions it does warn against over-reacting to so-called urban squalor and romanticizing the rural life. If services in urban areas are thoroughly inadequate, those in most rural areas are usually much worse.

The second reservation is that most of the criteria by which we judge housing conditions in poor countries are highly subjective and ethnocentric. While the poor of India would no doubt welcome flush toilets in their homes, it is unlikely that they view the lack of such a facility in the same cataclysmic way as would a European or North American. Families unaccustomed to such 'luxuries' often view their real needs differently. For many Third World poor, 'our' standards are often irrelevant because they have more urgent needs. To a hungry family food is of far greater importance than shelter, especially where the climate is dry and warm. The importance of this reservation will become clearer when we discuss architectural norms and public housing provision. The needs of the poor, or at least the ordering of their priorities, are frequently misunderstood by professionals, let alone by those of us who have lived most of our lives in the comfort of a developed country.

Thirdly, judgements about housing conditions must also take into account different cultural, social, and environmental conditions within Third World cities. Those who compare conditions in large and small cities need to recognize that the manifestations of poverty differ. In large cities poor housing conditions are likely to be represented in the high proportions of people living in one room and paying high rents. Jobs and services may be available in the central areas of those cities, but space is at a premium. In small cities the problems may be just the reverse.

Regional differences are also important in international comparisons. In certain parts of Asia renting is very common. In Chinese cities, almost every household rents accommodation from the state or from state enterprises. In Korea, 59 per cent of Seoul households were renting in 1987, in India 76 per cent of Calcutta households were renting in 1981, 61 per cent of those in Bombay, and 68 per cent of those in Madras (ISHOC, 1987; India, 1989); elsewhere in the continent ownership levels are much higher. In many African cities most families rent accommodation (O'Connor, 1983; Peil and Sada, 1981); 62 per cent of those in Kumasi in 1986, and 52 per cent of those in Rabat in 1981 (Tipple and Willis, 1989; Keles and Kano, 1987). The reasons for such different tenure structures are not entirely clear. In most large Latin American cities many rent homes because they cannot yet afford to own, but in Africa the reasons behind high tenancy levels may have less to do with income. Certainly, Peil (1976, 1981) and Muench (1978) note that renters in West Africa and Kampala are not poorer than owners; tenants do not purchase homes in the city because they have every intention of returning to the countryside, or because they prefer to invest surplus funds in a business, to avoid the responsibility of home ownership, or to maximize their residential mobility (Muench, 1978). Similarly renting may not constitute as exploitative a situation as that which used to exist in many developed countries. Frequently, as in Lagos or in most Colombian cities, owners live in the same dwelling and let rooms at relatively low rents (Peil, 1981; Edwards, 1982; Gilbert, 1983; Marris, 1979; 426). Similarly standards of physical construction should be viewed with caution. It is no coincidence that most pictures of bad housing conditions are taken in hot climates. To most European eyes a bamboo house is inferior to one built of brick; wood inferior to cement. Had the story of the three little pigs been written by an African, it might well have ended differently. In a hot climate bamboo and wood are entirely adequate construction materials and it may be the poor of more temperate brick-built cities, such as Quito, La Paz, or Buenos Aires, who suffer the 'worst' housing condition. As Peil (1976) suggests, it is often far more pleasant to live out of doors in tropical countries.

Local factors must be considered carefully, therefore, when making comparisons between living conditions across the Third World. At the same time the positive correlation that exists between the level of national wealth and the quality of housing means that there are higher proportions of poor urban dwellers in Asia or Africa than in Latin

America. In addition it is probable that the African and Asian urban poor live in worse conditions than do their Latin American counterparts. Standards of servicing, for example, are superior in Latin America to those of Africa or Asia. In Latin American cities the poor will often wait several years for legal electricity or water supplies but in the meantime will have established illegal links to the mains; there are few urban settlements in Latin America without light, relatively few without water. By contrast, the poor of most Asian or African cities may never receive services. Such variations are in large part an outcome of differences in the levels of national and urban prosperity. Cities such as São Paulo or Buenos Aires, which concentrate large proportions of the total wealth of countries which by African standards are immensely rich, will naturally possess better conditions than those of African capitals. While these superior services and resources may be disproportionately concentrated in middle- and upper-income housing areas, even the Latin American poor benefit to some degree from the higher level of national and urban resources. Clearly there are other sources of variation between cities. Societal organization is a vital consideration; in terms of service provision, if not in terms of living space, the poor of Havana live better than those of other Caribbean islands (Acosta and Hardoy, 1972). But, as many socialist leaders would be the first to admit, there are limits to what can be done with reduced financial resources.

Rationality among the Poor

The differences in housing conditions in different Third World cities are a function of differing levels of per capita income, the distribution of wealth, the rate of urban growth, and the form of societal organization. But they also reflect differences in the responses of the poor in each city. Such responses vary dramatically according to the poor's own expectations of their life chances, their own view, reasonable or untenable, of what kind of housing they want and the degree to which they are organized to improve their housing situation. It is difficult for the poor to escape their poverty given the economic and social situation in most Third World countries. But, within the limits we shall discuss, the poor's response to that poverty is rational, innovative, and nearly always more perceptive than often they are given credit for.

Perhaps no one played a greater role in drawing our attention to the rationality of the poor with respect to their housing situation than Abrams (1964), Mangin (1967), and Turner (1967, 1969). They demonstrated that the shanty, which was so often, and of course sometimes rightly, denigrated as the ultimate in penurious living conditions, was frequently the basis of an adequate shelter. Rather than merely being a shack without services, it was the foundation upon which the more fortunate, better off, or more innovative sought a way out of their poverty. Over time, spontaneous housing tended to improve as inhabitants built outside walls, extra rooms, a solid roof, and sometimes a second floor. *In favourable circumstances* the poor could produce substantial, spacious, and reasonably serviced homes.

These writers did more, however, than merely show that over time many poor families are able to consolidate their housing. They also demonstrated that the reaction of the poor to poverty was rational and that families recognized the most sensible ways of improving their living conditions. Such an argument was diametrically opposed to the conventional wisdom of the day which owed much to Oscar Lewis's (1966) concept of a 'culture of poverty' (see pages 169–75). At its crudest, this view encouraged the idea that the poor are poor because they are poor. Poor children eat badly, receive a poor education, and receive from their families and cultural peers a training that encourages them to accept their poverty as inevitable. As Portes (1972: 269) notes, the concept came to 'denote a situation in which people are trapped in a social environment characterized by apathy, fatalism, lack of aspirations, exclusive concern with immediate gratifications and frequent endorsement of delinquent behaviour'.

The 'culture of poverty' view of the poor persists to this day among many higher income groups. It persists, perhaps, because it is a highly convenient explanation to the wealthy; by implication poverty is the poor's own fault. In this sense it serves as 'a vehicle for interpreting the social reality in a form which serves the social interests of those in power' (Perlman, 1976: 247). But, convenient though it may be, it has little basis in reality. The poor respond sensibly and rationally to the choices and opportunities open to them in their housing situation. And while the poor undoubtedly contribute at times to their own poverty, the basic causes of that poverty are beyond their control. The poor are not a separate sub-society but act much like everyone else. In Perlman's (1976: 234) words: 'In short, they have the aspirations of the bourgeoisie, the perseverance of pioneers, and the values of patriots. What they do not have is an opportunity to fulfil their aspirations.'

Rejection of myths such as the 'culture of poverty' is gradually leading to important changes in housing policy. If the poor are considered to be incapable of helping themselves, then they have to be helped. In a housing context this tends to mean that only governments are capable of building satisfactory housing for the poor. By contrast, the major policy recommendation in Turner's work is that governments are best advised to help the poor to help themselves. Such a recommendation has the additional virtue that self-help housing often produces superior shelter to that produced by governments, if only because the poor understand more clearly the role that housing plays in their lives. Architects, by contrast, are too often concerned with their own self-image and by their frequently erroneous views of what the poor really want. Turner has consistently argued that architects believe too strongly in the idea that good housing is an end in itself. Shelter should not be judged only in terms of whether it has a good roof and adequate drainage or would satisfy the board of examiners of an architectural faculty. While no one doubts that in an ideal world most houses would be well designed and serviced, in conditions of poverty another criterion is more important. That criterion is whether housing suits the needs of particular poor families. Turner (1976) demonstrates the choices facing two Mexican families, comparing the situation faced by a family living in an architecturally satisfactory house with that of a family living in seemingly more squalid physical conditions. In architectural terms the former mason's house is superior since it is both modern and supplied with services. Unfortunately, as a result of moving from his previous residence in the shanty town, the elderly mason's unreliable main income is no longer supplemented by the earnings of a small shop serving tourists. The family's income, in fact, is quite inadequate to cover the costs of the modern house which they have long coveted. Around 55 per cent of the family's total income is now devoted to maintaining the house and paying for the services compared to only 5 per cent previously. Given the present circumstances of the family, the 'quality' of the house is an irrelevance; it constitutes oppressive, not good, housing.

By contrast, the 'supportive shack' occupied by the ragpicker, while offering nothing in the way of good architecture or services, does match the family income. In fact it provides just the support needed by the family at a time when the man's employment as a car sprayer became unprofitable. Since the shack is in the back garden of a godparent whose house provides the services used by the family, they do not suffer the inconveniences of many shanty dwellers. And because the family is young and healthy, there is a good chance that they will obtain superior housing in the future.

The point of Turner's comparison is not to justify bad housing but to demonstrate the futility of poor people living in shelter of high architectural standards when it does not match their needs and incomes. Good housing should not be designed on the basis of assumptions about what the poor's needs ought to be, but should provide the flexibility by which the poor can trade off one need against another. In an earlier paper (1972) he suggested that all families have three basic needs; security, identity, and opportunity. Each income group within a city tends to make a different trade-off between these three needs (see Figure 5.1). In the context of housing, the poor value proximity to unskilled jobs (opportunity) much more highly than either ownership (security) or high-quality standards of shelter (identity). By contrast, a middle-income family gives much higher priority to modern standards of shelter and freehold ownership than to proximity. The policy implications of this argument are that the poor's needs are usually badly understood by governments and the kinds of housing provided by them are therefore inadequate. Much better, in Turner's view, is to give the poor greater flexibility to design and construct their own housing. It is not a recommendation for every family to build its own home, but for some kind of intermediate position between the somewhat autocratic dictates of large-scale enterprises, especially government, and the anarchy of letting every family build its own house (Turner, 1976). Give individual families greater choice over their housing design and location and the match with their needs will be closer.

The major criticism of this view is not that the poor, or for that matter the rich, are incapable of deciding their own best interests, but that in most circumstances the choices they make are tightly constrained. For example, we may argue that families make a choice between renting accommodation that is close to work (the classic 'bridgeheader' location) and building their own home in a more peripheral location (the classic 'consolidator' location). No doubt those families who earn an adequate household income can decide wisely which of those two alternatives they prefer. But, of course, many tenant families do not have the choice. Their only possibility is to settle wherever they can obtain cheap rented housing (Gilbert, 1983). For the old, infirm, and very poor, the trade-off

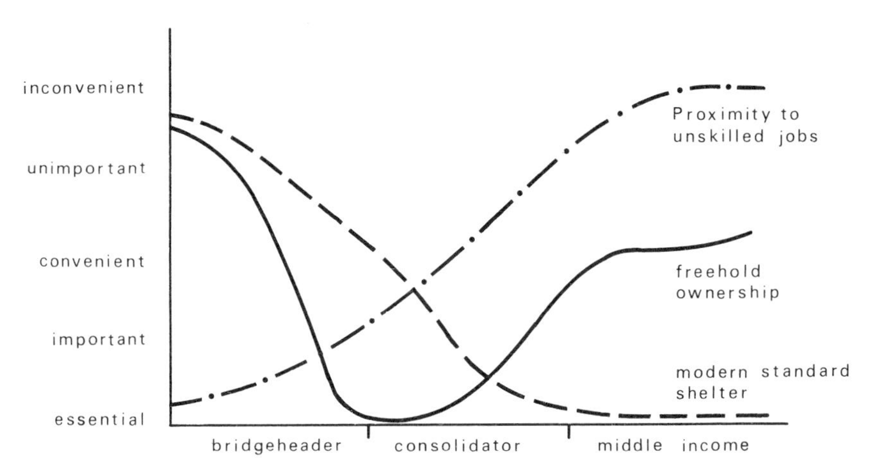

FIG. 5.1 The housing priorities of middle-income groups, consolidators, and bridge-headers
Source: J. F. C. Turner (1972)

is non-existent because it is determined by income level and the nature of the housing market in their city.

The same argument can be extended to cross-cultural comparisons. As we have seen, the balance between rental, self-help, and conventional housing varies greatly between cities, often within the same country, but notably between continents. To some extent the fact that more West African families rent homes in urban areas is a matter of choice (O'Connor, 1983). But, clearly, the availability or otherwise of adequate rental accommodation, the ease with which families can occupy land, the cost of building materials, the size of city, and therefore the transport problems involved in getting to work, are all factors determined by the organization of society rather than by the families themselves. The fact, therefore, that the poor in Lima invade land, those of Bogotá buy land, and those of Lagos rent accommodation, is only partly due to different household preferences. Much more important are the structural conditions that limit the choice.

What I am essentially arguing is that poor individuals are rational and make trade-offs which improve their welfare level. At the same time, their poverty and the conditions facing them in most Third World cities effectively limit their choices. Of course all our choices are limited by something, but in conditions such as those found in Calcutta or Kinshasa the choice is so limited as to be effectively compulsion. In such circumstances the poor do not choose location, but are pushed into any available accommodation. The only alternative in the long term is if the range of choice is enlarged by rising income levels or by changes in the housing or land markets; an argument to which I return.

The Nature of Self-Help Housing

The archetypal shanty hut on a hill is not the only, and indeed is often not the worst, form of housing in many Third World cities. At the same time it is the principal form of poor housing in the sense that most owners, squatters, and renters live in housing that began as some kind of self-help construction. But such an argument begs an important question: what is the critical element defining this form of low-income housing? Unfortunately, there seems to be no simple answer. Drakakis-Smith (1981; 1976*b*: 297) argues that 'the most acceptable definitions rest on the illegality of occupation of land, house or both'. Similarly Leeds (1969: 44) suggests that the 'only uniform identifying characteristics are their illegal and unordered origins by accretitive or organized invasion and, because of their origin, their continued juridically ambiguous status as settlements'. Such definitions are satisfactory in cases where land has been taken against the wishes of the owners, they include uneasily the frequent case of the poor purchasing the land on which they construct their dwellings. Here the right of 'ownership' is unquestioned although a legal title is rarely granted. This is because such land often lacks planning permission from the urban authorities because of its inadequate services, physical layout, ownership characteristics, or its location beyond the urban perimeter. Such settlements are well represented in Delhi (Bose, 1973), constitute the majority of low-income areas in Bogotá (Losada and Gómez, 1976), where they are known as pirate urbanizations, and cover substantial areas of Mexico City, where they are known as clandestine sub-divisions (Ward, 1976*a*). If Doebele (1975) is correct in his belief that more families in poor cities buy land than invade it, any definition of self-help housing must include this kind of semi-legal ownership. Similarly, any definition should include the common practice in parts of Africa and Melanesia to obtain permission from local officials or tribal chiefs to obtain access to communal land (Peil, 1976; Doebele, 1978; Baross, 1983).

In addition, it is important to remember that it is not only the poor who occupy land illegally. In San Salvador, for example, while few low-income settlements have been approved by the urban-planning agency, most settlements share their illegal situation with Colonia Escalón, the most luxurious of the upper-class residential suburbs (White, 1975).

Another common definition of self-help housing is that it has been constructed by the occupiers themselves. And although most people would agree that self-help is a vital ingredient in an adequate definition, few among the poor build their houses alone. Normally, they contract help from skilled neighbours who can lay the foundations, complete the more difficult brickwork, install the plumbing, or design the electrical system. Even Turner (1969: 525) notes that in Lima the poor may often have built only one-quarter of the dwelling themselves; the rest having been contracted out. While this estimate would seem high for Bogotá's low-income settlements, the point is sound.[2] Self-help settlements contain an important and sometimes a large component of skilled, even professional, labour (Moser, 1982).

Nor can we define self-help settlement in terms of tenure, because the inhabitants often include as many tenants as they do owners (Amis, 1983; Gilbert, 1983). In addition, while spontaneous settlers normally fall into the lower middle-income group in cities such as Bogotá (Valenzuela and Vernez, 1974), they may constitute a more middle-income group in the poorer cities of Africa or Asia.

I do not wish to make too much of this definitional problem. Nevertheless, it is essential to remember that different definitions reflect different philosophical approaches to the housing issue and that self-help housing takes a myriad forms. Generalization in such a situation is dangerous. This diversity also means that any of the pet terms to describe such housing, whether it be shanties, irregular settlements, squatter areas, or spontaneous dwellings, are often misleading. The squatter settlement, for example, is a misleading term to describe houses built on purchased land; the term 'shanty' is inappropriate for brick-built consolidated housing which might once have been a shanty. Spontaneous settlement is misleading in the sense that many such urban developments have been highly organized by their leaders to avoid eviction. Not only are many invasions planned carefully but they are often supported, and even initiated, by political groups, including those of the government. Perhaps the least objectionable adjectives for such settlements are 'spontaneous', 'irregular' and 'self-help'. 'Spontaneous settlement' raises something of the sense of innovation that the poor bring to their individual housing problems and also acts as a reminder that such housing, even if now solidly built, often began on the fringe of the law, sometimes after an invasion, and was usually built in some measure by the inhabitants themselves, when resources became available. 'Irregular' housing emphasizes the individualistic and transitional nature of the accommodation and how it often lies outside the formal legal framework. 'Self-help' housing is accurate in so far as few houses have been designed by architects and built ready for occupation by the new owners; the owners have usually designed or built much of the dwelling themselves. Nevertheless, I would not defend the use of any of these terms at all vehemently.

In general however, self-help settlements fall normally into two or more of the following categories: (1) most of the dwellings were built by the families which originally occupied or now occupy them; (2) the settlement as originally founded suffered from some degree of illegality or lacked planning permission; (3) when the settlement was first formed most forms of infrastructure and services were lacking and in many settlements services are still lacking; (4) the settlements are occupied by the poor, however defined. Obviously this definition is neither tight nor concise but it does eliminate most of the 'conventional' forms of settlement that exist in Third World cities. Within the category of self-help housing may be included as subtypes: *invasions* (of either public or private land, whether organized or incremental), where no purchase of the lot is involved; *pirate settlements*, where the land is purchased, but lacks planning permission; *rental settlements*, where houses are built on rented land; and *usufruct settlements*, where permission to use communal land has been granted by tribe, local government, or private owner (Baross, 1983). I am aware that there are many further variations and in certain cases such subtypes overlap.

The Consolidation of Self-Help Housing: Prospects and Barriers

In cities in which the poor are not threatened by eviction there is plentiful evidence to show that they are capable of improving their housing conditions. Many settlements which begin as unserviced collections of huts gradually achieve the status of ordinary suburbs of the city. Little by little the huts are transformed into solid dwelling units, the community and friendly politicians put pressure on the authorities until electricity, water, buses, drainage, schools, and health centres are provided.

For the populations of these consolidating settlements progress is slow but consistent. When money is available, they invest in improvements to their housing. When times are hard, they are not evicted because they do not pay rent. In times of inflation the price of food and transport may rise, but at least their investments in the house and the land are safe. To the successful consolidator the dwelling also offers a source of income either through renting or through turning the front room into a shop. *At their best*, spontaneous settlements offer a great deal of necessary flexibility to the poor. A large family which needs space can extend the structure at will, the small family which prefers high physical standards to space can achieve this goal. And where the community is not riven by political conflicts (see Chapter 7) the whole settlement benefits as a result of sustained pressure on the service agencies.

The question that needs to be raised, however, is whether the process of consolidation is possible for the majority or merely for a minority. What proportion of the Third World poor in fact make the transition, from bridgeheader to consolidator to middle-income dweller? To what extent do the proportions vary from city to city? Is consolidation mainly a phenomenon of the more prosperous Latin American city? Who actually are the consolidators? In short, what are the conditions under which successful consolidation can occur?

Turner (1967, 1969) emphasizes the importance of security of tenure in the consolidation process. Without a high level of confidence that they will be permitted to retain the land, no family will willingly invest time and money in consolidating their dwelling. But who or what ensures security? Where the poor buy land, the security generally follows, even where planning permission or a legal title is lacking (Varley, 1985). Where the poor rent land on which to build, as in parts of the West Indies (Clarke, 1974) or South Africa (Ellis *et al.*, 1977: 7), security depends more on the length of lease and upon being able to maintain rental payments. Where, as in many shanty towns of Montego Bay, there is no security, houses are designed so that they can be moved easily to another location (Eyre, 1972: 406). Where land has been obtained through invasion, security is still more problematic. Effectively it is the attitude of government authorities, together with the amount of political pressure that can be maintained, which establishes security. If the government changes, or its political power weakens, the settlement may be threatened. Security under such conditions is especially problematic where the right of tenure has little relationship with the letter of the law. In Caracas, for example, Pérez and Nikken (1982: 209) note that the law relating to building on other people's land has more relevance to nineteenth-century France than to Venezuela. Squatters are generally permitted to stay because 'action by the police against the squatters does not often commend itself and there are generally more reasons for political authority to side-step such activity than to undertake it.' A change of government, of course, may change the policy and with it tenure rights. Security is often as much a state of mind as a reality. And where, as so often occurs, the government decides to introduce an urban-renewal programme there may be little that can be done by the squatters to prevent it (see below).

Another critical factor determining the ability of the poor to construct and to consolidate spontaneous settlements is the availability of land. Clearly the nature of the land market and government policy are the key factors here. If land invasions are tolerated, it would appear reasonable that more families might gain land than where invasions are prohibited. In the latter case the availability of land depends upon the price of suitable land relative to the incomes of the poor. In some African and Melanesian societies communal land is made available by tribal chiefs or local governments so that the poor only face the problem of construction (Angel *et al.*, 1983). In other exceptional cases, such as the Republic of South Africa, access to land is controlled according to racial type. A black African wishing to work in Cape Town is compelled to live apart from his family in a workers' hostel or run a high risk of eviction from the few available squatter areas. He is effectively prohibited from obtaining land which will permit him to consolidate.

As cities become larger, and unbuilt land becomes scarcer, more and more families are compelled to buy land (Payne, 1982). Once dependent on the market, the availability of lots is determined by price. Unfortunately the rates at which land prices increase in Third World cities are generally high. Evers (1975: 782) notes that in Asian cities 'population increase and other social processes have intensified the pressure on urban land and have led, in the 1970s, to a wave of land speculation and spiralling land prices.' In an earlier period the effects of the petroleum economy in Venezuela led to the total land value of the central 5.4 million square metres of Caracas increasing four times from 1938 to 1951 (Lander and Funes, 1965: 322). Today land prices in the centres of large

Asian and Latin American cities are close to those in Central London and New York. While Walters (1978) is almost certainly correct that there is an upward bias in most estimates of land-price rises, there is no doubt that the profitability and low level of risk connected with land transactions, compared with other investment opportunities in Third World countries, attract large amounts of capital and push up prices. Land speculation is completely uncontrolled in most Third World cities and spectacular profits are possible for the shrewd investor. As Cornelius (1975: 34) reports, the dry Texcoco lake-bed in Mexico City 'was bought up in the 1940s for a few centavos per square hectare by a handful of entrepreneurs—including some army generals—who were later to become millionaires from their miniscule investment.' In a statistical analysis of the effects of land prices Grimes (1976) shows that costs are ensuring that public housing is built on the cheaper edges of cities and that many of the poor are being priced out of the land market. What makes matters worse is that government servicing and zoning policies frequently have the effect of valorizing land and increasing prices still further (Kowarick and Brant, 1978; Gilbert and Ward, 1978). And despite the efforts of some governments to hold down these rises, their interventions have more often accentuated land shortages and fuelled further increases (Payne, 1977; Soussan, 1982).

Price increases are also having a major impact on the ability to consolidate once land has been obtained. Costs of building materials are rising rapidly in many countries either because of monopoly practices or because of failures to increase production. The signs are that in Brazil, Mexico, and Colombia the costs of cement, glass, bricks, and steel have risen much more rapidly than the salaries of most low-income workers. Clearly such real price increases are bound to slow the consolidation process. In addition, as urbanization proceeds more and more materials are provided through the market (Burgess, 1978). There are fewer opportunities to cut down trees on the edge of the city or to make bricks from local clay deposits. As cities develop, building materials are provided by companies for sale. The process of self-help construction is thereby commercialized.

The prices of construction materials and land relative to wages and salaries are critical issues in determining the rate of spontaneous settlement growth and consolidation. As Ward (1978: 47–8) has shown for three settlements in Mexico City, 'once certain tenurial assurances are met, residential improvement at the household level is a product of the investment surplus that is created.' This suggests that consolidation can only take place among those groups who have reasonably well-paid jobs. This point has indeed been recognized by Turner (1969: 526), who argues that the 'vital differences between the semi-employed "bridge-header" and the more or less regular wage-earning "consolidator" ' are frequently overlooked. This failure to distinguish between the poor and the very poor leads to misleading generalizations about the poor and about the potential for spontaneous settlements to grow. Turner here, and elsewhere, clearly believes that the opportunities for consolidation are limited for the *very* poor. Squatters are an 'auto-selected community' who differ from poor non-squatters and who are able socially and economically to cope with the problems of spontaneous settlement (Turner, 1969: 513). The worrying implication of the argument is that within every city there will be a sizeable group of people who are unable to participate in the consolidation process. In particularly poor countries this group of non-consolidators may constitute a majority of the urban population. Such is the case in many cities of Africa and Asia (Dwyer, 1975: 203–4). The poor of Calcutta, Addis Ababa, or Jakarta may have little money with which to buy land or building materials. It may also be the case that the prospects of the majority becoming self-help consolidators are declining. This has clearly been the experience in most parts of Africa and Latin America as a result of the post-1980 economic crisis. In some Latin American cities, real incomes declined dramatically as a result of the combined effects of government policy, the world recession, and the debt crisis. In Lima, wages of private-sector manual workers fell by one quarter during 1988 and a further half in 1989; in Mexican cities the minimum wage fell by one half between 1980 and 1989 (UNECLAC, 1989). Added to this severe cut in incomes has been a dramatic rise in unemployment. Rates of unemployment during the middle 1980s had seldom been higher in most Latin American cities. As real incomes decline, the prospects for consolidation also deteriorate.

It may also be that the very process of urban growth is reducing the poor's chances of consolidating their housing. I have already noted that rises in land prices are being fuelled by urban expansion. But city growth may also accentuate other difficulties for the poor. For example, as cities grow physically larger, the journey to work will get longer. Already some workers in cities such as Lagos and São Paulo are travelling for three to four hours per day between centrally located work opportunities and peripheral homes (Ayeni, 1981; Kowarick and Campanario, 1986). Such journeys not only cut family budgets but also limit the time available for home consolidation. Of course such an outcome is not inevitable. Conscientious and effective planning could create conditions favouring employment decentralization, which would provide jobs close to homes. But such adjectives rarely describe the planning system in most Third World cities (see Chapter 8). Pessimistically, therefore, I must predict that the time and cost involved in travelling to work will pose an increasing burden on the metropolitan poor with an obvious effect on the pace of consolidation.

The changing physical size and shape of the city will also affect the desirability of certain residential areas. Such changes affect the poor almost as much as the rich, and may lead to deterioration in previously improving settlements. As Brett (1974: 186) puts it,

> many of the earlier consolidators may move out, renting or sub-letting their homes to poorer newcomers squeezed out of the city centre, or to provincial migrants fresh to the city. Where this happens, the selective out-migration of 'high achievers' will tend to lead to a downturn in the investment cycle, so that the peripheral settlements of today may become the slums of tomorrow.

If the poor increase in numbers and the better off start to move out, this may cause problems for the local environment. For the presence of higher-income consolidators would seem essential to the continued improvement of a spontaneous settlement. Such families contribute to the welfare of the rest of the settlement by creating a market for local stores, by providing casual employment, and by adding a more powerful voice in the petitioning for services (Doebele and Peattie, 1976). But in turn social differentiation within a spontaneous settlement raises a critical issue. While the richer families obviously contribute to the economy of the *barrio*, the poor may be an important source of income for the rich. For it is an undeniable fact that as settlements become older and consolidate, the proportions of renters increase; owners deliberately extend their houses to accommodate renters, thereby increasing their incomes. In Bogotá, where the author carried out household surveys in four self-help settlements, the proportion of renters in each *barrio* tended to rise with the age and the levels of servicing of the settlement (Table 5.1). A similar process of 'densification' of low-income settlements has been observed widely throughout Latin American cities (Riofrio, 1978; Kowarick and Campanario, 1986; Coulomb, 1985).

The implications of high levels of renting in self-help settlements are unclear. In Latin American cities many current owners were once renters, so that the presence of renters may be interpreted positively: they too will one day become owners. This interpretation would seem especially plausible in those cities in which the proportion of renters to owners is falling. On the other hand, where the proportion of renters is increasing it is probable that renters are finding more difficulty in obtaining land and building houses. In this case, rather than renting being a temporary stage prior to home ownership, it may be a permanent state. Such a trend may also signify that consolidation has become more difficult for those with land; in such circumstances the only way for owners to consolidate is to sub-let rooms. Unfortunately, few studies have considered the phenomenon of renting until recently and very few have studied the phenomenon over time (Amis, 1983; Edwards, 1981; Gilbert, 1983, 1987; Gilbert and Varley, 1991; Coulomb,

Table 5.1
Age, renting, and servicing of Bogotá's barrios

	Casablanca	Atenas	Britalia	San Antonio
Age of settlement (years)[a]	9.5	12.1	2.9	9.2
Percentage of households who are renters	42	43	28	43
Services and utilities score[b]	12	15	6	11

[a] On 1 Jan. 1979. Defined as the average date when the owners who were interviewed arrived in the settlement excluding those who bought from a third party.
[b] Points are awarded to a settlement according to its level of servicing: e.g. 2 points for settlements where virtually all households have a legal electricity supply, 2 points where both public and private telephones are installed, 1 point for a daily police patrol, 4 points for a settlement where three-quarters or more of the roads are paved. The maximum possible score for any settlement is 20 points.

Source: Gilbert and Ward (1985: 24).

1985; Malpezzi and Mayo, 1987; Tipple, 1988). Consequently, we have insufficient information to judge whether renting levels are increasing or whether the poor's chances to obtain property is increasing or decreasing.

Finally, the rate of settlement consolidation depends also upon the extent to which public agencies are able to provide infrastructure and services to the spontaneous communities. In some cities, notably metropolitan Latin America, electricity, water, and telephone companies are sometimes highly effective. On the other hand, in 1966 the homes of 1.7 million people in the Calcutta conurbation lacked water and the per capita supply of filtered water had declined by half between 1931 and 1965 (Dwyer, 1975: 215). To a considerable extent, service availability is linked to levels of per capita city income, but a number of other variables are important. Water supplies are critically affected by the environmental and topographical characteristics of the city; Mexico City and Caracas both face major shortages, whereas many cities such as Calcutta, Bangkok, or Jakarta, are almost awash with undrinkable water. Similarly, electricity services are dependent on available fuel supplies; cities located in countries lacking both coal and oil, and distant from rain-soaked mountains, face graver difficulties than others. But external features such as income levels, topography, and fuel supplies are not the only important determinants. The level of servicing for the poor depends also upon the efficiency and the allocation procedures of public utilities. In Mexico City and Valencia the allocation of services to the poor depends upon political patronage; where these links are effective, services are provided (Ward, 1986; Gilbert, 1981*b*). Universally, service variations are linked to income levels. In São Paulo services are available in middle- and upper-income *bairros* which can afford to pay the installation charges. In the poorer *bairros* community action is often required to install the services given the level of utility companies' charges. Indeed in some cities the poor even subsidize the services of the rich. In Cape Town 'the overcrowded, extensively used (and thus profitable) black [bus] services directly subsidise the white services which are sparsely used and which run at a loss' (Dewar, 1976: 16). In Port Moresby water, sewerage, and garbage collection was provided only to the rich who were long heavily subsidized owing to the low prices charged (Oram, 1976).

Are Conditions Improving or Worsening?

Having considered the parameters that limit the process of consolidation, it is interesting to ask whether general housing conditions in Third World cities are improving or deteriorating. To most readers this may seem a senseless question, for the balance of academic and planning opinion is clearly on the side of deterioration. A glance at the literature supports the belief that in most cities the situation is worsening. Abrams (1964: 7) argues that 'more than a million people . . . are homeless or live in housing that is described by the United Nations as a menace to health and an affront to human dignity. Worse still, in almost all the developing areas, housing conditions are steadily deteriorating.' With reference to Calcutta, generally recognized to be one of the world's housing disaster areas, and to the Indian situation generally, Rosser (1972: 180) notes that 'the housing shortage in urban areas has now reached staggering proportions, and grows annually worse as the rate of provision of new and satisfactory housing lags far behind the rate of population increase.' In China urban population growth generally has exceeded the rate of new home construction and in the fastest growing cities the housing shortage has undoubtedly worsened. In Sian the population 'increased more than threefold from 490,000 to 1.5 million during the period from 1950 to 1965, while the floor space of dwelling units only doubled' (Sen Dou Chang, 1968: 150). Even Turner (1967: 3) has argued that 'with squatter settlement growth rates of 12 per cent or more per annum in Mexico, Turkey, and the Philippines as well as in Peru and many other countries—double that of city growth as a whole—it is hardly exaggerating to say that city development is out of control.'

A common element in these arguments is that because larger absolute numbers and relative proportions of Third World urban populations are living in self-help settlements, general conditions must have worsened. Put as simply as this, the argument is unsatisfactory. If services such as water, electricity, and drainage are gradually extended to spontaneous settlements, if rooms are added and even a second floor constructed, then many dwellings will have more in common with conventional houses than with the flimsy shacks normally associated with the term 'spontaneous housing'. In such circumstances, therefore, a proportion, and possibly a considerable proportion, of the housing ought not to be classified with the poorest dwellings. To confuse self-help housing with slums and to assume that the quantitative increase in the former proves that a deterioration in housing standards has taken place, negates the useful work done by Abrams, Mangin, and Turner. If spontaneous settlement does improve through time, then perhaps different conclusions should be drawn. In any case, before the proliferation of spontaneous housing may be interpreted either favourably or unfavourably it is necessary to consider in detail the changing conditions of those dwellings. Are more dwellings provided with services than previously? How have changing land-use patterns affected the locations of squatter housing? Are the real costs of building these dwellings higher or lower than previously? Without such supporting evidence the proliferation of self-help settlements proves little beyond the fact that Third World cities are expanding rapidly and fail to provide conventional housing for all their inhabitants.

Even if it can be demonstrated that housing conditions have deteriorated, or for that matter improved, then accurate interpretation is still difficult. If, for instance, higher proportions of urban dwellers are living in bad housing conditions than previously, this may only be a symptom of the rapid movement of people from rural to urban areas. Since, as we have seen, poor quality self-help housing is virtually the only form of dwelling to be found in rural areas, the fact that it should spread in the urban areas over a particular time period may well reflect the movement of poor people to cities and the continuance of poverty in the society as a whole. Similarly a clear improvement in housing conditions needs to be interpreted carefully. It is common for housing improvements to have taken place in one or several cities at the expense of conditions elsewhere in the country. If, for example, governments have channelled investment resources into the major cities and neglected the smaller towns and rural areas, then housing improvements in the major cities are placed in a less favourable light.

Far too often, however, superficial arguments are used to support the idea that there is an urban crisis. For example, Abrams (1964: 51) argues that

> the less industrialized the country, the less apt it is to have a housing problem. The moment it begins to develop industrially, its housing problem burgeons . . . The moment the family moves from village to city, its members surrender the home that is usually their own, as well as the more ample space on which it

stands, the freedom from noise, smoke, traffic and danger, proximity to nature, and their place in community life.

One wonders whether Abrams had forgotten how bad rural conditions can be. For, despite the appalling conditions found in most Third World cities, more people have access to water, drainage, electricity, and health services than they do in rural areas. It is clear that drainage and fresh water are more necessary in crowded urban circumstances than in rural conditions, but in most rural areas few services are available. Since the quality of the construction is not dissimilar between urban and rural areas, this suggests that most urban families live rather better.

I am not trying to argue that there is no urban housing problem or to deny that in many cities, particularly during the recession of the 1980s, it may well have worsened. My aim is to demonstrate that most statements about the situation are based erroneously on at best one or two criteria. Because the priorities of the poor may be different from ours we need to take care. And even if we have objectively weighed the evidence and can state categorically that urban housing conditions have deteriorated, we still need to interpret the importance and the causes of this trend. In short, care in interpretation is the watchword. The tragedy is that despite such care, we can so often diagnose a decline in housing standards.

Government Responses

Numerous similarities exist in the ways in which governments have reacted to the housing situation in Third World cities. For many years they frequently demolished self-help settlements, seeking to build conventional housing units rather than assist the self-help process. They also produced planning regulations inappropriate to the housing conditions of the poor. Fortunately, such policies have become less common, most governments are now more tolerant of self-help housing. However, different governments have modified their policies at different rates. This is not surprising in so far as some governments had to face the rapid expansion of self-help settlement much earlier than did others; what was happening in Buenos Aires, Rio, or Mexico City in the 1940s was often of little concern to Lusaka, Lagos, or Dar es Salaam until the 1960s. Generalization about housing policy is also complicated by the fact that within the same country different attitudes have been manifest at different levels of the government hierarchy. Not infrequently the national government has maintained one policy, often one of benign neglect, while the real issues and policy decisions have been apparent at the local level, often in the form of slum demolition and residential segregation.

Despite these reservations, it is possible to make some generalizations about housing policy. At the national level the typical governmental response for many years was to ignore the country's housing situation and to restrict the growth of low-income housing in the main cities by limiting cityward migration. In parts of the British Empire population movement was controlled carefully through Pass Laws, a device intended to keep as much indigenous housing in the rural areas as possible (Peil, 1976; Tipple, 1976). Where native urban labour was required it was allocated employer-built rental housing (as in Zambia), or segregated into distinct residential areas (as in India) or forced into workers' hostels (as in the Republic of South Africa). While this attitude eased over time as the labour demands of industry and mining increased, the neglect component remained paramount in colonial policy for many years. Of course it was never seen in those terms. Rather it was rationalized in terms of racial differences in customs and needs or, more humanely, if equally unrealistically, in the belief that if economic growth and good administration were encouraged, the housing problem would eventually disappear. The later view persisted beyond independence; until recently in fact it was the learned opinion of most economists working in developing countries (Abrams, 1964). Only in the 1970s did an institution like the World Bank become actively interested in social infrastructure such as housing. Previously investment in housing was seen to be unproductive and most funds had been channelled into the growth-generating industrial, power, transport, and agricultural sectors. Most national plans, produced as they were by economists, failed even to mention housing until the 1960s. Only since doubts have been expressed about the ability of economic growth to remove poverty has the need for something to be done about the slum conditions of so many people been recognized. Certainly the realization that investment in housing could actually create income or that it could play a vital part in the economy of the poor has been very recent. Indeed, apart from one or two heretics (such as Currie, 1971), investment in housing was regarded as the classic means of slowing economic growth and of adding to the problems of the urban areas by attracting larger numbers of people from the rural areas.

Residential segregation

The local counterpart of national and international neglect was, and in modified form remains, a policy of deliberate segregation. In India the British planned their cities carefully; Delhi was 'built for two different worlds, the "European" and the "native"; for the ruler and for the ones who were ruled' (King, 1976; 263). Perhaps the word 'built' should be qualified, for only the European sections were planned and constructed by the colonial powers. Planning consisted of showing how towns should be organized, in building comfortable residential accommodation, wide roads, and open spaces for those who 'knew' how to use these facilities. The indigenous areas were left to look after themselves, probably on the assumption that nothing could be done to help them. The result in most cases was that pre-existing or new indigenous housing areas were allowed to develop in certain areas and new European centres constructed adjacent to them (Figure 5.2). Often the areas allocated to the native population were too restricted and led, under conditions of rapid in-migration, to very high housing densities. In Zaria the European residential area was clearly separated, grouped around a clubhouse with a race-track, a polo field, and a golf course. In Tunis there was an almost self-sufficient city outside the original indigenous centre, though in Cairo, while all Europeans lived together in a colonial quarter, they never constituted more than half of the population in that area (Abu Lughod, 1976: 35–6). In some cities the demarcation according to housing standard and race was carried to extraordinary lengths. In Singapore Sir Thomas Raffles delineated separate quarters for Europeans, Chinese, Indians, Malays, and Buginese; in Batavia, the capital of the Netherlands East Indies, there were at least fifteen distinct racial-tribal districts (Wee, 1972: 217; Evers, 1975: 780). After independence complete racial segregation began to disappear as indigenous groups began to move into the previously European districts. Ethnic segregation gradually diminished in cities such as Dakar, Abidjan, Harare, Rabat, Casablanca, and Tunis, and suddenly in Algiers (Abu-Lughod, 1980; Findlay and Paddison, 1986; Rule, 1989). Only in South Africa did complete racial segregation remain, although even there it began to break down after 1986 (Rule, 1989; Simon, 1989).

But if segregation on the basis of race and ethnicity has declined, separation of another kind has persisted and grown. In many African and Asian cities, divisions on the basis of race have given way to those based on class (Evers, 1977). As Abu-Lughod (1980: 253) puts it 'Moroccan cities never lost their basic patterns of spatial stratification, although the criterion of access changed from "ethnicity/caste" to a simpler "ability and desire" to pay.' Urban authorities no longer zone their cities according to race, but implicitly according to income and housing density. African and Asian cities are moving closer to the pattern, long apparent in Latin America, whereby income determines where people can live.

In case this form of segregation should fail to operate effectively, élite groups are additionally protected by zoning laws. As a consequence, beautifully planned élite *barrios*, the equal and better of the colonial townships, have emerged in all Latin American cities and are clearly

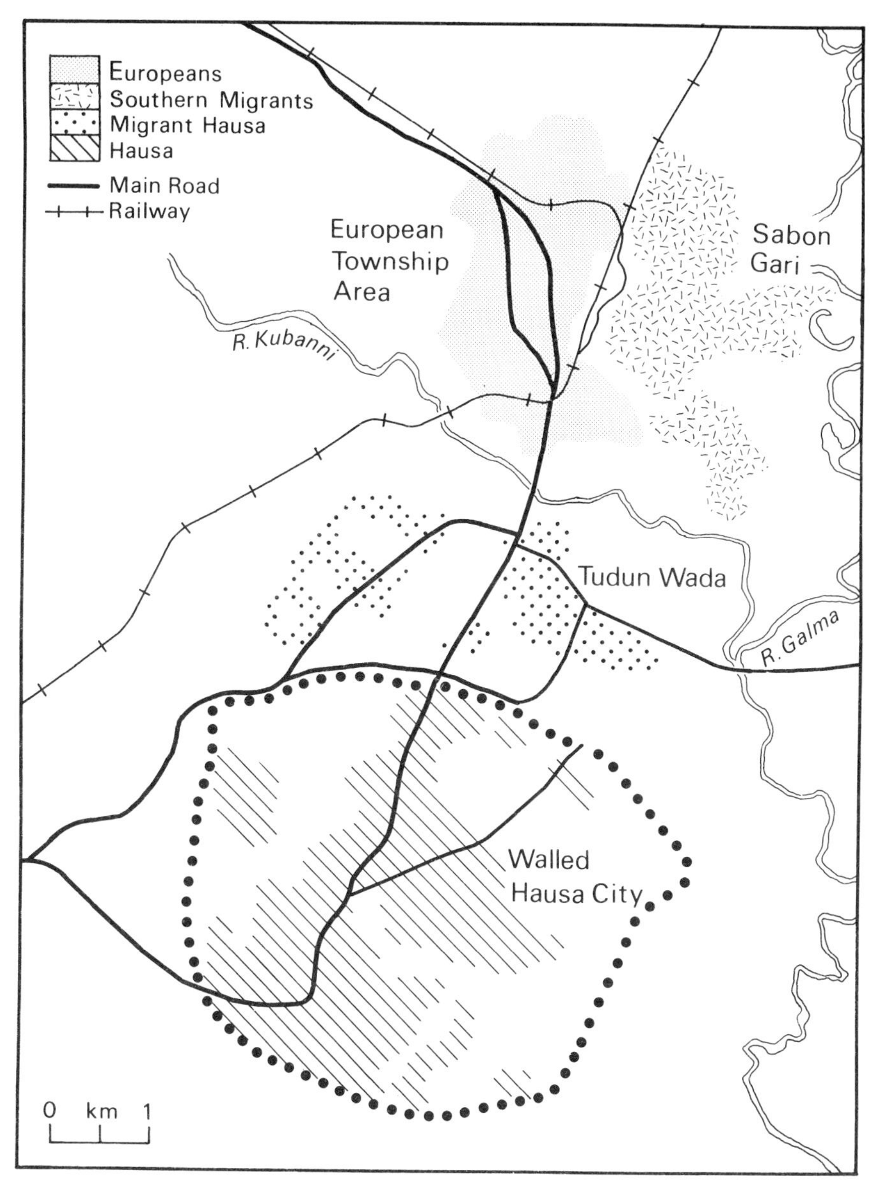

FIG. 5.2 Residential segregation in Zaria
Source: Schwerdtfeger (1972)

separated from the spontaneous settlements. A case can be made, in fact, that planning has become popular in Latin America mainly as a method of protecting élite *barrios* from the incursions of squatter settlers and the like (Amato, 1970; Violich, 1944). Today zoning, income, and government housing projects maintain residential segregation on the basis of class, not only in Latin America but throughout the non-socialist world.

Urban renewal and 'slum' removal policies

Normally linked to the policy of segregation has been one of demolishing undesirable housing. Indeed, the destruction of low-income housing is as old as the policy of separating populations according to their ethnic origins. In Singapore, for example, there are recorded cases of demolition occurring as early as 1840 (Wee, 1972: 220). Sometimes demolition was to discourage migration into the city, but more often it has been part of an urban renewal project or used to maintain the zoning laws. Unfortunately, urban renewal and demolition programmes have never been very effective in the sense of helping the displaced population. Too often the new housing has been unsuitable owing to its distance from work places, the regularity with which rents or mortgage payments have to be paid, and because of the limited house space provided. The most frequent reaction among the removed populations has been gradually to move back to areas similar to those from which they were removed. In the massive *favela* removal programme in Rio de Janeiro, large numbers of *favelados* sold their homes to higher income families or 'fiddled' the system in other ways (Valladares, 1978*b*: 18–23). In Nairobi most of the former squatters required to move to new sites in Kariobangi sold their lots to more prosperous applicants who moved in instead (Weisner, 1976). In Manila, of the 5,975 squatter families moved 35 kilometres to Sapang Palay in 1960, only 41 per cent remained in 1969 (Hollnsteiner, 1974: 313).

Such removal programmes have come under considerable attack in recent years as it has become obvious that the removed population seldom relish their new accommodation and that the reasons for the removal are often motivated less by interest in the conditions of the poor than in clearing land for prestige buildings or for speculative profit.[3] Frequently, indeed, demolishing housing is the worst of all possible strategies. As Abrams (1964: 126) long ago pointed out, 'in a housing famine there is nothing that slum clearance can accomplish that cannot be done more efficiently by an earthquake. The worst aspects of slum life are overcrowding and excessive shelter cost. Demolition without replacement intensifies overcrowding and increases shelter cost.' Unfortunately it seems that vested interests and/or ignorance on the part of planners have prevented this common sense from being heeded in many cities. In Nairobi the main squatter areas outside the Mathare Valley and Kibera were demolished in 1970 and President Kenyatta later defended the local council by arguing that he did not want Kenya's capital to turn into a shanty town (Stren, 1975: 272–3). In Cape Town blacks living in the now infamous squatter settlement of Crossroads have been regularly threatened with removal to the Transkei, 1,000 kilometres away, 'where they are legally entitled to be'.[4] Even the coloured population, the non-white racial group permitted to remain in Cape Town, is subject to urban-removal programmes (Smith, 1982). In its wisdom the government is in the process of building two new cities for coloureds at Mitchells Plain (27 kilometres by road to the south of the centre), and Atlantis (45 km north). While by Third World standards the accommodation being built is both cheap and of reasonable quality the fact remains that the new cities have few commercial or industrial employment opportunities, the proposed rail links to Cape Town are nowhere near completion, and it is less than evident that the poor actually want to move (Ellis *et al.*, 1977; Dewar and Ellis, 1979).

Government housing

The higher costs of transportation, rents, and services, and the disruption of informal social networks which previously had helped to sustain the low-income economy, have often been the main results of these removal plans. But many of the difficulties are rooted in the very notion of governments building housing for the poor in Third World cities. The universal error has been that governments have built at too high an architectural standard for the poor and without a clear understanding of the needs of the recipient population. The result has been that government housing has generally been too expensive, has offered very little flexibility in use, and has often been in unsuitable locations. Such public housing as has been built in San Salvador (Grimes, 1976), Bogotá (Laun, 1976), Rio de Janeiro (Portes, 1979; Valladares, 1978*b*), São Paulo (Batley, 1983), Mexico City (Cornelius, 1975), Lagos (Aradeon, 1979), Lusaka (Sanyal, 1981), Nairobi (Temple and Temple, 1980), Kuala Lumpur (Wegelin, 1977), Manila (Hollnsteiner, 1974), Tunis and Rabat (Findlay and Paddison, 1986), has mainly been occupied by middle-income people. This does not mean that the poor have not benefited at all, for often they have sold their rights to the new families. But building for middle-income families is not what many agencies believed, or at least pretended, that they were doing. On occasion the poor have been forced into government housing against their will and have suffered from conditions of 'oppressive housing', to use Turner's apt phrase. The *rancho* dwellers of Caracas forced by dictator Pérez Jiménez's national guard into the 'super blocks' are a classic example of the unpredictable consequences of such a policy. Not only did they treat the new accommodation badly and fail to pay their rents, but they also turned into a potent opposition force which helped to remove the dictator (Myers, 1978).

Indeed, a question mark must be placed against the desirability of governments building houses at all in most Third World cities. By necessity, governments are building houses which cannot be occupied by the poor because of their extraordinarily low incomes. Most commonly the accommodation is occupied by the upper- and lower-middle-income groups; in many countries it is government workers or those with appropriate political links who benefit from the accommodation. For these groups government housing is undoubtedly popular and waiting lists are long. Of course there may be advantages for the poor as a result of housing construction for higher-income groups. The total housing supply will increase and vacated middle-income accommodation may become available for the poor, thus easing housing pressure. But equally common is the situation whereby ownership of the vacated accommodation is retained by the original occupants and let at exorbitant rents. Perhaps the greatest danger in government housing projects is that they are provided with a substantial subsidy which benefits groups who are less in need. As Van Huyck (1968: 71) notes, in Calcutta most funds for subsidized housing come through indirect taxes from the poor. In short, the poorest families are called upon to pay the burden of rehousing the fortunate few who live in the public housing projects. This, perhaps, is the key issue for government housing agencies. If they build housing for the poor, it can be provided only with the aid of large subsidies. Since the countries are poor, few subsidies can be provided and in any case regressive taxation systems force the poor to pay for the housing. If the standard of housing is reduced, governments end up building one-room concrete shacks or apartments which all too often the poor could have improved on at much lower cost.

In certain circumstances this judgement may be too harsh. Clearly, reactions to living in multi-storey accommodation will depend upon previous housing experiences. It would seem that in Singapore, for example, the degree of overcrowding in the central areas was so great that the Chinese community adapted without major problems to apartment living. On the other hand, Malay housewives in the same city, who were more accustomed to village-style housing adapted less successfully; one comment on concrete floors was that it was like 'trying to make a home on the main road' (Wee, 1972: 227–8).

In general, however, the experience of governments building houses in Third World cities has been of dubious value. Most such housing has been either of high quality which has benefited few families or of such low quality that it has alienated the recipient families (Drakakis-Smith, 1981). With the exceptions of Singapore, Hong Kong, and Saudi Arabia, nowhere has a government building programme succeeded in stemming the tide of spontaneous settlement and improving housing conditions.

Some economists would argue that the principal reasons for failure are that too few government houses are built or that governments have failed to stimulate the private sector to build more (Currie, 1971). Nowhere is this better demonstrated than in socialist countries, such as China and Cuba, where the government has built little housing, the formal construction sector has been concerned mainly with building factories, offices, hospitals and schools, and self-help housing in urban areas has been actively discouraged. The result has been a severe housing shortage (Harms, 1982; Mathey, 1989; Fujima, 1987; Dwyer, 1987). Increasingly, economists are beginning to argue that expanding housing construction gives a major boost to employment opportunities, since construction is on the whole a labour-intensive industry. Variations on such programmes have been tried in Singapore, Hong Kong, Brazil, Colombia, and Mexico. In Colombia housing construction was the central plank of the National Plan between 1970 and 1974 and again between 1982 and 1986. Funds were generated through index-linked savings plans and the resultant construction boom undoubtedly helped the employment situation in Bogotá where most of the funds were invested. Whether in fact the plan helped the poor, since construction was concentrated on upper- and middle-income housing; or whether it stimulated the rapid inflation that ensued are more open questions (Robledo, 1985).

Where such a strategy has been employed, it has sometimes been associated with, but has not necessarily been the cause of, rapid economic growth. In four of the five countries in which some variation of this policy has been applied, the national economy expanded rapidly. On the other hand there is considerable danger of stimulating a speculative housing boom which further concentrates incomes, unless complementary urban taxation reforms accompany the introduction of the new strategy. In addition it is less than certain that where the accommodation is constructed for the poor it is suitable to their needs or, in places such as Brazil and Colombia, whether it benefits the poor at all. Clearly, more detailed evidence is required on this kind of economic strategy.

Rent controls

Another approach to improving conditions for the poor has been the institution of rent controls. Unfortunately the general experience seems to have been less than successful in the long term. In Mexico City 'rent controls, initially applied in 1942 to *vecindad* housing with a rent below 300 pesos per month and maintained ever since, have hastened the deterioration of the central city tenements, whose owners have little incentive to invest in their upkeep' (Cornelius, 1975: 27). While there can be little doubt that rent control discourages investment in rental housing, it is clear that rent controls are by no means the only factor discouraging it. Work in Latin America suggests that landlords ceased to invest in tenements as soon as other investment opportunities became available (Coulomb, 1985; Perló, 1979). The introduction of rent controls was a subsequent reaction, powerful economic interests had already moved their capital from this kind of housing to invest in suburban development or increasingly into the stock market, foreign capital markets or in currency speculation.

Once introduced rent controls reduce the profitability of rental investment but also protect, admittedly in poor housing conditions, the very poor. The beneficial effects of this legislation for some are apparent where rent controls are removed. This is clear from recent experience in Argentina and Uruguay (Cuenya, 1986); in Montevideo, decontrol in 1974 led to a major wave of evictions which increased the waiting list for public housing (Benton, 1987: 43). Generalizing on the basis of several urban experiences, Grimes (1976: 98) has argued that rent controls can only be effective in the short term and in combination with strict controls on prices and incomes. Employed in other circumstances they have either cut the rate of new construction and maintenance or have encouraged the development of illegal rationing systems such as the payment of 'key money'. This verdict has been re-emphasized by recent research undertaken by the World Bank (Malpezzi, 1989). Unfortunately, rents and landlord/tenant relations are such politically sensitive issues that it is difficult to introduce a satisfactory compromise which will both moderate rent rises and encourage investment in the rental housing stock (Malpezzi, 1989). Frequently, governments have chosen to maintain existing rent controls in central areas while ignoring the rental problem in the rest of the city.

Sites-and-services and upgrading programmes

In one sense there has been progress in government responses to the housing situation. During the seventies a consensus emerged that most capitalist Third World governments are incapable of building sufficient homes to remove spontaneous housing and that greater reliance must be placed on some kind of self-help policy. Major institutions such as the World Bank, government planning agencies, and many architects began to accept the advice of John Turner and others that more should be done to provide and service land and to leave the actual building to the people themselves (Linn, 1983). Two general policies can be classified under this general description, the one to upgrade existing settlements, the second to ease the development of new settlements (sites-and-services). The two policies obviously go hand in hand, promising to channel more resources directly to the poor and to give them greater security on the tenure of their land. At its best, a sites-and-services programme would offer a family a plot of land, gradual servicing, access to credit, advice on construction and materials, all at a price not beyond the family budget. The security afforded to the family would allow consolidation to take place and community action programmes to be introduced. Government agencies could supply more people at the same cost if people no longer occupied land on hillsides and river-valley bottoms which is expensive to service.

Self-help programmes have been taken up vigorously by numerous governments and, for a period, by the World Bank; Laquian (1977: 291) notes that in 1974 there were 'eighty proposed or completed schemes in 27 countries'. Similarly Grimes (1976: 20) reports that 'as of 1973 sites and services projects were a part of the national development plans for 13 countries. By the same year, Turkey, Chile, India, Pakistan, and Iraq had each completed more than 50,000 sites and services plots.' Site-and-service and squatter-upgrading programmes undoubtedly arrived on the international scene in the 1970s, and as Doebele and Peattie (1976: 9) note, represent 'one of the most important reforms in the housing policies of developing countries in the last decade'. At the same time, such a policy is not without many dangers: not least the risk that it may be turned into a universal 'answer' to the housing problem.

Certainly, as recent literature is at pains to point out, self-help is no panacea. That same literature records a catalogue of failures to put against the obvious common sense which underlies its adoption. Stren (1975: 270) describes how in Nairobi the government managed to spend less than one-fifth of its site-and-service project budget between 1969 and 1972, and how in Tanzania the government managed to build only 795 site-and-service units between 1969 and 1974, compared with the annual target of 5,000. Similarly slow progress was a characteristic of Papua New Guinea (Oram, 1976) and Bogotá (Gilbert, 1981*b*); and in Kuala Lumpur, where nine schemes were tried between 1965 and 1971, Wegelin (1977: 91) reports that 'the experience . . . has not been a very happy one.'

Most of these failures have been attributed to poor administration, due in part to lack of enthusiasm on the part of the local authorities. As the Chief Planner of Nairobi was reported to have once said: 'The Kenya government is still committed to the ideal that every urban family must have two rooms and a toilet and a kitchen and not everyone who matters is converted to a lowering of standards' (*Guardian*, 15 Dec. 1977). This reluctance to lower standards has been fairly general and has kept costs above those necessary to cater for most of the poor. Oram blames several failures in Port Moresby on over-strict building standards demanded of site owners. Stren (1975: 281) reports of one Nairobi site-and-service project: 'concrete slabs were provided for foundations. The cost of the slabs plus high standards of infrastructure put the project beyond the reach of most lower income families. The government was therefore obliged to heavily subsidize the project in order that low-income families could be settled in the plots and repay loans for building materials.' Rather than being accused of building, or condoning the building, of slums, government administrators have often prepared over-elaborate and hence over-costly programmes. In a sense, of course, this criticism opens sites-and-services projects up to the very problem that self-help schemes are trying to avoid. If costs rise, the only alternative is to provide a few families with large loans or subsidies or to spread funds thinly over a large number and gamble that the programme will still be viable.

In any case there is some doubt whether the very poor wish for higher standards if it means that they are to be burdened by debt repayments. Commenting on Zambian experience, Tipple (1976: 168) argues that a squatter resettlement scheme was unsuccessful because 'it was based on a false assumption that squatters would want to move to a serviced area and pay for the services provided rather than living in an unserviced area for no apparent cost.' Clearly responses will vary according to income level and to aspirations among the poor; very poor people are much more likely to settle for no services. It is likely, however, that more affluent spontaneous settlers would accept a serviced plot and related debt payments. Unfortunately, even where this is the case, bad administration all too often makes them wait too long for their lots (Oram, 1976) and services (Hollnsteiner, 1974: 315) or reduces the number of lots available.

Some degree of poor administration can be put down to teething troubles and inexperience. But in the case of self-help programmes this is a dangerous excuse. For one of the supposed beauties of such a policy is that it should be relatively free of bureaucracy. If in fact it demands extensive administration, then costs will rise and one of the main virtues of the programme is lost. Clearly, if administration cannot be kept to a minimum, the programmes will not work and planners must search for a new answer to the needs of the poor. Equally, if Third World governments are too incompetent to administer such conceptually simple programmes, the wisdom of sites-and-services and squatter-upgrading programmes must again be questioned.

In addition, self-help solutions have to answer two more profound criticisms. The first is that self-help is a cover for non-action by governments on critical issues such as urban reform, progressive taxation, and land speculation. As Wilsher and Righter (1975: 138) have expressed it,

> the doctrine of self-help is deeply attractive. It appeals to everyone's belief in human ability, neighbourliness, ambition and good sense . . . [however] it also, less nobly, encourages people to believe that there is nothing much to worry about, that the less interference that there is with natural forces the better, and that everything will work itself out in the long run.

A related, but still more sinister, accusation is that of Drakakis-Smith (1976*a*: 2): 'By satisfying some of the minimum housing desires of the urban poor, self-help projects can be said to maintain the status quo and as such have been eagerly adopted by urban power élites as multi-lateral funds have become available over the last few years.' Such a charge has been made specifically against the Chilean government of Eduardo Frei (1964–70). The introduction of the famous Operation Site project led to the development of paternalistic government policies aimed at winning votes and were a substitute for its intended attacks on the structural causes of the housing problem (Kusnetzoff, 1975: 292).

The second major criticism of self-help is that if it becomes widespread it will re-create the difficulties which it aims to conquer. Doebele and Peattie (1976: 6), for example, argue that sites-and-services schemes are likely to cream off the more affluent and innovative poor. Such creaming would leave the poor settlements without leaders who might have pressed government for more services and help. It would also lower internal demand for services and commercial activities in the poor settlements, 'breaking the ties with the neighbourhood store, the local food vendor, the seamstress, the firewood distributor, and all the dozens of other more marginal persons who depended on them for their own economic survival'. The same criticism is valid with respect to the effects of self-help on the land market, jobs, and city growth. Laquian (1977: 297) has criticized *early* site-and-service efforts because they were 'located in urban peripheries, requiring relocation of inner-city squatters and slum dwellers, resulting in economic, social and personal dislocations'. But this problem is inherent in any system that operates under an uncontrolled land market. Given the structure of land prices in Third World cities, only peripheral areas are available for low-cost programmes. And if self-help programmes should proliferate, the subsequent incorporation of a huge number of families into the conventional land market would further push up prices, relegating government site-and-service schemes to the cheapest most peripheral land. The major consequence of self-help schemes, without modifications in the land market, would merely shift the burden of finding cheap land from squatters or pirate urbanizers to the state. Without severe land taxes or direct government intervention in the land market, insurmountable problems would paralyse the self-help programme.

Similarly, site-and-service programmes would contribute only partially to helping the employment situation. Self-help would create more opportunities within low-income communities for skilled and semi-skilled labour. But without large-scale programmes aimed at creating well-paid jobs, too few people would be able to participate in even the cheapest self-help programme. Some observers would even argue that the incorporation of spontaneous settlements into the official sector through self-help schemes would open up the door to greater exploitation by the capitalist sector (Burgess, 1985). Such schemes might make it profitable for large-scale construction companies to provide prefabricated housing, thereby threatening still further job opportunities in activities such as bricklaying. It might also make the supply of building materials more profitable for large companies and undercut local small-scale manufacturers who now provide employment for many of the workers in the spontaneous settlements. Clearly, the validity of such an argument varies from city to city and opinion will differ according to ideological position. Nevertheless a fundamental issue is raised, which in my opinion is sound. To generalize self-help without reforming some of the basic inequalities of the Third World city will undermine the programme.

To be successful, therefore, sites-and-services and squatter-upgrading programmes need to be accompanied by structural reforms of the land market, taxation and zoning, and urban-planning policies. Without such reforms self-help programmes may help the less destitute but in no sense will the majority benefit, even in the more prosperous Latin American countries. Of course the more intelligent advocates of self-help recognize the limitations of these programmes in the absence of complementary reforms and they are clearly correct in their belief that self-help is more likely to help the poor in the absence of reforms than most other housing policies would. Sites-and-services and squatter-upgrading are no more than partial palliatives to the problems of the poor, and such programmes increase rather than obviate the need for reform. In fact, there seems little danger of such programmes becoming widely established. As a result of their experience even the World Bank has begun to withdraw from its former advocacy of this system. Evaluation of its first programmes was generally positive; although many of the less poor benefited from individual programmes, 'the great majority of beneficiaries is in the bottom half of the income distribution' (Keare and Parriss, 1982: viii). However, the major problem, and the one that would eventually reduce the Bank's commitment to this kind of programme, was the failure to recover costs in three out of the four programmes studied. Without cost recovery, replicability was endangered, especially when interest rates began to rise.

The Logic of Governmental Response

I have described how many governments have responded to the issue of poor housing. What we now seek to answer is why those policies have been adopted and why different governments have reacted in different ways to similar kinds of problems. In order to answer this question we need to know more about the state in Third World societies. Clearly, the form of the state varies with the organization of society; dictatorships often represent only a small élite group which manipulates the state in the cause of its interests; in true democracies the state may genuinely represent the masses and introduce policies which aim to improve housing and social conditions. Most countries fall somewhere between these extremes; the state is under pressure to represent a wide range of social groups, but tends to represent certain groups much more than others. In many capitalist countries the state represents primarily the interests of upper- and middle-income groups. But even dictatorships find it necessary to offer some benefits to the poorer sectors of society in order to gain a measure of legitimacy for their regime.

The role of housing and urban planning agencies in this context normally is to legitimate the policies of the state before the middle- and low-income groups. The extent to which such agencies are allowed to bring about real improvements for the poor will depend on the nature and the ideology of the state. In those countries in which serious taxation of the rich is absent, where land speculation is rampant and practised by government officials themselves, we can expect little success on the part of the housing and planning agencies. On the other hand, where resources are channelled into social infrastructure, where some effort is made to control the rise in land prices, and where serious attempts have been made to channel the growth of major cities in directions that will economize on the cost of services, something more than superficial improvements may take place. In any case the housing issue is always as much a political as a technical matter. For government agencies only receive large sums to engage in the housing battle when a favourable political decision has been taken. Similarly, politicians rather than technicians determine whether land speculation is being controlled, prices of building materials allowed to increase faster than average incomes, and infrastructure provided for the poor or for the rich.

Politicians, though, are always limited in their actions by social and economic realities. However Utopian the intentions of governments, a limit is posed by the resources of the nation. This is clearly demonstrated by the housing situation in urban China (Dwyer, 1987; Fujima, 1987). In the largest 192 cities, an inability or unwillingness to invest heavily in housing construction before 1979 led to serious urban overcrowding. As Badcock (1986: 149) notes 'the amount of floor space per person actually fell from a peak of 4.5 square metres in 1952 to 3.6 square metres in 1978'. While Communist China managed to redistribute a meagre national income, it is difficult to redistribute to the masses what it not being produced. Housing policy in Cuba has had no more success (Harms, 1982; Gutierrez *et al.*, 1984; Brundenius, 1981; Mesa-Lago, 1974). In addition, many Third World governments have been severely limited by their lack of technological and managerial competence. While I shall emphasize the political dimension in the subsequent discussions, these limits on action need to be remembered.

Having made these general points, I wish now to consider the major reasons for governments having adopted specific urban policies. Why were slum removal projects so popular for so long? Why has the price of government housing so consistently precluded access by the very poor? Why were there so few sites-and-services schemes during the 1960s and so many during the 1970s? Are universal explanations possible or do government policies vary too much to fit into convenient boxes?

Let me begin by considering why self-help housing is often demolished. A common argument in the literature is that planners are hostile to the idea of the poor constructing their own dwellings; such housing is badly designed, uses poor materials, and is difficult for governments to service. While this attitude is understandable, it is usually irrelevant to Third World urban conditions and is a product of the education received by architects and planners. Since most professionals are trained abroad or in schools whose curricula are based on those used in colleges in Europe or the United States, they imbibe the conventional wisdom of those societies. In ex-British colonies the influence of British architectural and planning schools continues to be strong (Dewar, 1976; Dwyer, 1975; Aradeon, 1978), and Rosser (1972) complains of the 'mental barrier' against the 'brute realities' of the Indian city which this influence creates. The typical reaction of architects is to recommend high-quality governmental construction. Many believe that standards will be lowered if the poor are encouraged or permitted to build their own housing. But, as Turner (1972: 148) correctly points out, 'the standards the objectors have in mind . . . are not something which can be achieved with available resources but, rather represent the objectors' own notion of what housing ought to be.' This notion is obviously an imported and Utopian idea of little relevance to reality. Planners see only the bad elements of self-help housing without seeing the good. Paper plans are seen to be the reality and the solutions of the people themselves and their problems become, in Grennels's (1972: 97) term, 'invisible'.

> This blindness is the result of a genuine desire to improve the living conditions of as many people as possible; a fixed idea of what constitutes 'good' housing; a recognition of severe limits on public and private commercial sector resources to attain these goals; an emphasis on standardization of design and production efficiency; and a consequent discounting of the role of the dweller in the provision of housing. The latter is based on assumptions that public participation is inefficient and time consuming, that people 'don't know what they want', or simply that trained technicians 'know better' about laymen's needs than they do.

There are many examples of the 'image' of the self-help settlement being so bad in the minds of professionals and politicians that it has led to instant demolition. Aradeon (1978: 1) recalls 'the military governor of Lagos state who, on arrival from an Australian visit, announced that his government was going to clear the slums of traditional Lagos and rebuild it with modern apartments and shopping centres'. Such views are both inconsistent and superficial and lead to inappropriate policy responses.

While such misguided attitudes have an important influence in the formulation of inappropriate responses, it would be unwise to overstress them. When, as happened in the Ford Foundation's International Urbanization Survey, most reliance is placed on greater and improved professional training, it diverts attention from solid political reasons for squatter areas or poor neighbourhoods being obliterated. In many countries demolition serves an important role in national political strategy and often serves to support particular social groups. In British Northern Rhodesia, for example, demolition maintained an important colonial policy: 'The urban areas were the milieu of the white population and legislation was enacted to keep it that way' (Tipple, 1976: 167). Such a policy is only in the process of being dismantled in South Africa (Wilson, 1972; Ellis *et al.*, 1977). For many years the state insisted that accommodation be built for the migrant labourer; spontaneous housing was prohibited in the cities although permitted in the rural areas. The policy against spontaneous housing in the cities was clearly not intended to help the poor black so much as to control the influx and to maintain white living standards. The transparency of South African housing policy is useful not only to warn us of the evils of apartheid but also to point to the way that governments manipulate housing policies.

The traditional Anglo-Saxon explanation of poor governmental performance is to blame it on short-sightedness on the part of the authorities and their advisers. All too frequently, however, the explanation lies elsewhere; it is in some group's interest that things continue the way they do. Many urban-renewal and squatter-relocation schemes clearly fit such an explanation. Consider the case when land close to the commercial business centre is occupied by low-income settlement. The technical explanation that it constitutes a health hazard or that the population would be better served by government-built accommodation elsewhere is simply too convenient. More often groups commanding political support persuade the planning authorities to act in this way and the project is legitimized in terms of the public interest. In Rio de Janeiro the *favela*-removal programme between 1962 and 1966 was supported on the grounds that the land was needed for mass-transit systems; in Mexico a similar programme was justified in terms of the construction of a metro; in Lagos 60,000 people were removed as part of a road-widening project; in São Paulo underground railway construction led to extensive displacement (Batley, 1982). Of course, where a project is genuinely in the public interest, any group physically blocking its completion should be moved. Unfortunately too many renewal schemes harm the poor and bring them few real benefits.

Not infrequently, therefore, technical criteria, often based on European or North American 'experience', are intoned to benefit higher-income groups. To some extent we can agree with those who argue that sensitive housing and planning policies depend less upon an awakening in professional attitudes than upon more political pressure from the poor. Collier (1976: 133) argues that the authorities in Lima have permitted land invasions because of pressure, albeit limited, from the poor. 'In other countries, the extensive eradication of settlements suggest that they are even more powerless in their relations with the government than in Peru . . .' Support for this contention comes from Ibadan where the political dominance of the central area population over the city council prevents slum clearance (Laquian, 1971: 65). Similarly '[in] India it appears from observation that the majority of low-income areas in the cities vote for the ruling Congress Party and that this has resulted in improvements to the physical environment of such areas even when this is in blatant contradiction to the government's own master plan' (Payne, 1977: 63). Political influence is clearly a vital ingredient in the orientation of housing policy.

Perhaps the role that political interests play within the capitalist system in determining governmental policy towards the poor is best shown by reversing my original question. Instead of asking why urban renewal schemes have been introduced and spontaneous settlements destroyed we should examine under what circumstances spontaneous housing has been permitted to develop. As broad hypotheses I can perhaps suggest that self-help settlements are allowed to develop when: (1) major political parties or governments require the political support of the poor; (2) land occupied by the poor does not directly threaten the principle of private ownership; (3) the operation of self-help settlements directly supports, or at least reduces the need for changing, the economic and social system.

We have already seen how in India and Nigeria partisan politics have prevented slum removal, but political interest can also lead to greater service provision and credit facilities. Drakakis-Smith (1976*a*: 12) argues that in Turkey, as the struggle for power between the two major political parties has become more evenly balanced, so the squatters have received *de facto* recognition of their occupation and have been provided with many facilities such as surfaced roads, electricity, and water connections. Similarly in Santiago land seizures rose from 13 in 1968 to 35 in 1969 and 103 in 1970 as an election approached (Cleaves, 1974). According to Kusnetzoff (1975: 294–5), 'These figures demonstrate that, however strong the Frei repression, growing workers' organizations, the support of opposition parties for land seizures, and the proximity of a presidential election were altogether instrumental in overcoming that repressive power and thereby the planning and control capacity of the government.' (See pp. 195–200 below for further discussion of invasions in Chile.) More recently, the rate of invasions has risen in São Paulo as a result of the return of democracy to that country and as a reaction to the difficult housing and economic situation in that city (Sachs, 1983; Kowarick and Campanario, 1986). The poor may also benefit, at least temporarily, under populist military governments. In Lima the military government of Odría (1948–56) began to give away land as a means of undermining the political support of APRA.[5] Collier (1976: 64) claims that the long-term effect of this ploy was to set in motion a 'sorcerer's apprentice dynamic' which obliged new governments, both elected and military, to give away land and to service already established settlements. Consequently Lima's invasions are a direct result of government encouragement rather than a threat to the Peruvian system, as is so often claimed.

While the poor as a group may benefit temporarily from populist programmes or from political power contests which demand their vote, it is more common for particular settlements to be singled out for special treatment. In many Latin Amerian cities service provision is a function of political patronage; *barrios* that promise to support a powerful political group or personality may receive telephones, water, or roads as part of the deal. Ray (1969) reports that in Venezuela many *barrios* are linked to particular political parties and benefit according to the fortunes of their parties. In Rio de Janeiro, according to Leeds (1969: 79), 'the larger the *favela*, the greater the flow of gifts or the more significant the gifts which can be commandeered: there are more votes to be delivered, more to buy . . . Thus, very large squatments . . . continue to grow, to improve, to wield political influence, while other considerably smaller *favelas* stagnate.'

In general, therefore, political interests are the critical element determining policy towards spontaneous settlements. How the poor as a group come out of the situation will depend on the degree of partisan political conflict as well as structural factors such as the prosperity of the country and the degree to which the poor can be helped without affecting middle- or high-income groups. These structural conditions are critical and lead us on to the second point.

The spontaneous settlers of Lima, Ankara, and Caracas have all been the beneficiaries of political competition. But in each of these cities a major factor has been the availability of public land around the city. Throughout the non-Communist Third World, governments generally have tended to defend private land but have permitted invasions of public land. Only in exceptional circumstances have invasions of private land taken place. Luis Echeverría, the President of Mexico from 1970 to 1976, encouraged the occupation of land belonging to the Magazine *Excelsior* in reprisal for its critical attitude to his administration. In Venezuela private owners have failed to discourage potential invaders when they have anticipated that the government might recompense them for their 'loss' (Pérez and Nikken, 1982; Gilbert, 1981*b*).[6] In Chile, in 1970, 'the situation was so topsy-turvy that owners of private land were encouraging *pobladores* to invade their land' (Cleaves, 1974: 301–2). Since land on the urban fringe could neither be farmed profitably nor sold privately for urban development, a landowner's best course was to sell invaded land to the government at an inflated price, freeing the owner from his legal responsibility to install urban services.

Much more common is the invasion of public land, especially where extensive areas of state land lie close to the urban area. Gradually, however, rapid urbanization is reducing the supply of accessible public land and leading to much less relaxed reactions to invasions. In Ankara it is now more common for the poor to buy land (Payne, 1984) and in Lima the last two years of the Velasco presidency (1974–5) saw a much stricter government response (Skinner, 1982). Governments in both cases now face the choice of cutting rates of city growth or preventing land invasion and thereby encouraging the purchase of land. In many Latin American cities, in fact, the poor purchase land in the same way as any other social group. In Bogotá, São Paulo, and Mexico City, many of the poor have purchased land in speculative subdivisions of fringe areas: in Bogotá over half of the city's homes have been built on such land (Doebele, 1975; Carroll, 1980; Gilbert and Ward, 1985). After paying a deposit, buyers are given three to four years to complete payment and are then given the title-deeds. The only important difference compared with that of the ordinary commercial market is that the land lacks planning permission and services. By most European standards it is strange that governments permit such extensive unregulated developments. On the other hand, these governments recognize that the poor need land and that if they compelled urbanizers to provide services it would force the poor out of the land market. The only alternatives to the process of pirate urbanization are higher levels of renting and, in the longer run, land invasion. While pirate urbanizations are often extremely expensive to service and encourage low-density patterns of urban growth, they provide land for the poor. More important still, they support the principle of private land ownership by reducing the need to invade and by turning large numbers of poor people into landowners. In Bogotá the state intervenes in pirate urbanizations only where the urbanizer is denounced by the population of the settlement (Gilbert, 1981*b*). Except for the over-grasping and corrupt, pirate urbanizers survive because government intervention would cause the state more problems than it would resolve. In addition, the system provides political interests with a source of patronage. Politicians seek planning permission, negotiate for services with the bureaucracy, and intercede with the pirate urbanizers. Less scrupulous politicians represent the interests of the pirate urbanizer before the authorities. In other cities, such as Nairobi, top politicians are themselves involved in sales of land and property to the poor and thereby have a direct interest in the maintenance of pirate urbanizations (Amis, 1983). Most commonly, however, the system of pirate urbanization survives because it offers the poor land without threatening the principle of private land ownership.

To some extent my second point overlaps with my third; the process of self-help settlement is permitted whenever it supports the existing social, political, and economic system. Whatever the fears and moral objections of élites to spontaneous housing, self-help keeps the Third World economy functioning. First, it lets the less poor into the housing market by keeping costs low, in a way moreover that does not threaten higher-income groups. Second, Marxists argue that the very cheapness of spontaneous housing allows the labour force to reproduce itself despite the low wages paid by modern industry (Pradilla, 1976; Burgess, 1978; Kowarick and Brant, 1978). Economic development occurs apace in dependent capitalist societies on the basis of cheap labour costs and the perpetuation of low-income housing helps to reduce the pressure for wage rises. While spontaneous settlements may offend middle- and upper-income groups, zoning regulations effectively segregate the different income groups. If spontaneous housing does not have to be seen too frequently, its advantages become manifest; it offers cheap labour to industry, a plentiful supply of servants, and the myriad of other cheap services available to Third World urban élites. Thirdly, the existence of spontaneous settlement, especially where there is an active process of consolidation, opens up profit opportunities for commercial and industrial companies. Glass, bricks, cement, tiles, and pipes are purchased in large quantities by the spontaneous settlers and provide a large market for construction material suppliers. In addition, spontaneous housing areas can help to channel the poor out of high-value central areas. As Collier (1976: 37) notes, 'out of sixty cases on which appropriate information was available, nearly 50 per cent in some way benefited public or private urban development or real estate interests. The ambiguities of land ownership in Lima have unquestionably served the wealthy as well as the poor.' Fourthly, partisan political interests are served by the process. As Pérez and Nikken (1982: 227) argue,

> squatting and the formation of *barrios* do not burden the state (or the party in power in local or national government) with any kind of obligation to provide land or services nor the task of sharing them. On the contrary, the State is granting a favour by not evicting the squatters for which those who promote the squat are paid in political power. The provision of public services implies a new round of favours.

My argument, therefore, is that while spontaneous settlement clearly brings problems for the state, it is generally helpful to the maintenance and reproduction of the social and economic order. In turn, this raises an important question. If it has been in the interest of the Third World capitalist system to encourage spontaneous settlement, why has the state sometimes demolished such settlements and engaged in urban renewal? If spontaneous settlements have been so useful to the rich, why have the World Bank and the United Nations had to work so hard to sell the idea of sites-and-services programmes? This, in turn, raises questions about the validity of some Marxist explanations of the Third World urban situation (Peattie, 1979). According to writers such as Burgess (1982), Harms (1982), Connolly (1982), and Portes and Walton (1981) self-help housing is functional to the system because it cheapens the cost of labour in the 'formal' sector. As Burgess (1982: 70–1) argues,

> In so far as housing is necessary for the maintenance and expanded reproduction of the labour-force, the dominant capitalist mode of production is satisfied to allow the self-production of such activities particularly when the absence of rents and the association of such housing with various forms of subsistence activities that extend family budgets will mean less pressure for wage increases.

While there is clearly some support for that case, there are also problems. First, the establishment of a cheap labour force is helpful to industry but it is also detrimental in so far as this group of population is largely excluded from participation in the domestic market. As such industry gains in terms of production costs only to lose in terms of sales. Second, serious political problems may arise if too many people are relegated to the self-help housing sector. If too many people are being excluded from the benefits of economic growth, if large areas of self-help settlement are not receiving infrastructure and services, a strong political reaction may develop hostile to the interests of the formal sector. While the capitalist sector requires some cheap labour, too much may cause the political system problems. In many countries, there may be just too many people living in the self-help settlements and 'it would be absurd to suggest that all those excluded from the modern sector play a functional role' (Sandbrook, 1982: 66–7).

In addition, Amis (1983) and Burgess (1982) argue that the dominant capitalist system increasingly penetrates the self-help settlement process. By purchasing land for reasons of speculation, by providing materials for self-help construction, and by increasingly dominating the retail market within such communities, the capitalist sector increases the rate of capital accumulation. The problem here is that such penetration, while increasing the profits of some, also increases the costs of labour reproduction. As a result, therefore, the principal 'function' of the self-help sector is undermined. A contradiction is involved here which needs to be carefully considered (Gilbert, 1986).

One possible resolution is to explain the contradiction in terms of the mutually conflicting interests of different élite groups (class fractions). The state is subject to pressure from representatives of various capitalist groups; its housing response will therefore constitute a balance between these demands. Thus national governments may respond to industrial interests, vaguely sympathetic to spontaneous settlement because of the cheap labour it offers, by failing to formulate a consistent housing policy or by supporting self-help programmes. Meanwhile local government policy may respond to pressure from élite residential groups and the construction industry whose best interests are served by demolition. Such local action would be favoured when the poor have occupied high-value land awaiting development, when they have blocked prestige public works programmes, and when the poor have established themselves close to middle- and upper-income housing areas, thus threatening to lower land and property values. In other circumstances local governments may accept spontaneous settlement because they are incapable of handling the problem; they cannot build proper houses to accommodate the poor, they cannot service the settlers, nor indeed can they muster the political power to remove them. Since in any case the process serves the interests of the rich and since the poor seem generally content with the actions they themselves have taken, the most sensible policy has been to leave bad alone.

At the national level, governments are increasingly admitting the 'blight' of spontaneous housing into their national development plans; development, after all, is now understood to include much more than economic growth. International pressure from the World Bank and other loan agencies in favour of spontaneous housing may not be entirely welcome, but it has been sweetened by hard currency loans. Further, urban problems can no longer be ignored by national governments because of the scale of the issue. Buenos Aires accommodates one-third of all Argentinians, Caraqueños make up one-fifth of the Venezuelan population, and Manilans represent one-eighth of all Philippinos. Since up to one half of the populations of these cities are living in self-help settlements, national governments are forced to intervene. The introduction of partial remedies such as sites-and-services projects can be expected to grow because Third World governments have no other solutions to offer. But the sorts of programmes that will be introduced are not those that were once recommended by the World Bank. For if national governments have been forced to intervene in the urban areas by political and economic realities, business interests have intervened because of the attraction of profits. A programme that would truly help the poor would involve attacks on land speculation, effective service provision, and firm, but sensitive, land-use planning which would reduce traffic congestion and unit costs of public service provision. But this is not the real world in a mixed economy and we are deluding both ourselves and the poor if we believe it. Sites-and-services schemes are both necessary and inevitable, but there can be no avoiding more radical changes which will redistribute the resources of the city and increase the rates of public service provision and job creation.

6
Social Organization in the City

Through the ages cities have elicited sharply contrasting responses. With the establishment of urban settlements humankind wrought the most conspicuous changes on the planet's surface and profoundly altered the social relationships that distinguish the human species. But after 6,000 years of urban experience we still appear ambivalent about our creation.

Whether the early cities were primarily seats of despotic rulers, repositories of holy shrines, or kingpins in trade which spiced food and embellished life, is a matter for speculation. To the philosophers of the Enlightenment the city represented virtue, but to later generations experiencing the Industrial Revolution the city appeared as a nest of vice and corruption. The *conquistadores* established their cities on the sites of the Aztec and Inca capitals they had destroyed, and they built their cathedrals with stones from the old temples, but their pride in their new creations eventually came to be shared by foreign immigrant and *mestizo* alike. Calcutta stood for over two centuries as the symbol of foreign intrusion, but it was in Calcutta that the Bengal Renaissance challenged the cultural hegemony of the West. The urban settlements of freed slaves established on the West African coast, Freetown in 1792, Monrovia in 1822, Libreville in 1848, came to symbolize the struggle for abolition. But on the same continent Johannesburg-Soweto stand as the foremost symbol of racial oppression today.

Our discussion of Third World cities also carries traces of ambivalence. The Third World city is characterized by its dependent position in the world capitalist system, but it is also a locus for strategies where more balanced development can be adopted and set into motion. I have argued that the city holds out real promises to the masses it attracts from rural areas, but I have also emphasized problems of urban unemployment, underemployment, and misemployment.

Here we shall focus on the characteristics of social interaction in the city. Assumptions as to what constitutes the urban way of life abound, and they underlie many of the judgements about the merits and evils of the city. But a good deal of research has enquired into urban social relations over the last half century and, as we shall see, some assumptions commonly held are no longer tenable.

The stage for the contemporary enquiry into urban life was set by

Wirth in 1938 in his classic essay 'Urbanism as a Way of Life'. He noted virtues of the city, but put far greater stress on what he saw as its dehumanizing aspects. As Wirth (1938: 1) summarized his argument:

For sociological purposes a city is a relatively large, dense, and permanent settlement of heterogenous individuals. Large numbers account for individual variability, the relative absence of intimate personal acquaintanceship, the segmentalization of human relations which are largely anonymous, superficial, and transitory, and associated characteristics. Density involves diversification and specialization, the coincidence of close physical contact and distant social relations, glaring contrasts, a complex pattern of segregation, the predominance of formal social control, and accentuated friction, among other phenomena. Heterogeneity tends to break down rigid social structures and to produce increased mobility, instability, and insecurity, and the affiliation of the individuals with a variety of intersecting and tangential social groups with a high rate of membership turnover. The pecuniary nexus tends to displace personal relations, and institutions tend to cater to mass rather than to individual requirements. The individual thus becomes effective only as he acts through organized groups.

A substantial body of research had demonstrated that generalizations like Wirth's are not warranted. At the very time when Wirth wrote, William F. Whyte (1981) studied a low-income Boston neighbourhood inhabited almost exclusively by Italian immigrants and their children. Middle-class observers looked upon the area as a slum, a formidable mass of confusion, a social chaos. Instead, Whyte found a highly organized and integrated social system; even young single men were integrated through the street-corner gang they had established. Subsequently Oscar Lewis carried out research among Mexico City residents who had come during the preceding half century from Tepoztlán, a nearby village he had studied previously. In 1952 Lewis (1970*b*: 424–5) concluded:

There is little evidence of disorganization and breakdown, of culture conflict, or of irreconcilable differences between generations . . . Family life remains strong in Mexico City. Family cohesiveness and extended family ties increase in the city, fewer cases of separation and divorce occur, no cases of abandoned mothers and children, no cases of persons living alone or of unrelated families living together. Household composition is similar to village patterns except that more extended families live together in the city. There is a general rise in the standard of living in the city, but dietary patterns do not change greatly. Religious life in the city becomes more Catholic and disciplined; however, men play a smaller religious role and contribute less money to the church in the city. The system of *compadrazgo* has undergone important changes, but remains strong. Although there is a greater reliance upon doctors and patent medicines to cure illness, city Tepoztecans still use village herbal cures and in cases of severe illness sometimes return to the village to be cured. Village ties remain strong, with much visiting back and forth.

Peasant and Urbanite

Many urban dwellers remain firmly rooted in the rural community in which they grew up. This is a widespread pattern in sub-Saharan Africa, much of Asia, and the Pacific, as we have seen in Chapter 3. Thus recent migrants who find themselves isolated in the urban setting—and they are the exception rather than the rule, as we shall see—may be lonely, but they are quite likely to feel secure in the knowledge that they continue to be members of the community they came from. Wives and children who had to be left behind, members of the extended family, and other village relationships continue to define a rural place as home for many.

The migration patterns distinguished in Chapter 3 are obviously related to the strength of the ties migrants maintain with their community of origin. Many migrants anticipate returning there. They continue to see themselves as members of a rural community, whether they want to be back in time for the next harvest, or plan to retire in the village after a lifetime of work in the city. The latter pattern prevailed in the early 1960s in what was then Eastern Nigeria. Urban dwellers invariably stressed that they were strangers to the city. Most urban dwellers identified with their community of origin, felt that they belonged there, affirmed their allegiance. In 1987, the next generation's commitment to the rural home was just as strong. They visited, contributed to development efforts, built houses, intended to retire, and wanted to be buried—'at home' (Gugler, 1971, 1991).

Significant ties with their rural areas of origin are not uncommon even among permanent migrants. Such a commitment to the extended family and the village is reported from Meerut, a city in North India, where Vatuk (1972) studied first- and second-generation migrants holding white-collar jobs. Most consider their real home to be in the village and say that they are living 'outside' in the city, or 'in service', i.e. at their place of employment. Frequent contact with the village and a sense of belonging to the village home carry on over several generations. Couples return for a visit to the husband's place of origin during holidays, women visit their natal homes with their children. The exchange of money, goods, and services between rural and urban segments of the extended family is not only normatively prescribed but common in practice, particularly if the rural residents are the parents of the head of the urban household. However, with few exceptions, these urbanites have no intention of ever returning to live in the village, nor will their children settle there. True, many retain joint ownership of rural land and homes, but such property apparently has little material value for these white-collar urbanites; it is not a significant source of present or potential income. Indeed, urban residents who have a claim to family property are reluctant to see the property divided because they continue to place high value upon the ideal of family 'jointness' (Vatuk, 1972: 131–41, 194).

Bruner (1972, 1973) reports a similar pattern among the Toba Batak, a Christian minority in Indonesia. They originate from Sumatra, and many of those residing in Medan, one of the two principal cities on the island, go to their home village for short visits to discuss lineage affairs, to look after property, or to participate in village ceremonies. However, very few return to stay, even after retirement. Batak in distant Jakarta—a ten-day boat trip away—send money home, help the children of rural relatives to attend school in the city, and keep an active interest in home affairs. The urban Batak do not lose their place in rural society. Indeed, many continue to own a house and rice-fields in their village, even after two or three generations in the city. However, as in Meerut, ownership of village property eventually comes to have symbolic rather than economic value.

The commitment many migrants have to their community of origin may be taken to suggest that they remain peasants at heart, that they do not become urbanites. Short-term migration in particular encourages such an interpretation. But even when migrants come for only a brief stay—no longer the typical pattern, as we have seen in Chapter 3—it does not follow that they continue to behave in rural ways while in town. Though migrants are used to rural modes of behaviour and frequently hold rural values, they also have varying degress of familiarity with urban conventions of behaviour and ways of thinking. They have learned about urban conditions in school and from visiting or returned migrants. Some have been in towns before to sell rural products, to make purchases or obtain services, or just as guests of kin or friends.

As soon as migrants arrive in town, they have to adopt behaviour that will allow them to pursue their goals effectively. The point was made forcefully by Gluckman (1960: 57) when he dismissed an earlier perspective that saw African urban workers as 'tribesmen' in his classic dictum: 'An African townsman is a townsman, an African miner is a miner.' That is, the migrant's behaviour is defined by the immediate context. Yet such a model of 'situational change' captures only one aspect of the migrant's adaptation. There is also a drawn-out process: migrants continue to modify their behaviour as they gain urban experience, as they undergo 'biographic change'.[1]

Language use is a conspicuous and important factor in adaptation. Some migrants need only to make a few additions to their vocabulary, or to modify their pronunciation. Others have no mastery, or only limited mastery, of the city's lingua franca. Many migrants in Asian and African cities have to switch to the national or regional language in common use. The transformation is particularly striking in those Latin American cities in which American Indian migrants come to be seen as *mestizos* as they learn Spanish, abandon their rural dress and hair style, and modify their food habits. The switch in language illustrates the propositions of both the situational and the biographic change model: as soon as migrants arrive in town they will need to employ what little they know of the lingua franca; as they stay on, their language skills will develop.

Adopting urban patterns of behaviour does not mean forgetting how things were done at home. Working-life migrants will continue to behave in urban or rural ways as the situation demands. Indeed, they have to be both peasants and townsmen in order to act successfully in the 'dual system' they have established (Gugler, 1971). Learning and acquiring new norms of behaviour through urban socialization, some individuals grow away from their rural ways, but most maintain the ability to relate according to rural norms, whether in the town or in their home areas. In other words, becoming urban implies extending oneself to the urban culture, but does not require a commensurate rejection or loss.

Kin and Home People

Most migrants move to a city where they expect to be received by relatives or friends. They will be offered shelter and food for a while, they will be introduced to the urban environment, and efforts will be made to find them an opportunity to earn their living. This pattern of initial urban association encourages persons from the same village, region, ethnic group to form residential clusters. Allocation or housing by public authorities or employers may inhibit such clustering. But even when residentially dispersed, people of common origin frequently maintain close ties.

The family has been commonly assumed to change from an extended form to the nuclear family in the course of industrialization. In fact preindustrial patterns varied widely. The nuclear family, comprising parents and their unmarried children, was the norm in many pre-industrial societies. And even where the extended family—married siblings co-resident with their parents—was the ideal, it was realized only by a minority, for example by some but by no means all wealthy families in China (H. D. R. Baker, 1979: 1–25). In contemporary cities joint families, i.e. married siblings maintaining a common household, are the exception. Stem families, that is three-generation households, are more common. In Calcutta, a survey found about one in five households to be constituted by stem families, about one in eight by joint families (Chakrabarti, 1985). A study of urban China reports 50 per cent of men and 59 per cent of women over the age of 60 living with a married son or, rather rarely, a married daughter (Whyte and Parish, 1984: 157, 144). If stem families are founded on the elders' control over resources in peasant societies, they may be thought to be more dependent on a culture of filial responsibility in the urban context. Still, urban elders frequently control precious housing and assist with child-care and domestic work. Such assistance is crucial for mothers of young children who work away from home and lack access to adequate child-care facilities.

While joint-family households are exceedingly rare in contemporary cities, relations with kin typically are important.[2] Rugh (1984: 244) reports from Cairo that families compensate for the lack of broader sets of convenient kin by placing greater reliance on nuclear family members and those relatives who happen to be close at hand. Even friends, office colleagues, and neighbours may be treated as sister, brother, or other kin. The practice is most common among first-generation urban dwellers who are poor and particularly dependent on support networks. But even in China, where the state provides a subsistence floor for most, people still call on kin when the rare emergency does occur and money or assistance with a sick person is needed. With the increase in unemployment in the late 1970s, kin have been called on to help with job contacts. Kin working in shops or elsewhere in the commercial network help find scarce goods. And when kin live in the nearby countryside, there may be frequent visiting back and forth—the urbanite bringing gifts of consumer goods that cannot be found in the countryside and the villager bringing gifts of food not available in cities (Whyte and Parish, 1984: 335–6).

Unlike the typical rural setting, the city gives a measure of choice how closely to associate with various relatives, whether to discontinue relationships with some of them altogether. In Cairo, the established middle class which has long since made its adjustments to city living, weighs the costs and benefits of relationships, and the competition for educational achievement and for status tends to cut it off from those below (Rugh, 1984: 244). A. L. Epstein (1981) describes the pattern in the 1950s on the Copperbelt in what has since become Zambia. The corporate kin group of the village is transformed into a social network of urban kin, a network maintained and developed selectively. At the same time the range of recognized kin may be extended. A minimum level of support may be expected from all people sharing a common origin.

Butterworth (1972) describes a group of men in Mexico City who migrated from Tilantongo, an isolated Mixtec community 300 miles away. Almost every weekend they meet at the house of a member who is the undisputed leader of the group. A member in need of aid turns to him first, but responsibility is corporate. For a member who had become an invalid, the group mustered considerable financial resources and spent a great deal of time and effort to gain access to public agencies. Repayment of such assistance is not expected, even loans go mostly unpaid, but continued affiliation with the group and willingness to help other members are implied.

The migrants share information about strategies for coping with problems in the city and introduce new arrivals to the complexities, but the most popular topic of conversation is their *tierra*, Tilantongo. They discuss the current state of affairs in the community and deplore the decline that has taken place. They critically analyse the needs of the community and possible ways to meet them, hatch schemes to get rid of the reactionary incumbents in the community offices, and weigh suggestions to be made at the monthly meeting of the formal organization of Tilantongo migrants in Mexico City. Their leader makes regular calls on the president of the organization of Mixtec migrants in Mexico City, a former deputy in the Mexican legislature with influence in government circles. Considerable efforts are directed towards getting him and officials from various government commissions to accept invitations to *fiestas* at the leader's home. Once present, the officials will be plied with liquor and a sumptuous barbecue *a la Mixteca* in an attempt to extract promises of aid to Tilantongo.

The solidarity of common origin, strengthened by feelings of mutual obligation, and the shared long-term interest in the future of their home community, is complemented by sentiment. These men rarely associate socially with anyone other than migrants from Tilantongo. As they drink together, they recall their childhood in the village, the way things have—and have not—changed:

> After the men have been drinking for a while, someone invariably brings forth a guitar to accompany sentimental songs about their beloved *tierra*. A favorite is the 'Cancion Mixteca'. ('How far away am I from the land where I was born | Immense nostalgia fills my thoughts . . .') Hardly a dry eye remains as the melody concludes: 'I would like to cry | I would like to die | of sentiment.' With tears streaming down his cheeks, one of the men is likely to stand and shout, 'I'm from Tilantongo!' (Butterworth, 1972: 139)

Formal organizations of 'home people' are prominent in parts of West Africa. Some ethnic unions go to considerable lengths to have all from home join in. They meet regularly, elicit intensive participation, and serve a wide range of explicit and implicit purposes. Members relax in one another's company, they evolve common responses to the urban milieu, provide assistance in personal crises, settle their disputes within the union, and are frequently involved in furthering the development of their home community (Gugler and Flanagan, 1978: 81–8).

Three variables affect the pattern of association among migrants. First there are the links with home people in the city and the ties to the common home. These tend to be mutually reinforcing as each enhances communication and social control in the alternate context. Mayer (1971: 283–93) emphasizes a second variable, the interaction between cultural traits and the structure of social relationships. Immigrants who have a traditionalist outlook will tend towards encapsulation in a group of like-minded home people who uphold shared rules of behaviour. Where Mayer's analysis represents cultural background as the prime determinant of patterns of association in the urban setting, Banton (1973) has focused attention on a third variable: opposition among social groupings in town. He suggests that the social density that characterizes the village is encouraged among urban groups both by the degree of discontinuity between the rural and the urban system and by the extent and strength of structural opposition in the urban system. I shall address ethnic alignments in political conflict in the next chapter.

Life-Style Alternatives

Contrary to the assumption that urban life is characterized by the absence of meaningful personal relationships, we have seen how many rural–urban migrants, far from being uprooted, maintain strong ties with their community of origin and establish in the city new communities based on common origin. Other urban dwellers, first-generation migrants as well as urban-born, feel securely anchored in networks of kinship and friendship. In a survey in Kanpur, India, an industrial centre of over one million inhabitants, nearly everyone reported seeing friends in the city more than once a week, and most characterized these friendships as intimate. Furthermore, a large majority had relatives in the city, and many described their relationships as intimate or extending to mutual aid.[3] The survey covered three neighbourhoods which differed in type of location, length of settlement, and socio-economic status, but the patterns reported were quite similar across these neighbourhoods. In each of them the great majority of residents appeared to have intimate ties in the city. And within each there was a good deal of visiting and exchanging favours (Chandra, 1977: 89–122, 189).

Religious groupings play an important integrative role for some urbanites. Roberts (1978: 145) attributes the development of Pentecostal and other Protestant sects in predominantly Catholic Guatemala to the attempts of those who seek a basis for social relationships. In the two low-income neighbourhoods he studied, members of the sects were often those without kin in the city and included women separated from husbands or whose husbands were alcoholics. Catholic voluntary groups performed similar functions for other low-income residents, providing the opportunity for single women with children, for example, to interact with others who could help them to find work or to obtain benefits from social welfare agencies.

Many an urbanite establishes a quite close-knit network with kin or home people, or through involvement in a religious or political group.[4] The ideal type of the close-knit network is provided by the isolated village community where everybody knows everybody else. In urban settings such a pattern is approximated in those exceptional cases in which a group of people are in one another's exclusive company not only during leisure-time activities, but also at work. Unlike the village, the city offers an alternative form of integration. An individual may have a wide range of meaningful relationships with people who do not know one another. Such loose-knit networks are specific to the city, a pattern impossible to implement in a rural context. Close-knit networks tend to be composed of like-minded associates who enforce conformity with the rules of behaviour prescribed by the group. The members of a loose-knit network, in contrast, tend to have a more open outlook; each one can take advantage of the choices the urban setting offers and has greater leeway in deciding with whom to associate and which cultural pattern to adopt. The moral pressures exerted by associates in the heterogeneous urban setting are frequently inconsistent, and changing one's behaviour, beliefs, norms, and values, while straining relations with some associates, need not lead to general ostracism.[5]

Gans's (1962) classic rejoinder to Wirth exposed the basic fallacy of a perspective that sees social relationships as determined by the urban environment, i.e. of ecological determinism. Reviewing urban neighbourhood studies in the United States, Gans showed conclusively that only a minority of the urban population lives in isolation, and that a considerable variety of life-styles is to be found across neighbourhoods. From this observation the central argument followed: to the extent that urban dwellers can choose their location within this heterogeneous environment, they also, more or less deliberately, opt for a life-style.

The shortcomings of ecological determinism are well illustrated by a study of crime in Kampala which focuses on two low-income neighbourhoods (Clinard and Abbott, 1973: 142–65). Kisenyi was well known as a high-crime area, while Namuwongo had a better reputation. Indeed, rates of crimes reported to the police and of arrests were considerably higher in Kisenyi for violent crimes and even more so for property crimes. The study proposed an interpretation in terms of differences in community integration and in the residents' perceptions of their communities. It failed to address the fact that for most immigrants in Kampala there was an element of choice in which neighbourhood to settle. Kisenyi was located near the bus station, major markets, and the business centre; it had the greatest concentration of prostitutes in the city, a wide selection of bars, places in which illegally brewed beer was sold, gambling and dancing establishments, and drugs. Namuwongo, in contrast, was situated outside the city limits and bordered upon the industrial sector. We would expect there to be a measure of self-selection among the people who respectively settled in two so diverse neighbourhoods.

Even within the same neighbourhood different categories of people may pursue distinct life-styles, a point strikingly demonstrated by Gans (1962) for the inner city in the United States. Anthony and Elizabeth Leeds (1970) similarly emphasize the heterogeneous composition of the *favelas* on the hills of Rio de Janeiro. Some of their inhabitants are unable to earn a living and barely survive on hand-outs; they frequently die young. Following Gans, we may call them the trapped, for they have nowhere else to go. Others have experienced a crisis which left them no alternative but to seek refuge in the *favela*; they are going through a period of stress, but there is the prospect that they will surmount their problems and move out. A third category of *favela* residents live there by choice in order to economize; they could afford regular housing but are attracted by the opportunity to pay little or no rent and perhaps the opportunity to raise fruit, vegetables, pigs, or chickens. Finally a few are well off—the Leedses encountered an accountant, a watchmaker, and a retired teacher; they have a taste for the freedom the *favela* offers from conventional constraints, and for the social recognition and prestige they enjoy among their co-residents. Lomnitz (1978) describes a category of relatively affluent residents who are bound to shanty towns in Mexico City for their very livelihood. Small entrepreneurs need close contact with relatives and neighbours who provide them with cheap labour or custom. Brokers similarly have to live in the shanty town, for example the leader of a construction gang who recruits labour, the jobber who puts out sewing to women in the neighbourhood, the local political boss.

Apart from these various motivations, residential heterogeneity is fostered through processes of change. Where a neighbourhood is invaded—for instance, when the original squatters sell out to more affluent people once their settlement has become legalized—there will be high heterogeneity during the transition period. Conversely, a measure of heterogeneity arises when people remain in a neighbourhood in spite of changes in their economic fortunes or household composition—for instance, because they feel attached to the locality or they continue to have significant social ties in the neighbourhood, or because housing shortages make it difficult to find a satisfactory alternative, or moving out means giving up rent-controlled housing.

Neighbourhood relationships are important for most urban dwellers, but they are only part of the urbanite's social network. Few urban dwellers are bounded by their neighbourhood, most have social relationships which reach farther afield. Many migrants, as we have seen, remain integrated 'back home'. Where a migrant finds himself among strangers, he may well lead the life of an exile, affirming his continued membership in a home community. Whether migrant or urban-born, most urban dwellers have significant ties in other parts of the city and frequently beyond. The maintenance of such geographically extended social networks is facilitated by fast transport becoming more easily and more cheaply available, and better means of communication. Modern technology has dramatically expanded the human environment.

The attempt to explain the behaviour of people in terms of their immediate environment thus has to take into account three specifically urban phenomena: life-styles vary to a considerable extent across the urban agglomeration, and most urbanites have a measure of choice where to locate; a considerable variety of life-styles is found within some neighbourhoods; and urban dwellers, to the extent that they can take advantage of modern transport and communication, are not bounded by the neighbourhood. Conversely, certain categories of urban dwellers can be seen to be constrained by their immediate environment. Many are forced into an environment not of their choosing, for example those who out of economic necessity work in the gold mines of South Africa and have to live in the dormitories provided by their employers. Others have no satisfactory options within their immediate environment, they are outsiders in their neighbourhood, for example a lone minority household in an otherwise homogeneous neighbourhood. Finally, the environment is narrowly circumscribed for some, either because they cannot afford efficient transportation and communication or because they are homebound—some for practical reasons, such as mothers caring for children, the young, invalids, others because of cultural norms: Janet Bauer (1985) characterizes women in Teheran as 'locked into space' even prior to the establishment of the Islamic Republic. Still, most urban dwellers have a measure of choice where to live and with whom to associate.

The choices it provides distinguish the city from rural areas. There is not one urban life-style, distinct from a rural way of life, but a variety of life-styles unknown in the village community. Some urbanites lead encapsulated lives, nearly as if they were in a village community. Others strike out, associate with like-minded persons, separate when they no longer agree, become individualists. The bigger the city, the better the chance for even the most unusual mind to find others so inclined. The city allows the unconventional, those labelled 'deviants' in the society at large, to establish social relationships with those of their ilk, and to develop viable social roles. In the city adherents of a new religion, protagonists of a new political idea, carriers of a new fashion can aggregate in sufficient numbers to support one another. Cities are centres of innovation because it is in cities that innovators can constitute a critical mass (C. S. Fischer, 1984: 37).

The varied opportunities the city offers dissidents to associate among themselves, leaving others in the dark about their thoughts, and indeed about their activities, present a serious problem for totalitarian regimes. Such regimes invariably attempt to control urban populations through grass-roots organizations. In China, residents' committees exert close control over neighbourhoods while working closely with the local police station. Work units monitor and sanction the behaviour of their members not only at work but beyond, for instance informal sanctions may be used to deal with what are termed 'life-style problems': excessive drinking, gambling, reading forbidden literature, sexual infidelity. The combined efforts of residents' committees and work units leave almost no place to hide, or in the Chinese phrase, no 'dead corners'. By all accounts, crime, prostitution, and drug abuse are quite low, whether compared to the era before the Communist victory or to other societies. Still, in the wake of the Cultural Revolution hundreds of thousands of rusticated middle-school leavers returned to the cities without authorization, urban crime rates rose, and even youth gangs appeared (Whyte and Parish, 1984: 231–73). In Cuba, the functions of the Committees for the Defence of the Revolution have fluctuated over the years, but vigilance was considered a prime concern much of the time. In the early 1960s, especially around the time of the Bay of Pigs Invasion, concern focused on local collaborators with foreign intervention. Vigilance has continued to be directed against political deviance as well as common crime (Domínguez, 1978: 261–7; Butterworth, 1980: 105–18).

Mental Stress and Crime

Part of the negative image of the city is based on the assumption that mental stress is more characteristic of the city than of rural areas. The evidence from Third World countries is scanty, and less than clear-cut. The Harvard Project on the Social and Cultural Aspects of Development provides data on men aged 18 to 32 in Argentina, Chile, India, Nigeria, and Pakistan (Inkeles and Smith, 1970). The survey asked about such psychosomatic symptoms as difficulty in sleeping, nervousness, headaches, or frightening dreams; more were reported among those with longer urban residence in four of the five countries, but the relationship was statistically significant only for Argentina and Pakistan. Similarly, when urban non-industrial workers were matched with cultivators on variables such as education and ethnic membership, the workers reported more psychosomatic symptoms of stress in four countries (no data are available for Argentina), and the differences were statistically significant in Nigeria and Pakistan. However, when long-time factory workers were matched with cultivators, while the workers appeared less well adapted in four countries, the only statistically significant difference appeared in India, where they reported less stress.

A similar study in Kenya compared Abaluyia women who shared their lives between Nairobi and their rural home communities in Western Kenya, Gikuyu market women in Nairobi, and rural Gikuyu women (Weisner and Abbott, 1977). The rural Gikuyu reported substantially higher stress scores than either of the other groups, and the relationship held when education and age were controlled for; however, the differences were not statistically significant. Certainly, to date, research has failed to provide the consistent results that would sustain the popular stereotype that urban life is more stressful than rural life.[6]

Measurements of psychological adjustment have to be qualified as brave attempts at best. The data base is similarly precarious when it comes to addressing another salient facet of the negative image of the city: the belief that crime and vice are rampant there, in contrast to supposedly more idyllic rural areas. Usually all we have to go by are statistics of crimes and arrests recorded by the police. Clearly not all crime is reported, nor are all criminals arrested. Of particular concern for our purposes is the fact that the proportion of crimes reported and of criminals arrested varies by type of community.

Violent crime in Third World countries does not appear to be more prevalent in urban than in rural areas. The most comprehensive data available compare the national homicide rate and the rate for a major city in eleven countries (table 6.1). The countries divide about evenly between those in which the national rate is higher, and those in which it is lower than the city rate. While every caution as to the reliability of crime statistics is in order, the major bias to be expected in poor countries—that rural crimes are more likely to go unreported, and that rural criminals are less likely to be brought into the national system of justice—suggests that rural crime rates are understated compared with urban rates. We conclude that the rural-urban dichotomy does not constitute a promising explanatory variable for the rate of homicide in the Third World.[7]

In contrast to violent crime, property crime as well as victimless crime universally appear to be more common in urban than in rural areas. Four types of explanation compete: the disorganization argument that may be identified with some of Wirth's writing, the anomie thesis

Table 6.1

Homicide rates for eleven countries and major cities within them, 1960s[a]

Country (period)	National rate[b]	Major city	City rate[b]
Guyana (1966–70)	6.18	Georgetown	5.21
India (1966–70)	2.72	Bombay	2.85
Kenya (1964–8)	5.67	Nairobi	5.27
Mexico (1962, 1966, 1967, 1972)	13.24	Mexico City	13.34
Panama (1966–70)	11.07	Panama City	4.96
Philippines (1966–70)	7.98	Manila	23.86
Sri Lanka (1966–70)	6.09	Colombo	5.59
Sudan (1961–4, 1968)	5.67	Khartoum	30.25
Trinidad and Tobago (1966–70)	14.00	Port of Spain	15.31
Turkey (1966–70)	9.65	Istanbul	4.84
Zimbabwe (1966–70)	5.33	Harare	7.20

[a] The national rate includes the rate of the major city, hence the comparison given here underestimates the difference between the major city and the rest of the country. The reader is cautioned against making comparisons of homicide-rate levels across nations or across cities. The definition and reporting of homicide varies considerably among countries, and the indicator used here, while identical for each country and city in it, is not consistent across countries.

[b] Average annual number of homicides per 100,000 inhabitants.

Source: Archer and Gartner (1984: 105–7).

advanced by Merton (1938), structural interpretations, and the compositional proposition related to Gans's approach.

The disorganization argument assumes that urban dwellers are no longer effectively integrated into a community: no longer subject to informal social controls over their behaviour, without a firm commitment to community values, they are easily attracted by the promise of quick gains, seduced by the lure of vice. In fact, as we have seen, many urban dwellers are well integrated. Still some, especially young adults, are quite footloose. Accountable to none, they are ready to try their hand at crime.[8]

The contrast between rich and poor is glaring in the cities. Rural–urban migrants tend to rest content that they have improved their position. They are preoccupied with establishing themselves securely in urban employment and housing. And their ambitions for social mobility in the urban context are cast in terms of hope for their children's future. However, many of the urban-born look beyond their own status group and feel severely deprived. Limited opportunities for social mobility thwart their aspirations, and they experience anomie. A career in crime appears as the only way out.

Structural interpretations emphasize the difference in objective conditions between rural and urban areas, and indeed among urban neighbourhoods. First, theft and burglary are facilitated in anonymous urban settings where the stranger goes unnoticed.[9] Second, rural dwellers are usually assured of subsistence, but some urban destitute have no alternative but to steal or rob for survival. Sometimes they organize for the purpose as in the bands of *gamines* in Bogotá. Third, the city's underworld is sufficiently large for a division of labour among professionals, indeed for organized crime. Fourth, the greater number of customers brings forth a wide variety of illegal services to cater to their wishes, whether they want sex, drugs, or gambling.

The compositional proposition shifts the focus to the characteristics of the people who come to live in different types of agglomerations. Where young men predominate among migrants, more or less outright prostitution is likely to be encouraged. Customers for illegal services visit to avail themselves of the greater variety of such services in the city. Criminal elements are attracted by better opportunities in the bigger cities. Some may find it advisable to depart from the village or small town where they are too well known for comfort; the city offers the opportunity to shed one's past—and to hide present activities.

The Subculture of the Poor

High crime rates mar the attractions of the city, but it is the urban slum that confronts us with the worst failure of humankind's urban endeavour. The contrast between rich and poor within one city—found everywhere but particularly striking in much of the Third World—dramatically exposes man's insensitivity to the plight of fellow man. However, a closer look at rural realities, at the condition of Untouchables in India, at the degradation of Indians on the Amazon, suggests that this is not a specifically urban phenomenon. Still, disregard for others is facilitated by cultural distance, or rather the perception of such distance, and the juxtaposition of locals and various migrant groups in the city readily provides bases for cultural distinctions to be made.

Middle-class observers tend to perceive urban poverty in distorted terms which are encapsulated in the notion of the slum. This is how Perlman (1976: 13), in an account reminiscent of Whyte's (1981: xv–xvi) famous introduction to his study of a Boston 'slum', contrasts the outsider's perception of a squatter settlement in Rio de Janeiro with an insider's view:

From outside, the typical favela seems a filthy, congested human antheap. Women walk back and forth with huge metal cans of water on their heads or cluster at the communal water supply washing clothes. Men hang around the local bars chatting or playing cards, seemingly with nothing better to do. Naked children play in the dirt and mud. The houses look precarious at best, thrown together out of discarded scraps. Open sewers create a terrible stench, especially on hot, still days. Dust and dirt fly everywhere on windy days, and mud cascades down past the huts on rainy ones.

Things look very different from inside, however. Houses are built with a keen eye to comfort and efficiency, given the climate and available materials. Much care is evident in the arrangement of furniture and the neat cleanliness of each room. Houses often boast colorfully painted doors and shutters, and flowers or plants on the window sill. Cherished objects are displayed with love and pride. Most men and women rise early and work hard all day. Often these women seen doing laundry are earning their living that way, and many of the men in bars are waiting for the work-shift to begin. Children, although often not in school, appear on the whole to be bright, alert, and generally healthy. Their parents . . . place high value on giving them as much education as possible. Also unapparent to the casual observer, there is a remarkable degree of social cohesion and mutual trust and a complex internal social organization, involving numerous clubs and voluntary associations.

We have emphasized, against ecological determinism, the choices the city offers. But the range of choices available to urbanites varies widely, depending on the power they wield in the political arena and in the market-place. Lack of leverage to affect the political process, and lack of the economic means to compete effectively in the market, severely circumscribe the choices open to many urban dwellers. Are their lives determined by their political position and their material condition? The notion of a culture of poverty focuses on similarities among the urban poor in different societies, but emphasizes that the behaviour and values of the poor are not determined by their circumstances, rather they constitute a culturally evolved response.

The concept of a culture of poverty was introduced by Oscar Lewis, an anthropologist with considerable research experience among American Indians, in India, in Cuba, and with Puerto Ricans, both in Puerto Rico and in New York City, but who is best known for his work in Mexico. Lewis (1959: 16) first proposed the concept in his account of the life of five families in Mexico City. He expanded and modified his arguments over the years, and I shall base my discussion on his last statement, published in 1970, the year he died.

The rather catchy phrase, 'culture of poverty', designates common cultural elements found among poor people in different societies. The approach focuses on cultural traits, i.e. patterns of behaviour and values, specific to the poor in a given society. These traits do not constitute a separate culture, but rather a variation on the national culture, a subculture. Lewis argued that these subcultures observed in different societies have a common core: the absence of childhood as a specially prolonged and protected stage in the life-cycle, free unions or consensual marriages, a trend toward female- or mother-centred families, and a strong predisposition to authoritarianism.

Critics first of all point out that many other elements listed by Lewis are not cultural traits, but rather part of the objective conditions of poverty, for example the lack of effective participation and integration of the poor in the major institutions of the society at large, the poor housing conditions and crowding they have to contend with. Furthermore, while Lewis interpreted the subculture of the poor as both an adaptation and a reaction to their position in a class-stratified, highly individuated, capitalistic society, many of his critics tend to see the cultural traits of the poor as by and large determined by the economic and political reality they face, by structural constraints. The very similarities across countries are taken as an indication that the poor have little scope for innovation, that the constraints they face are so severe as to narrow the range of possible responses. The most profound disagreement in theory, and serious concern in praxis, arises over Lewis's (1970*a*: 69) contention:[10]

> The culture of poverty, however, is not only an adaptation to a set of objective conditions of the larger society. Once it comes into existence, it tends to perpetuate itself from generation to generation because of its effect on the children. By the time slum children are six or seven years old, they usually have absorbed the basic values and attitudes of their subculture and are not psychologically geared to take full advantage of changing conditions or increased opportunities which may occur in their lifetime.

If Lewis subscribed to early childhood determinism, he simultaneously affirmed that any movement, be it religious, pacifist, or revolutionary, that organizes and gives hope to the poor and effectively promotes solidarity and a sense of identification with larger groups destroys the psychological and social core of the culture of poverty. He ventured the proposition that the culture of poverty does not exist in socialist countries, and specifically commented on a slum in Havana which he first visited in 1946:[11]

> After the Castro Revolution I made my second trip to Cuba [for five days in 1961] as a correspondent for a major magazine, and I revisited the same slum and some of the same families. The physical aspect of the slum had changed very little, except for a beautiful new nursery school. It was clear that the people were still desperately poor, but I found much less of the feelings of despair, apathy, and hopelessness which are so diagnostic of urban slums in the culture of poverty. They expressed great confidence in their leaders and hope for a better life in the future. The slum itself was now highly organized, with block committees, educational committees, party committees. The people had a new sense of power and importance. They were armed and were given a doctrine which glorified the lower class as the hope of humanity. (Lewis, 1970*a*: 75)

Lewis returned to Cuba in February 1969 with several collaborators to carry out a three-year research project. A good deal of information had been collected by the time research came to an abrupt halt in June 1970. The project included a study of a housing development of 100 units in which residents from the very slum Lewis had visited previously, Las Yaguas, had been resettled in 1963. While the research was still under way, he wrote to a colleague: 'It is . . . clear that many of the traits of the culture of poverty persist in the housing project. I believe I was overly optimistic in some of my earlier evaluations about the disappearance of the culture of poverty under socialism. However, there seems to me no doubt that the Cuban Revolution has abolished the conditions which gave rise to the culture of poverty' (Lewis *et al.*, 1978: 526, n. 1).

Butterworth (1980), who co-ordinated research on the housing development for four months in 1970, provides an account which is limited in so far as it is based on incomplete records from the suddenly interrupted study.[12] He emphasizes the tangibles that the Revolution brought to the former slum dwellers: secure jobs, a sufficient and balanced diet, excellent health-care, and improved housing. A substantial proportion of the men were skilled or semi-skilled workers, and a majority of these were young men who had acquired their skills since the triumph of the Revolution. Unlike the situation in Las Yaguas, where the father was sometimes an inadequate provider and thus may have been a marginal member of the household, he was now virtually assured of employment and a steady income. The result was a more stable household. Only 5 of 71 family households were made up of single women with their children.

If there had been major improvements, serious problems remained. A quarter of the children between six and fourteen years of age had either dropped out of school or never entered. Eight homes were recognized as black-market centres. While gambling in any form was forbidden, at least six homes served as gambling centres. Rum was illegally distilled and sold. There was trafficking in marijuana. Most striking was the low level of integration into mass organizations and campaigns. Such was the case for the campaigns to perform voluntary work in agriculture, for the Federation of Cuban Women, and for the neighbourhood organization, the Committees for the Defence of the Revolution (CDR). The three CDRs had experienced a limited degree of success for a year or so after their founding in 1964 but had gradually become inactive. Cursory investigations suggested that the CDRs in the six other housing projects where Las Yaguas residents had been resettled had followed the same pattern.

Butterworth points to both internal and external factors to explain the persistence of such problems. There had been a good deal of inertia in the CDRs, and personal feuds and animosities inhibited co-operation. We may further surmise that those involved in illegal activities had reason to resist the effective operation of a government-controlled neighbourhood organization. Indeed, much conflict was created within the settlement because of the attempt to enforce rules from the outside in an environment characterized by a relatively high incidence of deviance. One might be led to conclude that a populace reared in the culture of poverty largely persisted in its ways. However, Butterworth also notes that the residents still carried the stigma of having come from Las Yaguas; indeed, the housing development was referred to as 'Las Yaguas made of cement'.[13] They were looked down upon by the people in a middle-class neighbourhood close by.[14] Their children in the primary school had to put up not only with belittlement from the other pupils but with a school director who maintained that over 90 per cent of the children from 'the Las Yaguas block'—but not one of the other pupils—had been diagnosed as mentally retarded, or severely disturbed psychologically. The local People's Court held its trials outside the housing development, and the judges, none from the settlement, expressed a middle-class morality. Officials at the next higher level of the CDR organization had become disheartened at the lack of progress. As tasks were not attended to, they had eventually stopped all communication and co-operation with the local CDRs.[15]

Lewis's general argument, if not his initial perception of reality in post-revolutionary Cuba, may be labelled 'hard-culture'—in contrast to a 'soft-culture' approach arguing that changed circumstances elicit quite rapid cultural responses (Hannerz, 1969: 193–5).[16] If adults can modify their behaviour, if not their values, in response to changing conditions, intergenerational changes in behaviour and revision of values tend to be more far-reaching. Adolescents measure the cultural traits evolved by previous generations, proffered by their parents and their teachers, against the economic and political situation they confront, or more precisely, against their perception of that situation. As each new generation evaluates its cultural heritage against the conditions it faces, it collectively evolves its own patterns of behaviour and values. Changing perceptions of reality provoke the reworking of the available cultural inventory. Such a model of cultural change offers a better understanding of social process than either a hard-culture approach or structural determinism.

The critique of the culture of poverty concept was fuelled by concern that a hard-culture approach can be used to justify the continued neglect of the poor. Criticism was also inspired by a rejection of the rather bleak image of lower-class life Lewis presented. Negative evaluations of the behaviour patterns of the poor come about in two ways. Most poor are aware of middle-class values. While often living by different standards, many aspire to middle-class life. And middle-class observers tend to perceive lower-class behaviour as inferior. An appreciation of the conditions the poor face helps to understand the gap between ideal and actual behaviour, and goes some way toward putting into question such judgements made by outsiders. But a balanced perspective requires more: it has to be based on an appreciation of the positive elements of the subculture of the poor. Recurrent patterns of mutual aid and of solidary action among the poor merit special attention from such a perspective. Hollnsteiner-Racelis (1988: 232–3) reports from a neighbourhood in the lower-class Tondo section of Manila:

> Being poor forces a closeness beyond mere sociability, for crises arise frequently enough to encourage strong patterns of neighbouring. Mutual aid consists largely of contributions of food, money, or service upon the death of a household member or the happier celebration of a baptism or marriage. It surfaces again in the borrowing and lending of household items and money, maintaining surveillance over a neighbour's house or children while the mother runs an errand, notifying one another of job openings and (particularly for adolescents) support in the event of a gang fight with rivals from other blocks.

Lomnitz (1977, 1988) argues that the urban poor in Latin America find their ultimate source of livelihood in market exchange, but cannot survive individually: the market fails to provide any security, and the poor are not in a position to accumulate savings. They survive by complementing market exchange with a system based on resources of kinship and friendship: a system that follows the rules of reciprocity, imbedded in a fabric of continuing social relationships.

Lomnitz describes the pattern of reciprocity in Cerrada del Cóndor, a small shanty town in Mexico City. Here the prevalent rural patterns of individualism and mistrust are superseded by powerful tendencies toward integration, mutual assistance, and co-operation. Recent arrivals are housed, sheltered, and fed by their relatives in the shanty town; the men are taught a trade and oriented toward available urban jobs, in direct competition with their city kin. The migrants thus become integrated into local networks of reciprocity. Clusters of neighbours practise continuous exchanges of goods and services on an equal footing. They are made up of three or four—less frequently two, five, or six—nuclear families, and nearly all nuclear families belong to a cluster. Ideally, each cluster is composed of neighbours related through kinship, but a third of the clusters were partly or totally based on friendship. Ties within the clusters are reinforced through godfather relationships (*compadrazgo*) and through drinking companionship among men (*cuatismo*). The exchange is underpinned by a strong ideology of assistance:

> The duty of assistance is endowed with every positive moral quality; it is the ethical justification for network relations. Any direct or indirect refusal of help within a network is judged in the harshest possible terms and gives rise to disparaging gossip. People are constantly watching for signs of change in the

economic status of all members of the network. Envy and gossip are the twin mechanisms used for keeping the others in line. Any show of selfishness or excessive desire for privacy will set the grapevine buzzing. There will be righteous comments, and eventually someone will find a way to set the errant person straight. (Lomnitz, 1988: 259)

Patterns of mutual aid are common among the urban poor and appear as an effective adaptation to their circumstances. A more profound change in their condition requires solidary action. Indeed, in many Third World cities great numbers of the apparently powerless have grasped a measure of power through collective action: they have organized as squatters, they have gone on strike, they have taken over city streets. To these collective responses to poverty I shall turn in the next chapter.

Determinants of Urban Social Organization

We began our discussion of social organization with Wirth's classic essay that argues that urbanism constitutes a distinct way of life, a way of life determined by the demographic characteristics of the city—the fact that the city is a relatively large, dense, and permanent settlement of heterogeneous individuals. Wirth proceeded to posit a lack of social integration as the root cause of urban problems. We emphasized, in contrast, the substantial evidence that urban dwellers are quite well integrated: some remain rooted in a rural community of origin, many establish new communities of common origin or shared faith in the cities, most are integrated in networks of kin and friends. And, indeed, the evidence on mental stress and homicide (in the Third World) fails to support common assumptions about the detrimental consequences of the supposed disorganization of the city. And if property crime and victimless crime are more common in urban than in rural areas, social disorganization provides only one among several explanations.

The city is characterized not by a particular urban life-style, but rather by the life-style alternatives it offers. Urban dwellers, in sharp contrast to peasants, tend to have options. Their choices are circumscribed by political position and material condition rather than demographic variables. And even those with the least choice, the urban poor, can be seen to form their own response to the conditions they confront. Culture holds a central place: it shapes the responses of urbanites to their environment, even as it is modified in these responses. And kinship, the forms which it takes, the purposes it serves, appears as a core variable in these cultural patterns.

Whyte and Parish (1984: 358) present a Chinese model of urbanism, especially as it was developed during the Cultural Revolution. Among the basic structural elements of this model are strict migration controls and minimal urbanization, in spite of considerable economic development; a penetrating system of residential and work-unit organization; a highly developed bureaucractic allocation system; an emphasis on production rather than consumption; a relatively egalitarian pattern of distribution; a rejection of schools as the basic mechanism for sorting talent; much stress on citizen involvement in public health, social control, and other realms; and rigid taboos on all forms of dress, expression, ritual life, and communication that do not conform to the official ideology.

Whyte and Parish report that these structural elements contributed to a number of distinctive social features. High stability in jobs and residences, involvement and familiarity with neighbours and work-mates, minimal differentiation of consumption patterns and life-styles, low divorce, high female work-participation, and rapid changes in fertility and religious customs stand out. Whyte and Parish (1984: 368) observe how entrenched some urban characteristics, such as bureaucratism and the prestige hierarchy, proved to be. They further emphasize that when elements of social organization are effectively changed, the results are not always as expected: the pursuit of equality through class struggle, and an emphasis on class labels, alienated many and failed to eliminate awareness of occupational rank and privilege; the pursuit of comradely relations produced more interpersonal knowledge and concern but also a certain interpersonal wariness, and weak attachment to middle-level community and work organizations. And several reforms produced contradictory results: the pursuit of equality through class struggle, for example, destroyed much of the former unity of purpose of the goal of national strength and growth. Since the death of Mao Zedong in 1976 there have been major reversals on policies affecting urban social organization, changes expressly directed to deal with some of these unexpected consequences.

The Chinese experience thus focuses attention on a nation's political economy and level of economic development as a third variable to be considered, along with the cultural patterns of a society and the demographic characteristics of the city, in any attempt to understand urban social organization.

7

Patterns of Political Integration and Conflict

A wide gap in income and wealth, power and status separates the élite from the mass of the population in most Third World countries, and the middle class is usually quite small. The majority of the urban population have a standard of living so low that it is inconceivable to the average citizen of an industrialized country. While most urbanites are better off than the rural masses, there are some who have no shelter, who can barely clothe themselves. Malnutrition is common. For many the quest for food, for themselves and for their children, is a daily struggle for survival.

The urban masses cope with their condition in a variety of ways. In the last chapter we saw how the urban poor secure the help of kin, neighbours, and friends in the crises which frequently threaten those living close to subsistence. But government officials loom large as a threat or a resource. Police officers and tax collectors have to be evaded, trade licences and building permits are required, government employment or welfare assistance has to be applied for, public services may be secured.

Government leaders are less than responsive to the needs of the masses. They are usually preoccupied with securing the support of the armed forces and promoting the investment of indigenous and foreign capital. They are constrained by multinational corporations and foreign governments which provide investments, buy exports, supply raw materials and spare parts, and wield the ultimate threat of subversion and military intervention. In this concert of the powerful, the voices of the mass of the population usually remain muted. Many regimes have done away with elections altogether. In other countries they are a farce: in Togo, the general won 99.7 per cent of the votes in a 98 per cent turnout for the presidential elections in 1986.

Even where elections offer a choice, and the ballot is secret, one party is frequently well entrenched. Since it appears assured of re-election, such a party may receive widespread electoral support because areas voting for opposition candidates face neglect from the government or even outright retribution. Patron–client relationships such as characterize the Partido Revolucionario Institucional in Mexico act in a less crass manner. Once clients have established a working relationship with a patron, they have good reason to vote for, and to persuade their followers to vote for, the local list which renews their patron's mandate. Voting in opposition candidates would entail depriving oneself of local representatives who have access to government agencies.

In other countries, parties effectively compete for electoral support, but they are not mass-based. They become active in low-income neighbourhoods primarily at election time, offering limited material rewards in exchange for votes. Anthony and Elizabeth Leeds (1976) describe such a pattern in Brazil, and I shall return to their account. Wirsing (1976) reports on municipal elections in Nagpur, India, in 1969. Many candidates who provided patronage in the past were well positioned to secure long-term benefits for their constituents, but as election day approached short-term tactics came to predominate. The voters' choice was restricted through the kidnapping of political rivals, paid withdrawals of opposing candidates, and the setting up of bogus candidacies. Voters were intimidated, for example through administrators controlling the employment or the housing privileges of low-ranking government employees. And votes were bought:

> For many of the urban poor, the campaign period is a time when the market value of their support appears to soar and when the clever seller may turn the value of his vote or the votes of his followers to good advantage. The poor are wined and dined, wooed with gifts and bribed with cash. There are free haircuts, free saris, free mutton dinners, free entertainment, and plenty of free intoxicants. (Wirsing, 1976: 195)

If the circumstances vary, the outcome is similar throughout most of the Third World: the masses are rarely able to make significant inputs through an electoral process. This is particularly the case for rural populations. Where they do take part in elections, local élites frequently control their vote through patronage or outright coercion. Powell (1980), in a comprehensive review of rural voting patterns, concludes that almost everywhere, just below the surface, there is a 'rural mafia' system. The rural masses are thus usually without a voice and left with but three options. They may rebel when their land is taken away from them, taxes are raised, or the prices paid for their crops are lowered, but such upheavals tend to be isolated. Alternatively, peasants can withdraw from the market and revert to subsistence farming, jeopardizing food supplies for the cities and export crops. Either strategy may force limited concession, but neither can constitute an effective threat to the élites in the cities.

The most common response of the rural masses to the neglect, or even outright exploitation they face, is to vote with their feet. They form a large stream of rural–urban migrants who cannot be absorbed into the urban economy productively. They put severe pressure on urban resources. They swell the ranks of the urban masses strategically poised at the centres of local, regional, and national decision-making. They people the nightmares of conservatives and the daydreams of the radical Left:

All over the world, often long in advance of effective industrialization, the unskilled poor are streaming away from subsistence agriculture to exchange the squalor of rural poverty for the even deeper miseries of the shanty-towns, *favelas*, and *bidonvilles* that, year by year, grow inexorably on the fringes of the developing cities.

And they, too, are the core of local despair and disaffection—filling the *Jeunesses* movements of the Congo, swelling the urban mobs of Rio, voting Communist in the ghastly alleys of Calcutta, everywhere undermining the all too frail structure of public order and thus retarding the economic development that can alone help their plight. Unchecked, disregarded, left to grow and fester, there is here enough explosive material to produce in the world at large the pattern of a bitter class conflict finding to an increasing degree a racial bias, erupting in guerrilla warfare, and threatening, ultimately, the security even of the comfortable West. (Ward, 1964: 191–2)

This statement, by a Third World specialist soon to be appointed to the Albert Schweitzer Chair of International Economic Development at Columbia University in New York, was published in August 1964.[1] Earlier that year the military had toppled President João Goulart of Brazil after he had launched a campaign for broad structural and political reforms. It established a regime of repression. During the following decade there was little evidence of the urban masses threatening a regime anywhere in the Third World. Only in 1978 did mass demonstrations and strikes challenge the rule of the Shah of Iran. That same year the towns of Nicaragua rose to the call of the Sandinistas to topple Anastasio Somoza. Up to that point governments everywhere had shown themselves in control. That more and more countries came under military rule seemed to escape the attention of those social scientists who now swung to another extreme position: urban masses were described as conservative in outlook, ignorant of the political system, and apathetic in public affairs.[2]

Any generalizations about the political stance of the masses are bound to be misleading.[3] The urban masses are neither radical nor apathetic. Their behaviour, the attitudes that underlie it, and ultimately even their values have to be understood in terms of the economic, social, and political realities they face. The masses are frequently hostile to the more powerful and affluent. In varying degrees they understand the forces that create and perpetuate the inequality they so acutely experience. However, they have to get on with the daily struggle for survival. They evolve a design for living that takes the existing situation as a given to be coped with, only rarely defining it as a contingency to be challenged

(Portes and Walton, 1976: 72).[4] The recurrent theme in the interplay between the masses and those who control the state and dominate economic decisions is that the power of the masses is usually narrowly circumscribed. The quest to break these constraints is the preoccupation of a few visionaries; only rarely are historical constellations such as to offer them an opportunity to lead the masses in an effective challenge to the established order.

In subsequent sections we will discuss ethnic, religious, and caste identities that usually divide the masses, but at other times serve to focus opposition to a regime. We will explore three forms of mass action—squatter movements, labour movements, and street movements—that at times impact on the urban political arena. And we will conclude this chapter with an analysis of the rare cases where popular movements have overthrown a regime—I will argue that contemporary revolutions are invariably and necessarily urban-based. But first we will address the widespread patterns of patronage that muffle the voice of the masses.

The Politics of Co-optation

In any country public officials allocate a variety of resources, and enforce a range of sanctions. The proportion of resources they control tends to be large in Third World countries. The government is a major employer; in many places such employment extends beyond public administration to public or semi-public corporations in charge of major sectors of the economy. Governments frequently control prices, for example for agricultural products that have to be sold to government agencies, or for food sold in urban markets. They allocate scarce resources such as credit and foreign exchange. They attempt to collect an assortment of indirect and direct taxes. A wide range of activities, from building permits to trade licences, are usually subject to government regulation. The scope for sanctions is extended in those many Third World countries in which civil rights are limited or non-existent. Penal sanctions can be particularly severe where prison conditions are harsh and the death sentence is common.

Third World governments are frequently ineffectual in their exercise of control and sanctions.[5] Myrdal coined the phrase 'soft state' to express the fact that laws and government regulations tend to be flouted in Third World countries:

> The term 'soft state' is understood to comprise all the various types of social indiscipline which manifest themselves by: deficiencies in legislation and in particular law observance and enforcement, a widespread disobedience by public officials on various levels to rules and directives handed down to them,

and often their collusion with powerful persons and groups of persons whose conduct they should regulate. Within the concept of the soft state belongs also corruption . . . These several patterns of behaviour are interrelated in the sense that they permit or even provoke each other in circular causation having cumulative effects. (Myrdal, 1970: 208)

The soft state reinforces inequality. Those who control state power, or who have the economic resources to gain access to it, find opportunities for large-scale gains and evade sanctions. For the mass of the population the allocation of resources appears quite arbitrary. There are only a few jobs for the many workers with limited qualifications. Schooling, subsidized housing, adequate medical care, and trade licences are available only for some of the many who require them. Impersonal rules meant to govern the allocation of such resources remain a dead letter. The mass of the population has little leverage in dealing with the élite and its representatives who control these resources, whether it be managers in private firms or government officials. Not only do they have few legal claims, but those claims commonly are unavailing. Such a situation may be challenged through collective action, and in subsequent sections of this chapter I shall explore three forms such action may take. Or individuals may try to deal with the system as best they can. Cultivating a patron—a person in a position to further one's career and to assist in crises—is a promising approach.[6]

Four characteristics of clientelism stand out. The patron–client relationship is a relationship between two individuals, a dyad. The relationship is reciprocal: in exchange for the patron's help, the client gives political support and contributes to the patron's status. This exchange is not based on legal or contractual requirements but is an informal understanding; where the two have a legal or contractual relationship, such as employer and employee, the patron–client exchange introduces additional elements. Finally, the relationship is profoundly unequal: not only does the patron have greater power, more economic resources, and higher status, but he usually has numerous clients and the leverage any one of them can exert on the patron is therefore narrowly circumscribed.

Patron–client relationships may be seen as addenda to formal institutions (Landé, 1977). The fulfilment of institutionalized requirements does not establish the special patron–client bond. To take our example, meeting what are established expectations of employees in general cannot constitute the basis for patronage. Where the employer favours some of his employees, for instance giving assurance of permanent employment, and they reciprocate by performing beyond what is expected of employees in general, the patron–client addendum to the institutionalized employer–employee relationship is established. The point highlights the fact that the scope for clientelism is a function of the lack of impersonal rules for allocating resources. To pursue our example, if job tenure were institutionalized, if it were a function of qualifications, performance, and/or seniority, if it were not a special favour at the whim of the employer, then it could not be a source of patronage.

Complementary to the instrumental content of patron–client relationships, there are usually affective elements. Kinship, common origin, shared experiences, can be the source of genuine affective ties; mutual appreciation constitutes an effective reward. Sometimes a fictive kinship relationship is established. The *compadrazgo* system in Latin America and the Philippines is frequently used in such fashion. The client asks his actual or potential patron to act as godfather at his child's baptism, and the two men thereby become *compadres*.

The owner of a small firm who is a patron to some of his workers, or the mayor of a small municipality, has ultimate control over the resources he provides to his clients. In contrast, many patrons are themselves clients to higher-placed patrons. Thus, in large urban systems the poor man's patron frequently finds himself in the role of broker: he has to obtain many of the benefits for his clients through a patron of his own, and what his clients offer him in return—for example their votes—goes to a higher-placed patron with whom they do not interact directly. There is a pyramidal structure, where those at the intermediate levels are both patron and client, they are mediators between levels of social organization in a complex order (Michaelson, 1976). Patron–client relationships do not necessarily have to be repeated at the different levels of interaction, but typically clientelist patterns prevail throughout the hierarchy of social strata. The lack of impersonal rules and collective action characterizes the entire society, and clientelism is an all-pervading part of its political culture.

Clientelism takes different forms, but it has a long tradition in many societies. In pre-colonial Mali, patron–client relationships were found between urban merchants and peasants, and between long-established citizens of a town and both newcomers to the town and peasants. N. S. Hopkins (1972: 142–5) describes how the system evolved in the years following the imposition of colonial rule. It was now three-tiered, involving petitioner, grantor, and an intermediate broker. The relationship between the petitioner and the broker was often seen as the traditional link between merchant and farmer. After independence the brokers were usually people who held positions in the ruling party. Grantors might hold party positions, but the essential qualification was that they were in administrative positions with some degree of control over admissions to schools, recruitment for jobs, assignments of building lots, and so on. In a detailed study of Mushin, a recently established suburb of Lagos, Barnes (1986) shows that most political participation is based on patron–client relationships. A clientele that can be mobilized for various forms of action—voting, influencing, disturbing—is the basis for power, and this power in turn is used to strengthen and expand client support.

In Lebanon the origin, and some of the persisting peculiarities of patronage, can be traced to a feudal past. To a large measure, much of the country's socio-political history may be viewed as the history of various groups and communities seeking to secure patronage: client groups in search of protection, security, and vital benefits, and patrons seeking to extend the scope of their clientage (Khalaf, 1977). With the advent of electoral politics, success at parliamentary elections became the prerequisite for gaining access to governmental patronage. The political system was dominated by locally powerful leaders called *zu'ama* (plural of *za'im*). In the cities, they developed sophisticated machines to recruit a clientele. Michael Johnson (1986, 1988) describes the clientelist system in the Muslim Sunni quarters of Beirut before the outbreak of civil war in 1975. *Zu'ama* were not elected on the basis of a programme, but on their ability to provide their clientele with services:

> The *za'im* maintained his support in two important ways: first by being regularly returned to office, so that he could influence the administration and continuously provide his clients with governmental services; and secondly, by being a successful business man, so that he could use his commercial and financial contacts to give his clients employment, contracts, and capital. Depending on the wealth and influence of his clients, the *za'im* provided public works contracts, governmental concessions, employment in the government and private sectors, promotion within the professions and civil service, free or cheap education and medical treatment in government or charitable institutions, and even protection from the law. In order to survive electoral defeats and periods in opposition, the *za'im* had to be rich, or have access to other people's wealth, so as to buy the support of the electorate as well as the acquiescence of ministers and officials responsible for particular governmental services. Although election to the Assembly was of considerable advantage, it was not always essential, and a *za'im* could survive temporary periods of opposition and political weakness by using the credit he had built up in the past. (Johnson, 1988: 310)

The clients were expected to offer in exchange consistent political loyalty. It usually took the form of voting for the *za'im* and his allies in parliamentary elections. The clientele could be required to support the *za'im* in other political conflicts, and even be expected to take up arms in disputes with other *zu'ama*. Clients publicly demonstrated their loyalty: on feast days, they visited their *za'im* to wish him the compliments of the season; and when a *za'im* returned from a journey, his supporters usually turned out to welcome him home. Such occasions often involved the supporters holding a great reception, during which the *za'im*'s armed retainers would fire their machine-guns into the air to express loyalty, political strength, and jubilation.

The most important part of the *za'im*'s organization was a core of strong-arm neighbourhood leaders, the *qabadayat* (plural of *qabaday*). The *qabaday* recruited a following on the basis of his reputation as a man of the people, as a helper of the weak and the poor, as a protector of the quarter and its inhabitants, and, most important, as a man who was prepared to defend his claims to leadership by force. Typically, the *qabaday* was a criminal involved in protection rackets, gun-running, hashish smuggling, or similar activities. The *za'im* provided him with protection from the police and the courts in return for his political loyalty and services. The *qabaday* recruited supporters for the *za'im*, assisted the *za'im* in the allocation of patronage to his supporters by vetting their loyalty and political reliability, and turned them out for mass demonstrations of support. Most important, the *qabadayat* acted as 'election keys' and could usually ensure that clients voted in the way they were told. The *zu'ama* maintained control because they were in a position to play arbiter: to shield their followers, and in particular the *qabadayat*, from the authorities, or to call in the police to punish the recalcitrant.

The use of the *qabadayat* fragmented the electorate and prevented the emergence of self-conscious social categories linked horizontally, such as interest groups or classes. By forcing clients to approach him through their quarter or family *qabaday*, the *za'im* encouraged the individual client to see himself as a member of a particular quarter or family grouping, and discouraged the formation of other social groupings that might have posed a real threat to the *status quo*. In some cases *zu'ama* actually created artificial quarter and family identifications. The electorate was further fragmented because most favours remained in the gift of the *zu'ama* and went to individuals.

The clientelist system worked quite well in Lebanon until the outbreak of civil war in 1975. However, the power of the *zu'ama* was such that there were few arbitrating mechanisms to control conflicts among them. Costly feuds arose. Such feuds contributed to the outbreak of a full-scale civil war in 1958. But the system was flexible enough for enemies to become reconciled. In the 1970s, however, unemployment rose, and it became increasingly obvious that the Muslim community—and especially the Shiite sect—was underprivileged compared with the Christians. With the development of social and communal conflict, the clientelist structures came under increasing pressure and eventually collapsed. The *zu'ama* largely controlled both the allocation of state resources and the state's coercive potential, and their position was predicated on a state providing resources, and holding the means of coercion in the police and the army. The collapse of the state in the civil war in 1975 deprived the *zu'ama* of their base. They lost control over their *qabadayat*. Some *qabadayat* emerged as prominent militia leaders over the course of successive rounds of fighting.

Co-optation provides the foundation for a small élite's power in many other countries. It manifests itself in various ways:

> Again, the weakness of the state allowing the emergence of patrons, may have different forms. It has been said, for instance, that whereas the Lebanese state is an association of patrons, the Tunisian state is a machine for the making and unmaking of patrons. There may be a kind of spectrum, ranging from a state which is an association for the protection of a pre-existing patron class, via a state which plays off patrons against each other, to a state which creates and destroys them through temporary allocation of political positions. (Gellner, 1977: 5)

While patrons usurped much of the power of the independent state established in Lebanon in 1943, India presents an intermediate position. The state has considerable power, but major public resources are expressly shifted to voluntary associations. Two key arenas are education and housing: most schools, from pre-primary to collegiate, are run by associations heavily dependent on public financing; and co-operative housing societies allocate land and credit. These associations are often set up along linguistic or caste lines. Many local politicians are active in them. Obtaining and dispensing government patronage, they maximize electoral support. Thus a third of the candidates for municipal councillor in Nagpur in 1969 held executive positions in co-operatives, and a quarter did so in education societies (Wirsing, 1973; Michaelson, 1979).

In Mexico political power is concentrated at the apex of the state, in the hands of the president and his close advisers. The Mexican government legitimizes itself on the basis of a revolution, justifying social change in the name of revolution. Yet the government consistently flouts the officially proclaimed revolutionary goals. Nevertheless, in the regular elections the Institutionalized Revolutionary Party, ruling since 1928 under a variety of 'revolutionary' names, has not been threatened by any of the other parties until quite recently. Co-optation throughout the political and administrative system effectively channelled and scaled down demands, and paralyzed nearly all potential opposition (Hellman, 1983: 135–63). In a study of three low-income areas in Mexico City, Eckstein (1988*a*: 78–102; 1988*b*) found a range of social and economic associations, in addition to political and administrative groups. On their own, residents had organized to deal with the municipal government. They had also joined national groups that deal with municipal authorities. Some of these groups publicly profess a concern with social welfare and have grass-roots organizations. However, all these efforts had met with limited success. Eckstein (1988*a*: 87) explains:

The informal processes that inhibit the political effectiveness of organized residents are rooted in various national institutional arrangements: mainly in the hierarchical nature of national inter- and intragroup relations, the class structure, the personalistic style of Mexican politics, the multiplicity of groups operating nationally along with the containment of overt élite competition, and the government's threat to apply force. They reflect social structural forces closely associated with the state, but not necessarily deliberate efforts of functionaries to regulate local poor. For these reasons political stability can be maintained with little cost, and the state can nationally advance the interests of capital without using brute repression against a major base of political support.

Local leaders are constrained to conform with national political 'rules of the game'. Otherwise they can neither personally advance in politics nor secure benefits for their constituents. The prospects of removal from office also compel local subordinates to conform with the expectations (or their perception of the expectations) of higher-ranking functionaries. Thereby, 'appropriate' local concerns are delimited and heads of local groups are encouraged to be subservient, although they are not necessarily accordingly rewarded.[7]

Eckstein reports that local leaders usually established institutional ties with national political-administrative groups. They felt that their political and economic mobility would be thereby enhanced, and that they would secure social and urban services for their followers. On occasion they fashioned covert arrangements with non-local functionaries because they felt that constituents or potential constituents were antagonistic to openly associating with the ruling party. Mexico's urban and rural poor thus could not use the ostensibly democratic political institutions to advance their own interests effectively. And the Mexican urban poor did not appear any better off than their counterparts in Latin American countries that were non-democratic at the time.

When Eckstein revisited the three neighbourhoods in 1987, she found that local group life had atrophied in two of them. The civic groups of the early 1970s were now defunct. No new independent, or quasi-autonomous, civic groups had emerged in their place. Government-Party hegemony had prevented such movements from arising. Residents no longer saw any reason to organize. As a consequence, co-optation no longer had the regulatory effect of earlier years. Thus, the means for bringing out the vote for the ruling party had disappeared. In one of the neighbourhoods, the party's share of the vote had halved between the last two elections, and about a quarter of voters had abstained even though voting is obligatory in Mexico (Eckstein, 1988*a*: 253–8).

In the inner-city neighbourhood, however, local groups had asserted themselves and pressured the state to accommodate to them. The victims of the 1985 earthquake took to the streets *en masse* making public their demands. Alone or with victims from other sections of the city, they marched to the Plaza of the Constitution to present petitions to the mayor; to the Zócalo, the central square of Mexico City; and to Los Pinos, the home of the President. In March 1986 about 50,000 earthquake victims from the neighbourhood and elsewhere threatened to camp out in the Aztec stadium, where the world soccer tournament was scheduled to begin two months later. The earthquake victims in the neighbourhood succeeded in getting the government to expropriate the property of absentee landlords, and build new housing to be owned by the former tenants. They pressured the government to build them provisional housing nearby as well, so that they could oversee the construction and maintain any local business they had. When they considered the temporary housing unsafe they set it on fire—and made the government construct safer housing in its place.

The disruptive tactics of the earthquake victims succeeded in specific circumstances: the earthquake received extensive international and national media coverage; foreign agencies, Mexicans from well-to-do neighbourhoods, professionals, political parties, and the Catholic Church provided assistance. As a consequence, the neighbourhood was not socially or politically isolated, as low-income communities typically are. When the earthquake victims fought official indifference, the Mexican state eventually responded: it responded in a manner that gave protesting groups a stake in the reform programme. Residents defied the public order in ways that did not attack the state frontally. And they did so with the support of better-situated individuals and groups. The government in turn negotiated with the movement and made major concessions. However, it required the movement to work with, and not against, the state. The government thus co-opted opposition through reform until order was restored (Eckstein, 1988*a*: 263–78; 1990). The Mexican opposition made a strong showing in the 1988 elections. It remains to be seen how much longer the Institutionalized Revolutionary Party can control the state. Whatever the future may hold, the Party succeeded, through the manipulation of clientelist networks, to hold on to power for more than two generations.

Cultural Elements in Political Alignments

Glaring contrasts between rich and poor characterize most Third World countries. The primary alignments, however, are frequently in terms of shared identities of a different order. Three modes of distinguishing 'we' from 'they' stand out: region of origin, religion, and caste.[8] The salient characteristics of such identities are articulated in juxtaposition to the identities of other groups. Such antagonisms are not only present in individual consciousness and enacted in more or less public settings, they are also stored in the unconscious (Saberwal, 1986, 65–73). In much of Africa, and in major parts of Asia, most migrants in the cities and their children identify themselves and others in terms of region of origin. To refer to such a pattern as 'tribalism' is unfortunate, because of the pejorative connotations the term 'tribe' has acquired. It is also misleading. Such identities in the urban setting bear a rather tenuous relationship to traditional societies and their culture. The past provides the raw materials, but ethnic identities are fashioned in the confrontations of the urban arena.

Adherents of two or even three world religions live within the same city in parts of Africa and Asia. At times they have clashed with each other, most dramatically Hindu and Muslim in South Asia. But conflict can also crystallize around internal division within a religion, such as the Muslim sects in Syria. Caste is of central importance in Hindu religion. For the majority of Indians it remains the primary referent. Determined by birth, and supposedly immutable, caste can change its guise according to the situation.

Recognizing common origin is the basis for political alignments in many parts of the world time and again. In the cities of sub-Saharan Africa, the great majority of adults are first-generation urban dwellers: they spent their early childhood in a culturally more homogeneous rural environment. In town a whole series of identities of origin are available to them: their extended family, their home village, the village group to which it belongs, their 'subtribe', their 'tribe', their 'supertribe', their nation, their race. Some identities represent traditional bonds and shared culture, others are new. I refer to such identities or origin and descent as 'ethnic'. New arrivals may discard some of these identities, others they may not yet be ready to recognize as their own. Which ethnic identity they embrace varies according to their situation, in terms of the categories of people to whom they oppose themselves. Banton (1965: 145) observed that the migrant to Freetown, Sierra Leone, was involved in a series of oppositions: African versus European, tribesman versus Creole, Temne versus, say, Mende. The three categories—African, tribesman, Temne—were in a straight hierarchy, each one was socially relevant only when a higher-order opposition did not enter. Such a series of identities of origin may be seen as a nested hierarchy (Leeds, 1973). A set of concentric circles provides a graphic model of the more narrowly or more largely defined ethnic groups (Gugler, 1975*a*).

Members of an ethnic group frequently monopolize major political and/or economic opportunities. Two processes are at work. First, an ethnic group builds up an important lead. Thus in many countries Christian missions focused their work on particular regions, giving those people a head start in Western education. When cash crops were introduced, they favoured particular regions, such as country suitable for cocoa-growing. And the military officers who rule many countries today, frequently come from ethnic groups, the 'martial tribes', preferentially or even exclusively recruited into colonial armies.

Second, once some members of an ethnic group are in a privileged position, they wield considerable influence over the opportunities open to others. When their patronage goes to kinsmen, fellow villagers, or any 'brother', an entire group can be seen to enjoy privilege, excluding others. Thus Marwaris from Rajasthan have come to play a leading role in trade and industry throughout much of India (Timberg, 1978). And Hausa control the long-distance trade in cattle in West Africa.

The perceived coincidence of ethnic difference and economic opposition may prompt a reaffirmation of cultural distinctiveness and culminate in an ethnic renaissance. Abner Cohen described such a process among the Hausa people in Ibadan, Nigeria, concluding:

> The situation will be entirely different if the new class cleavages will overlap with tribal groupings, so that within the new system the privileged will tend to be identified with one ethnic group and the under-privileged with another ethnic group. In this situation cultural differences between the two groups will become entrenched, consolidated, and strengthened in order to express the struggle between the two interest groups across the new class lines. Old customs will tend to persist, but within the newly emerging social system they will assume new values and new social significance. A great deal of social change will take place, but it will tend to be effected through the rearrangement of traditional cultural items, rather than through the development of new cultural items, or, more significantly, rather than the borrowing of cultural items from the other tribal groups. Thus to the casual observer it will look as if there is here stagnation, conservatism, or a return to the past, when in fact we are confronted with a new social system in which men articulate their *new roles* in terms of traditional ethnic idioms. (Cohen, 1969: 194)

drawing

Language is frequently the key element in delineating common origin.[9] Common language not only simplifies communication, it is the medium of ideology. In Nigeria, the Ibo Union was established in Lagos in 1936. A Pan-Yoruba cultural society was founded in London in 1945. Both played a principal role in enlisting popular support for two major Nigerian parties after the Second World War, parties which were soon identified with the Igbo (the current spelling) and Yoruba people respectively. A third party came to represent Northern Nigerian interests. These divisions shaped civilian politics until 1966, underlay the civil war that visited widespread suffering from 1967 until 1970, and continue to be a salient feature of Nigerian politics. In the past neither the Igbo nor the Yoruba actually constituted a political unit. Still, these new groups follow lines of cultural affinity, and in particular of shared language. The various dialects of the Igbo people are clearly distinct from the languages of their neighbours. At the same time, there is such variation within Igbo, that some dialects are barely intelligible to speakers of others. Significantly, major efforts were made to standardize Igbo speech and writing. In addition, certain forms of traditional culture, such as music and dance, were reaffirmed, and a new pride in Igbo identity cultivated.

In some cities conflicts arise between indigenous peoples and migrants. In Nigeria this has occurred in Kano (Paden, 1973), Ibadan (Sklar, 1963: 289–320), and Lagos (Pauline Baker, 1974), urban centres established in pre-colonial times. However, such conflicts are the exception in sub-Saharan Africa. Because of the recent origin of most cities and their explosive growth, the urban-born usually are a small minority. Even in India, with a greater proportion of cities many centuries old, with colonial cities dating back to the nineteenth century, and comparatively slow urban growth, the lines of ethnic conflict are usually drawn between migrant groups.

People from a city's hinterland and migrants from more distant lands are frequently juxtaposed in conflict. 'Nativism' is a label which describes such confrontations between 'sons of the soil' and 'outsiders'. The 'native' protagonists are sons of the soil in that they were born in the city's hinterland and are newcomers to the city. They typically maintain that their needs have precedence over the needs of outsiders who hail from distant lands and/or speak a different language. In federations such as India and Nigeria, state boundaries reflect cultural diversity and at the same time provide a ready framework for the assertion of subnational rights. Natives hold that they are committed to the advancement of their region, whereas outsiders are perceived to take an exploitative approach, oriented towards the development of their region of origin. Nativist demands are usually fuelled by resentment against outsiders. They are seen as having a head start, and controlling major opportunities. As Weiner (1978: 293) concluded for India: 'Nativism tends to be associated with a blockage to social mobility for the native population by a culturally distinguishable migrant population.'

Bombay is one of India's most cosmopolitan cities. Its migrants constitute a wide variety of religious, cultural, and linguistic communities. They were competing for political power and educational opportunities already in the nineteenth century (Dobbin, 1972: 217–46). Still, Bombay is one of the few Indian cities where the city has become a dominant element in the residents' identity. Here they call themselves Bombaywallahs: signifying that they belong to a culturally and linguistically heterogeneous city. Once the state of Maharashtra was established, Marathi speakers attempted to turn Bombay, the state's premier city, into a Marathi city. They held that the city belonged to Maharashtra, that native Marathi speakers had special rights, and that Marathi had to be used at least as a second language. The Marathi nativists explicitly rejected the concept of the Bombaywallah, in favour of ethnic identities of origin. Their party, the Shiv Sena, articulated their demand that jobs in the city should not go to immigrants from other Indian states. Within two years of its founding in 1966, the Shiv Sena became the largest opposition party in the municipal elections, supported by a majority of the Marathi-speaking people in the city. The Shiv Sena did not gain power in the state government, but the governing Congress Party adopted many of its stands: pressuring private employers to recruit Marathi-speaking people over other migrants, giving preference to local people for employment in the state government, and tacitly supporting signs in Marathi on public and private places. Such political pressures were re-enforced by physical attacks against business men who were not co-operative (Katzenstein, 1979; Weiner, 1978).

Language constitutes a central factor in ethnic conflict in countries in which indigenous languages compete for dominance. Not only does language delineate ethnic groups and serve to affirm ethnic identity, but language policy is of vital concern to the various groups. The political decision of which language or languages to use at various levels of the educational system affects educational opportunity. The choice of official language affects access to the bureaucracy and, of particular importance, circumscribes career opportunities there. The language policies adopted by private firms have similar consequences. In contrast, language policy has rarely become such an issue in sub-Saharan Africa. While all countries in the region are multilingual, nearly all are dependent on the lingua franca introduced by the colonial power. Decisions about the use of such a lingua franca have class rather than ethnic implications: the children of the educated élite have a decisive advantage in educational systems where a foreign language is the language of instruction from an early age.

Alignments of shared origin, religion, or caste usually obscure stratification at the national level and defuse class conflict. At other times, those in power are perceived to represent an ethnic minority or a caste and become isolated. Nor is religion invariably the opium of the masses. In Latin America, the ecclesial base communities inspired by 'liberation theology' have nurtured dissent, undermining the legitimacy of established structures and leaders. They have laid the foundation for new kinds of leaders and solidarities. The new theology stresses issues of justice and equality, calling for a 'preferential option for the poor'. Religious values thus provide justification for poor people to mobilize against immoral social structures. Religious symbols come to signify the just struggle. Injustices once claimed to be divinely ordained are now portrayed as social constructs that can be changed. Ecclesial base communities have fuelled revolutionary movements in Nicaragua and El Salvador. They have organized mass mobilizations for democratization in Brazil, Chile, and Haiti. They have supported protests against austerity programmes in the Dominican Republic (Eckstein, 1989).

Religious leaders and their institutions enjoy a moral authority that sets a high price on repression. The Shiite clergy in Iran spearheaded the movement to overthrow the Shah. In Malaysia, the intellectual leaders of the Islamic revitalization movement have formulated radical critiques of multinational corporations. They have informed the consciousness of worker-members, providing them with a lens for recognizing their situation as exploitative and a political idiom to articulate this *exploitasi* (Ong, 1990). In Northern Nigeria, Islam speaks to the experience of proletarianization to develop class consciousness. In a survey of unskilled workers in three factories in Kano, Lubeck (1986: 262–303) found that those who had spent more years in Koranic school were more likely to express support for factory trade unions and to favour workers' political activity. Members of Islamic brotherhoods similarly distinguished themselves from non-members. In addition, they were more likely than non-members to hold that their pay was unfair and that Islam approves strikes for a fair reason. Among these urban workers, those most integrated into Muslim organizations—and presumably most knowledgeable about Islamic values—were thus most likely to interpret these values as supporting labour militancy.

The Struggle for Land

In many Third World cities the most conspicuous political action of the urban masses is the illegal occupation of land. The urban dweller who can barely make ends meet will try to avoid paying rent. For some this means no more than space in the open street to lie down and sleep. But more common is securing a spot where some kind of shelter can be constructed without paying for the land.[10] Squatting may occur unobtrusively: an individual, a family, or a small group, establish themselves undetected or, when found out, are tolerated by the authorities. More typically, they are liable to forcible eviction by police. In a number of Third World countries, however, the illegal occupation of land has been organized on a scale so large as to persuade the authorities initially to condone it and eventually to grant the squatters legal title. (On the problems of categorizing different types of low-income housing and the use of terms like 'squatter', see the section on 'The Nature of Self-Help Housing' in Chapter 5 above.)

Large-scale seizures of land are carried out rapidly, and usually at night, so as to present the authorities with a *fait accompli*. This is reflected in some of the designations used for squatter settlements: in Mexico they are known as *barrios paracaidistas*, i.e. 'parachute settlements'; in Turkey as *geçekondu*, i.e. 'built overnight'. Such invasions are major operations which require careful planning. A group of people prepares in secret. A site is selected and surveyed where the likelihood of eviction is low. The site's ownership is particularly important: eviction is less likely from land owned by the government or the church, from property in dispute, and from sites that hold little interest for anybody but the poor, for instance land subject to flooding, waste dumps, steep hillsides. A public holiday, or the visit of a foreign dignitary, is chosen as an invasion date when the authorities will be reluctant to use force. Finally, contacts are established to secure support from political or religious leaders, and to ensure that sympathetic journalists will be present to denounce police excesses. On the agreed night, the chosen site is occupied by people numbering in hundreds and sometimes more than a thousand, plots are marked out, and rudimentary shelters put up. Usually the squatters project an image of moderation: they fly the national flag on their new huts, naming the settlement after a religious or government figure.[11]

The specific form squatting takes varies with the political context. In Peru various military and civilian governments since 1948 have tolerated land invasions, though specific policies toward squatters fluctuated (Collier, 1976). Anthony and Elizabeth Leeds (1976) attribute this liberal approach to the fact that the Alianza Popular Revolucionaria Americana (APRA) had a mass base, primarily in rural and urban trade unions. All political action, whether undertaken by any of the other parties, by coalitions, by the bureaucracy, by a civilian or a military executive, or by the Catholic Church, had to cope with APRA's popularity and its covert organizational links in the bureaucracy, the military, the legislature, and elsewhere. In order to circumscribe APRA's popular support, even quite conservative groups had no alternative but to pursue policies *vis-à-vis* the squatters that ranged from accommodation to the allocation of substantial resources.

In Brazil the political context is quite different. Parties compete in state and national elections—they did so even during the military regime—but none of them has a mass base:

> Thus, in Brazil, are found an array of parties, none of which has a mass base in the sense of effective, formally organized, local, self-expressive constituencies acting within the operational norms of the party. All the parties both intentionally and unintentionally (perhaps!) follow policies, make choices, and act in such a way that such a mass base cannot or will not develop. Specifically, whether deliberately or not, they act to maintain links to the constituencies through patronal ties alone and even encourage their clients in the masses to operate as lower level or sub-patrons to *their* clients, voters in general . . .
>
> This structure of party organization, of course, defines the rules of the game

that must be played *among* parties. For example aside from each party's elite-class interest *not* to allow an organized mass-base to develop, it is also *constrained* from organizing such a base (which, in the ideal, it could hopefully keep under control) by the threat of the other parties' also organizing a mass-base—indeed, by the threat of creating a Chilean kind of party situation, a quite intolerable idea in Brazil. The peculiar congeries of circumstances which, in Peru, allowed an APRA to spring up as a unique mass-based party did not occur in Brazil. Preventive action (e.g., such as after the coup of 1964, the military's smashing the unions, low-ranking military, student, and left-wing clerical movements) is taken expeditiously when such a set of circumstances seems to be developing—usually by a broad elite-party coalition of peculiar bed-fellows who fall out with each other shortly afterwards. Brazil, then, from the point of view of elite-bounded parties, is characterized by a pulsating tension between controlledly mobilizing for votes a mass excluded from real participation and intensifying its exclusion sometimes to the point of almost universal repression. One major feature of the procedure is the maintenance of that fragile bond, the paternalistic politician, who can so easily, too, be withdrawn as a contact route for the proletariats. (Leeds and Leeds, 1976: 205)

In this context *favelas* are, for the most part, settled by accretion. The parties deliberately avoid the establishment of formal political organizations at the local level. Instead, the local scene is dominated by informal cliques. Clients of the party organization, they furnish an ambience of pro-party feeling which produces votes. They are not to bring into being an organization that might have an independent basis of power.

The repressive character of Brazilian politics is reflected in major campaigns to eradicate squatter settlements. Following the military coup in 1964, and particularly between 1969 and 1972, the *favelas* of Rio de Janeiro suffered from sweeping evictions. It is estimated that 80 out of 283 *favelas* were totally or partially destroyed, and that 140,000 people were rehoused.[12] Opposition was quelled forcefully. In 1966, in the *favela* Jardim America, the police confronted about 2,000 people, many of them children. They had received word on the previous day that their homes were to be demolished. To speed up the process and discourage resistance, gunshots were fired randomly into the crowd, and those who seemed to resist were beaten. In 1968, the Federation of Favela Associations in Guanabara held a congress which resolved upon 'rejection of any removal, and the condemnation of the human and financial waste and of the social problems resulting from removal'. The federation immediately mobilized to prevent action against the *favela* Ilha das Dragas. Together with the local Residents' Association it instructed every household to refuse to give information to those in charge of preparing the resettlement. Soon after, the leaders of the federation and of the local association were arrested. They were held incommunicado for days. And they were threatened with severe consequences if there were further opposition. Open protest by the federation was effectively ended. When, in the following year, 7,000 residents of Praia do Pinto refused to be relocated, the *favela* 'accidentally' caught fire. Although residents and neighbours called the fire department, no fire-engines came. Orders had evidently been issued that no help was to be sent. Most families were unable to salvage their few possessions. By morning almost everything had been destroyed. Ominously, the *favela's* leaders disappeared. In spite of such repression, by 1977 one million people, one out of every five inhabitants of Rio, were estimated to live in *favelas*, more not just in absolute numbers, but in relative terms as well, than in 1968 when the major efforts towards eradication were initiated (Perlman, 1976: 200–7; Valladares, 1978*b*).

Chile was characterized, until the military takeover in 1973, by fierce electoral competition among several mass-based political parties. For the most part they were closely connected with sectors of the well-established trade union movement. Squatter settlement took two forms: *callampas*, through accretive occupation of a particular site; *campamentos*, through organized invasions. The inhabitants of the *callampas*, like their counterparts in Brazil, manipulated the competition among the parties, establishing patronal ties with bureaucracies and trade unions. If the affiliations of *callampas* were heterogeneous, *campamentos* were invariably led by a political party. Indeed, the original invasion was typically encouraged by a party and organized by its leaders.

The extent of the organized illegal occupation of land and its sponsorship was a function of the national political configuration. There was a dramatic change even during the reformist presidency of Eduardo Frei of the Christian Democratic Party. In 1965 a few land invasions sponsored by the Left were violently squashed. With the municipal elections of 1967 there was a sudden rise in invasions, and the government responded again by repression. However, in 1969–70, with the presidential elections approaching, government resistance broke down. Early in 1969 the authorities still opposed invasions forcibly; in March several squatters were killed in Puerto Montt. A large invasion in January 1970 initially met with strong resistance, but it was ultimately successful. As the presidential campaign got under way, repression diminished. Following the public outcry at the treatment of squatters in Puente Alto in July 1970, the government abandoned all recourse to force. When Salvador Allende, leader of the Left-coalition Unidad Popular, was elected president on 4 September 1970, the period until his inauguration on 3 November was marked by indecision on the part of the authorities. This encouraged many new invasions, leading to the establishment of *campamentos* of a variety of political tendencies. A total of 23 invasions was reported in 1969, but the movement exploded to 220 invasions in 1970. The movement reached its peak in 1971 with 560 invasions: once Unidad Popular had been established in authority, it was the Christian Democratic Party rather than the Left that encouraged land invasions, and even extended the movement to occupying recently built apartments (Collectif Chili, 1972; Castells, 1983: 200). (See p. 149 above for discussion of the incidence of land invasions in Santiago.)

The different strategies of government and opposition parties can be seen to arise from the responsibility of those in authority to mainain law and order, that is the respect of private property and the enforcing of planning, building, and health regulations. However, those in power have a wide array of resources to attract the vote of the poorly housed. The government allocates low-cost housing and site-and-services plots; it controls the issue of land titles to squatters; it grants planning and building permits; it establishes and maintains urban infrastructure; it may even offer building materials to squatters and employ them to build their own housing, as was the case in Chile under the Unidad Popular government. The opposition, in contrast, cannot offer any of these resources. To sponsor the illegal occupation of unused land or buildings is an alternative approach towards gaining popular support. If such actions undermine the authority of the ruling party, so much the better.

The politics of squatter settlements vary from country to country, and change within a country over time. They are also articulated differently by various squatter settlements in a country at a particular political juncture. In Mexico, the Institutionalized Revolutionary Party used to be highly effective in incorporating squatters. The squatter settlements on the periphery of big cities were characterized by strong community organization. They were under the tight control of leaders who acted as intermediaries between the squatters and administration officials. Squatters were allowed to stay on the land they had occupied, and were eventually provided with urban services. The Party in turn was assured of their votes and support. In the 1970s different patterns emerged in a few squatter settlements out of the junction between the interests of the squatters, the reformist policy of the administration of Luis Echeverría Alvarez, and the commitment of intellectuals radicalized by the repression of the 1968 student movement. The radical Left tried to organize and politicize some squatter settlements, connecting their demands to the establishment of permanent bases of revolutionary action and propaganda within these settlements. They were often unable to overcome the squatters' fears of reprisals and the Party's solid political organization. When the radicals did succeed in organizing a squatter settlement as a revolutionary community, the state resorted to violent repression. Nevertheless, a few settlements resisted police repression and survived by maintaining a high level of organization and political mobilization.

The radicals had their greatest success in Monterrey, in northern

Mexico. There students led land invasions and established settlements expressly designed to oppose the administration from 1971 onward. In 1976 the Popular Front for Land and Liberty was established, bringing together 31 settlements with approximately 50,000 squatters, 16 tenants' associations, 3 *ejido* organizations and 3 workers' unions.[13] They refused to have their illegal land occupation legalized by the government: economically, legalization would have meant high payments for a long time; ideologically, the movement asserted the people's natural right to land; politically, legalization would have undermined the movement by creating a specific relationship between individual squatters and the administration. The squatters similarly emphasized self-reliance in the provision of services. They established a meat co-operative, a co-operative providing construction materials, a petrol station, a super-market, a chicken farm, a clothing plant, and a shoe factory. Collectively they built schools, health facilities, and civic centres. Water, sewerage, and electricity were provided by illegal connections to the city systems. Repeated seizures of buses persuaded the bus company to service the settlements. Honour and Justice Committees passed judgment on conflicts in the Monterrey settlements. They established an internal security organization to maintain public order. A prison was constructed for those who had committed minor offenses. Alcohol and prostitution were prohibited. Those who collaborated with the government were expelled from the settlements, while the police were not allowed inside.

The success of the radical squatter movement in Monterrey has to be understood in terms of a particular political constellation. Monterrey is an industrial city dominated by a local bourgeoisie that stands in opposition to the ruling party. The Monterrey bourgeoisie openly opposed President Echeverría's reformism and launched a major attack against the state governor he had appointed. The governor reacted by taking a populist line. The squatter movement was thus in a position to play the national and the state government off against the city administration. These were exceptional circumstances, and the Monterrey movement remained the only urban movement of such size and character in Mexico. Its leaders had only a local constituency and could not make a national impact. Even within Monterrey, the squatters were isolated. Nearly all had recently arrived from rural areas. The cohesion of the movement was enhanced by family ties among squatters, by *compadrazgo* and friendship, by shared origin. But these recent immigrants, unable to secure jobs in the large industrial plants, had little contact with the industrial workers to whom the paternalistic local bourgeoisie provided relatively high wages and substantial benefits, including housing.

The political constellation was altered when the more conservative regime succeeding Echeverría appointed a new governor to the state in

1979. Between 1973 and 1978, 29 invasions were successfully carried out, and only 2 failed. Now, however, all attempts to establish new settlements were repressed. Not one out of 42 invasions attempted between 1979 and 1984 succeeded. Radical trade unions solidary with Land and Liberty were repressed as well. And a land-title programme was launched to divide the squatter movements. Most of the squatter settlements in Monterrey were legalized. The Land and Liberty movement, representing 14 per cent of the squatter families in Monterrey, was left quite isolated. The movement was divided by leadership conflicts, and by 1981 most of the co-operatives had ceased to function. When the state offered to deal with the squatters, part of the movement split between those maintaining the oppositional stance and others dealing with the authorities (Castells, 1983: 194–9; Vellinga, 1989).

In Chile, a survey of 25 *campamentos* in 1971, the first year of the Unidad Popular government, showed major variations in form and extent of political activity. The key variables appeared to be, on the one hand, the social composition of a given *campamento*, i.e. the extent to which common workers, élite workers,[14] and lumpenproletariat were represented; and on the other hand, the political line pursued by its outside sponsors (Collectif Chili, 1972). Presumably there was some interplay between these two variables because different parties would tend to sponsor different social strata in their attempts to secure land and amenities.

Squatter settlements composed primarily of blue-collar workers in low-paying industries and sponsored by the Movimiento de Izquierda Revolucionaria (MIR) were the most politicized and radical. The radical commitment of MIR was clearly established. A student organization founded in 1965, it had remained undergound during the Frei period, and acted as a leftist opposition to the Allende government for the first two years before joining it. Common workers were most receptive to radical mobilization due to their interests and their political culture. They had more to gain from revolutionary change than élite workers; and, unlike the lumpenproletariat, they had an experience of organized political struggle in the trade unions.

Radical transformation for squatters meant both, alternative institutions within the settlements, and revolutionary changes in the national political structure. Internally, the most dramatic change was the emergence of autonomous judicial control in a few *campamentos*. With it came new definitions of what was considered unacceptable behaviour, and a search for integrative forms of punishment. Non-participation at meetings or poor direction of an assembly called for an accounting. Particular attention was directed toward resolving conflicts within families. Drunkenness was a target of social control: those who arrived drunk at a *campamento's* entrance would be detained, and in some, alcohol was banned altogether. Sanctions ranged from self-criticism, study of revolutionary texts, physical punishment, internment, and fines, to expulsion from the *campamento*. But reprimands and the settlement of quarrels and reconciliation of those involved were most common (Collectif Chili, 1972). Handelman (1975: 40, 42) emphasized the significance of these developments:

> Because the *campamento* is somewhat removed from the existing socio-political order and lacks basic municipal services, its inhabitants develop their own civic institutions to deal with a variety of daily problems—maintenance of law and order, criminal justice, housing, local administration, and the like. For the most part the police, welfare agencies, courts, and government bureaucracies of both Frei and Allende's civilian administrations did not operate within the *campamentos*. Consequently, the marginal communities created new alternative institutional forms in some of these areas that offered the basis for greater mass participation and political consciousness in the society at large. Thus, the settlements may well have contained the seeds of a new, mass-based social order . . .
>
> Had the Chilean military coup not terminated such *campamento* organizations, their political implications might ultimately have been quite significant. In effect, some *campamentos* had created a small-scale state within a state. Locally controlled courts, work brigades, and other administrative agencies offered an opportunity for active political participation that is normally far beyond the reach of the urban poor in the developing world.

The radicalized *campamentos* made demands above and beyond the initial quest for land. They occupied hospitals to demand health services, they occupied administrative offices to protest delays, they participated in mass demonstrations (Collectif Chili, 1972). When the truckers went on strike in 1972, the *pobladores* opened up the highways and organized supplies. They established committees to control distribution and forged links with factories, eventually obtaining food directly from the rural areas. In 1973, when the awareness spread that a military coup was imminent, the squatter settlements prepared to resist it.[15]

On 11 September 1973 the military acted in concert. The few loyalists within the armed forces were rapidly neutralized. The Moneda, the presidential palace, defended by a handful of the faithful with sub-machine-guns, succumbed to an assault by tanks and planes. Salvador Allende, who led the resistance at the palace, perished. With him died the promise that socialism could be established in Latin America in a peaceful transition, through the electoral process. Pockets of resistance at some factories were crushed by military might. Mass arrests, systematic torture, and random killings ushered in a regime of terror. In what came to be known as the *Pinochetazo*, more than 2,000 people were killed. Many more were imprisoned or went into exile. A brutal military regime had imposed its variety of law and order.

The Force of Organized Labour

The classic response to the oppression of the masses is the call for the workers of the world to unite to create a just order. Such solidarity must appear as a far-distant goal at a time when workers in the industrialized countries nervously eye the industrialization process elsewhere, opposing investments in Third World countries and imports of manufactured goods from there. Workers in Third World countries have to rely by and large on themselves. They depend on their own resources to ameliorate their condition or to radically transform their society. They may chant 'The Internationale', but the chorus in distant countries is rarely heard.

In contrast, the resources and opportunities available to business interests and governments are less circumscribed by national boundaries. Multinational corporations can pick and choose a location, securing advantageous conditions for their investments when they play host governments off against each other. Most governments can count on external support to counter any popular threats to their rule, whether it be developing the mass media they control or equipping and training the army and police. The Brazilian military government sent torture experts to instruct Pinochet's men in Chile.

Generations of workers have laboured in the textile mills of India, down the gold mines of South Africa, on the docks of Brazil since the nineteenth century. By now nearly all Third World countries have a history of worker protest.[16] Knowledge is travelling of revolutionary changes wrought in some countries, of reforms obtained by workers in many others. Rising levels of education foster critical awareness of historical precedent, developments around the globe, and their own condition. The wide gap in living conditions separating the labouring masses from local élites, foreign advisers, and tourists is less and less taken for granted. Still, the emergence of a broad, class-conscious movement of workers is very much the exception in Third World countries. Such movements are usually aborted by governments responsive to pressures from workers in strategic positions. Time and again potentially powerful sectors of the labour-force are induced to behave as vested-interest groups, concerned to preserve and improve their privileges rather than practise solidarity with the great numbers of less-privileged workers.[17]

Industrial action in Third World countries tends to have immediate political repercussions. The government is invariably a major employer.

Its bureaucracy absorbs a substantial proportion of the labour-force. In addition, public or semi-public corporations often have full or partial control of utilities, of mining, and of major industries. Furthermore, the government frequently plays a part in setting wages, and establishing and regulating working conditions, in the private sector. In this situation wage-earners direct their grievances towards the government: they are expressed in a political form, and the state cannot appear in the role of arbitrator between wage-earners and their employers.

Any country's economy is dependent on its working force. However, the leverage different sectors of the labour force can exert varies greatly. On the one hand, unskilled workers are easily replaced by eager recruits from the unemployed and underemployed. On the other hand, many skills are in short supply. At the same time, underdeveloped economies are heavily dependent on the effective operation of a few key sectors. A strike by railroad or dock workers brings the entire economy to a halt in many a country. A work stoppage by mineworkers can threaten a foreign-exchange crisis. Those who work such crucial economic resources, and whose skills are not easily replaced, can thus exert very real political power.

The workers in large-scale enterprises constitute a compact mass difficult to manipulate—unlike small firms where workers are usually closely controlled by a paternalistic employer. The very numbers of workers in a big plant can become menacing, as in the *gherao,* i.e. workers blockading managers in an office. And threats of mass dismissals lack plausibility, because large numbers of skilled workers are difficult to replace at short notice.[18] Further, conflict frequently arises in more than one plant at once. Many issues—minimum wages, compulsory contributions to retirement funds, social-security benefits—transcend the boundaries of any single firm. They affect workers in general and bring them together in opposition to political authorities. Once substantial numbers of workers are mobilized, there is a potential for mass riots in which urban poor are likely to join.

The task of organizing industrial workers is made easier because they are concentrated in the factory, industrial estates, and residential communities. In Karachi, labour protest has been based on both place of employment and residence. In 1961 the decision to call a general strike was reached at a meeting of about 10,000 workers on a hilltop between the main industrial area and the neighbouring workers' colonies. The 1972 strike showed a high degree of co-ordination between industrial and residential areas. The workers' colonies, where the police hesitated to venture, were a haven for the strike leaders. The factories were efficiently *gheraoed,* often by work-shifts: the shift going home would bring the news from the industrial front and the shift going to the sites would take back directives from the leadership (Shaheed, 1979).

Sandbrook and Arn (1977: 57) conclude, from their survey of political attitudes in two residential areas in the Accra-Tema conurbation, that the emergence of a working-class orientation is encouraged among workers who live in 'occupational communities'. Communities where—owing to a concentration of the similarly employed, their insulation from moderating outside influences, and their particular schedules occasioned by shift-work—workmates interact both on and off the job, creating and reinforcing common images of the world.

Labour organizers face the threats of repression, and the blandishments of co-optation. Trade-union leaders are prominent among political prisoners in Third World countries; they have suffered torture and met violent death. Many have compromised their vision of a more just society. Some have done so simply to reap the benefits promised them individually; others chose to compromise to obtain benefits for limited sectors of labour. The classic case occurred during the Mexican Revolution when the 'Red Battalions' of urban workers supported Venustiano Carranza, the representative of the middle class, against the disinherited Indians of rural Mexico led by Emiliano Zapata. Urban workers secured substantial benefits in Uruguay when they gave their loyalty to José Batlle at the beginning of this century, in Brazil when they were wooed by Getúlio Vargas from 1937 on, in Argentina when they provided the mass following of Juan and Eva Perón from the mid-1940s.

The rule of the Institutionalized Revolutionary Party in Mexico presents a case study in co-optation, as we have seen. During the presidency of Lázaro Cárdenas, in the 1930s, organized labour was incorporated into the Party as one of four sectors, the others being the peasant sector, the military, and a confederation of 'popular organizations'. Thus much of Mexican labour was organized in unions affiliated, through union confederations, with the ruling party. According to the standard account, state control of the union bureaucracy constrained rank-and-file militancy, strikes were few and predictable, wages rose less than productivity, and organized labour could be counted on as a fundamental pillar of support for the policies of the regime. Roxborough (1984: 34), however, argues that corporatist control over labour in Mexico has been both weaker and more uncertain than commonly assumed. He predicts that the long-run trend is to further erode corporatist mechanisms.

Roxborough's study of unions in the Mexican automobile industry started out with the assumption that the new 'independent' unions established in the 1970s would operate differently from the old-style 'official' unions tied to the ruling party. He found that the three independent unions were, indeed, among the most strike-prone in the industry. However, two of the official unions were also strike-prone. The common characteristic of these five unions, which had gone on strike repeatedly between 1970 and 1980, was democratic process within the union. These unions, unlike all but one of the four oligarchically run unions, had full-time union officers paid for by the company, and a greater degree of control over work-loads and line speeds. They tended to be more supportive of their members when conflicts went to the labour courts and to have obtained somewhat faster wage growth. Union democracy characterized the three independent unions established by movements breaking away from the official union confederation affiliated with the ruling party. The success of these three breakaway movements may be explained in turn by the fact that they operated in plants established in regions of the country where the official confederation was relatively weak. A fourth militant union had achieved a compromise whereby it became independent in substance while remaining within the official confederation. The origins of democracy in the fifth militant union could be traced to its establishment during the pro-labour regime of Cárdenas. Roxborough suggests that the political liberalization that began in the 1970s makes the suppression of union dissidence increasingly hazardous, greatly expanding the scope for manœuvre of trade-union militants (Roxborough, 1984: 155–63, 174).

Tacit collusion among government, big business, and trade unions is common where organized labour has not established a political alliance with the ruling élite. In capital-intensive industries the share of labour costs is low. Employers are concerned to minimize labour turnover and absenteeism among trained workers. They want to avoid strikes that entail huge losses in output and may lead to sabotage of valuable equipment and violence against managerial personnel. Employers, especially if they represent foreign or minority capital, also need to be careful about their public image. Governments in turn are acutely aware of the crucial contribution certain sectors of labour make to the economy. And they are made uneasy by the disruptive potential of massed urban labour. Elements of urban labour thus wield power than can be translated into wages and benefits, working conditions, and job security governed by legislation and/or collective bargaining. I have already referred to them, in Chapter 4, as a 'protected' work-force. They appear privileged *vis-à-vis* the rest of urban labour in nearly every Third World country.[19]

The internal stratification of labour is usually codified by social-security legislation which is partial and unequal. In Latin America, only a small portion of the economically active population was covered by such legislation as late as the 1960s. Countries such as Brazil, Colombia, Guatemala, and Mexico provided coverage for less than 30 per cent of the economically active population. At the other extreme Cuba provided total coverage, and Chile about 70 per cent coverage (Rosenberg and Malloy, 1978).[20] The history of social-security legislation in the region is instructive:

> Social security coverage in general evolved on a piecemeal, group-by-group basis. The usual sequence in the evolution was: the military and public functionaries; workers in critical infrastructure activities; workers in important urban services; and industrial workers. By and large, the quality of coverage was positively correlated with the sequence of coverage. Both the sequence and quality of coverage were determined by the power of groups to pose a threat to the existing sociopolitical systems and the administrative logic of the contractual type of social insurance schemes developed within the region. As a result, the great mass of rural workers and urban marginals were either ignored or received coverage of an inferior quality. The upshot was the incremental evolution of social security systems that were both highly fragmented and unequally stratified in terms of the quality of programs. Even among late adopters, which had more unified administrative structures, coverage was extended on a piecemeal and unequal group-by-group basis.
>
> The corporatist structures of 'representational participation' were an important factor accounting for the fragmentation and inequality of the social security systems. These structures, which were often a part of a general corporatist approach to labor relations, reflected the goal of established elites to undercut the emergence of a broad, class-conscious movement of workers. In the social security area these structures both encouraged and permitted discrete groupings to pursue their own particularistic advantage at the expense of other groups. (Rosenberg and Malloy, 1978: 168)

The degree of protection enjoyed by sectors of the labour-force varies from country to country. According to Roberts (1978: 134), state regulation of the economy in Brazil depressed wages to such an extent that there was relatively little difference between the wages of unskilled and semi-skilled workers in the large-scale sector and those in the small-scale sector. He attributes this to the influence of a strong internal and foreign-linked bourgeoisie. Roberts describes Peru, in contrast, as an 'enclave' country where the national bourgeoisie was weak. The state under nationalist pressures imposed strict labour regulations on the predominantly foreign-controlled large-scale enterprises. The workers in such firms enjoyed much better conditions than their counterparts in the informal sector. The parallel with the different government response to squatter settlements in these two countries, discussed in the preceding section, is striking.

In countries with electoral politics, trade unions tend to be attached to political parties. Indeed, multi-party competition is often mirrored in a proliferation of trade unions hostile to one another. Although the connection with political parties resembles the tendency of squatter movements to come under party control, industrial workers have greater leverage than squatters. Their unions, if allowed to operate freely, are not easily manipulated. In India, unions are started by party functionaries. Once established, they are linked to parties both through trade-union federations—which are informally tied to parties through their leaders—and directly through important union officials who are members of the sponsoring party and union cadres who are also the local party cadres. Every union is referred to by the name of the party to which it owes allegiance. Union and party collaborate closely in strikes, elections, agitations, and meetings. Intra-party struggles and shifting party alliances are reflected in conflict within unions and changes in union coalitions. Through prolonged membership the rank and file develop loyalties to the sponsoring party. Yet the unions have a logic and momentum of their own. The political commitment of the workers, and even of the union leaders holding high offices in the party, is strongly influenced by their perception of union interests. Thus when the Congress Party and its national trade-union federation denounced the *gherao* movement as unconstitutional, the unions associated with them in different parts of West Bengal nevertheless organized *gherao* and other violent protest movements. Rather than adhering to the more conservative ideology of the Congress Party, they responded to the establishment of a leftist United Front government in the state which had cut them off from sources of patronage. In contrast, when the major Marxist union moved to support the United Front government, stopped all agitations, and began collaborating with management, it lost much of its support (Sengupta, 1977; Chatterji, 1980: 147–58; E. A. Ramaswamy, 1973, 1984, 1988).

The primacy of union over party is demonstrated by the case of the Chilean mine workers. They supported the Left while it was in opposition, but when the Unidad Popular government came to power they increasingly transferred their electoral support toward the parties of the centre and the right. Zapata (1975) infers that the miners' political radicalism was not ideological, but was instead connected to a search for political support for economic goals. This pattern was highlighted when the workers at the El Teniente copper mine went on strike against the Allende government in April 1973. Substantial numbers stayed on strike until June, even though the 'workers' government' was already severely threatened by pressures from foreign powers, Chilean capital interests, sectors of the middle class, and the military.

Shared economic goals may unite workers in their demands in spite of profound ethnic divisions. A. L. Epstein's (1958: 234–40) principle of 'situational selection' is pertinent. He used it to explain developments, in colonial days, on the Copperbelt in what has since become Zambia. The African migrant workers gave allegiance to ethnic leaders in some situations, but in other situations had moved away completely from representation on an ethnic basis: in a situation of conflict with the mining company they presented a common front. In deeply divided Lebanon, the trade unions preserved their unity and effectively defended the interests of workers even in the middle of the civil war (Hanf, 1988).

The repressive character of many Third World regimes leaves little scope for dissent. Workers in key industries tend to be among the few groups able to articulate opposition. In South Africa, trade unions representing the disenfranchised majority have played an important part in the confrontation with the racist regime for nearly two decades. In early 1973 a series of illegal strikes erupted in the Durban area. Over the course of the year 90,000 workers went on strike, despite government repression and dismissals. In the aftermath, unions based upon effective shop-floor organization in the factory and regular involvement of the rank and file were established. By 1979 the regime was forced to grant official black trade-union rights—for the first time in South Africa. Many of the new unions obtained substantial gains for their members. Some ventured beyond the factory gates to co-ordinate protest campaigns in the townships (Friedman, 1987; Mitchell and Russell, 1987).

Drake (1988) shows how in the 1970s trade unions in Chile, Argentina, and Uruguay survived the onslaught of military regimes intent on destroying them. In Brazil, the military who had usurped power in 1964 ultimately failed in their strategy of incorporating organized labour. They arrested leading unionists, placed many unions under the control of the Ministry of Labour, and imposed strict limits on union activity. Almost all strikes were made illegal. But the Metalworkers' Union of São Bernardo do Campo, representing the workers in the automobile plants in the Greater São Paulo region, came to be the most serious opposition to the military regime. In 1978 it initiated strikes which spread to other cities and to other industries, reaching even schools, hospitals, and banks. An estimated 280,000 workers in over 250 firms stopped work during the first four months of the movement. The government deposed the leaders of the Metalworkers' Union in 1980, and imprisoned them for a while. However, the only interim executive the government could appoint owed allegiance to the deposed leaders. During a strike at Ford in 1981, the firm negotiated with Luis Inácio da Silva, the former president of the union popularly known as Lula, even though he had no official position whatsoever. When elections for the union executive were eventually held, the slate supported by the deposed leaders, who were not allowed to stand, overwhelmingly defeated an alternate slate (Humphrey, 1982: 24, 160–207, 229). Lula gained such national prominence as to come close to being elected president of Brazil in 1989.

The strike is the classic, potentially powerful weapon of labour. Workers can also protest by less directly confronting the powers that be. Foot-dragging, slow-downs, absenteeism, pilferage, and sabotage undermine productivity. They may lead employers and/or governments to address the grievances of workers. Thus Cuban workers have, on occasion, quite effectively expressed resentment by poor on-the-job performance and high absenteeism. Their low level of compliance compelled the regime to modify policies implemented in the late 1960s. At the time, workers had been asked to work long hours—without additional pay in the case of the sugar harvest. Meanwhile, the government stressed exports and investments in basic industries, and living standards plunged. Asked to work for 'moral' reasons—out of a commitment to Communism—workers resented the demands on their time and the deterioration in their living standards. In the absence of institutional channels to voice their discontent, they expressed it through poor work performance. The regime, confronted with a sharp decline in productivity, not only reintroduced material incentives and expanded the supply of consumer goods, but also granted labour greater participatory rights in enterprise, union, and government decision making (Eckstein, 1989).

In Chile, the massive protest against the Pinochet regime that emerged in 1983 was led by the Copper Workers' Confederation. Initially it called for a national strike. When important union locals refused to participate, the confederation decided to build on the growing discontent among the population at large and called for a 'national protest'. On the day of the protest, there were strikes, widespread absenteeism, work slow-downs, and demonstrations at work-places. Assemblies and demonstrations took place at the universities. Younger children stayed away from school. In the city centre and on main thoroughfares drivers honked their horns and people staged brief demonstrations. In middle- and lower-class neighbourhoods alike, residents boycotted stores; at night they turned out their lights, banging pots and pans. Some shanty towns erected barricades. The union leadership continued to play a central role in the convocation of subsequent 'national protests' that were called almost monthly. While the protests failed to dislodge the regime, they obliged the regime to make a number of economic and political concessions (Garretón, 1989).

The Chilean 'national protests' demonstrate the wide range of possibilities of expressing opposition. Where an opposition manages to establish itself on public places and in city streets, it can become a veritable challenge to those in power.

Street Power: Demonstrations, Riots, and Insurrections

Most Third World regimes narrowly circumscribe the scope for public debate. The opposition is denied genuine elections, the right of assembly, and access to the media. Challenges to the regime are frequently made by organizing large-scale meetings in public places, or marches through city streets. Massed demonstrators who are prepared to stand their ground, or indeed to press against police or soldiers, present a regime with difficult choices. Violent repression may elicit unfavourable international reactions, but such reactions usually generate more heat than action. Of greater concern to the regime must be that the victims of repression may become martyrs overnight and rouse as yet acquiescent groups to join the opposition. For civilian regimes there is the threat that once the military have been called on to repress a mass movement, they will decide to take power into their own hands. Withdrawal from the confrontation suggests that the regime is weak. It can spell the destruction and looting of public and private property, and may physically threaten regime supporters in their very offices. Control of city streets means power, and a movement that establishes itself on the main squares and thoroughfares of the capital city constitutes a threat to government.

Las Madres de la Plaza de Mayo, the Mothers of the Plaza de Mayo, innovated an approach to public space that played a major role in undermining the military regime in Argentina. Las Madres was composed exclusively of women. Most of them were mothers of children who had been kidnapped and disappeared. They had met in antechambers and public offices while attempting to obtain information about their children. The women decided to meet in the Plaza de Mayo, the square in the heart of downtown Buenos Aires that faces the presidential palace, to publicize the plight of their children. Fourteen women met there on 30 April 1977. They decided to return the following week, and a pattern of regular meetings was established. Initially the women sat on benches, or huddled in groups, exchanging information and discussing the names of women who might be willing to join them. The police told them that they could not loiter in the Plaza. They decided to walk silently for half an hour around a small pyramid erected in the Plaza to commemorate the beginnings of the independence movement.

Marching silently around a national symbol in a downtown location bustling with people, including tourists, became a weekly ritual. The group marched every Thursday at 3.30 p.m. By June 1977 it had grown to one hundred. As the repression subsided and the military regime weakened, the ranks of the organization swelled. By 1982 Las Madres claimed a membership of 2,500. The women attracted local and international attention to the *desaparecidos*: the large number of people who had disappeared at the hand of agents of a regime that refused to acknowledge its responsibility. At first the women used Catholic symbols: for example a carpenter's nail on their back in memory of Christ's sacrifice. Later they began to wear white kerchiefs on which they wrote the names of their kidnapped children. They carried pictures of their children around their necks or in their hands. As the military loosened their grip, especially after the disastrous Malvinas campaign, the women's marches spearheaded the vast mobilizations that hastened the return of civilian rule to Argentina. By the time the electoral campaign began, Las Madres had successfully transformed the *desaparecidos* into an issue that no political party could ignore or could afford to negotiate (Navarro, 1989).

The occupation of Beijing's Tiananmen Square in 1989 dramatically demonstrated the symbolic importance of physical control over urban space. Large numbers of dissidents occupying the most prestigious location in the nation's capital for several months, a leadership unable to receive its most important foreign visitor in many years, Mikhail Gorbachev, in proper fashion, undermined the legitimacy of party and government. Whatever route the regime chose to dislodge the protesters, the costs were bound to be substantial to those committed to the status quo. Either concessions had to be made to the dissident movement— tacticians can argue about the relationship between the concessions necessary to regain control and their timing, strategists can ponder how concessions might engender new demands. Or the dissidents had to be repressed: the option eventually chosen. That choice cost about a thousand lives. And the regime's legitimacy was profoundly compromised both in the country and in the international arena.

The predicament of the Chinese leadership was aggravated by the assiduous coverage of the events by foreign media sympathetic to the dissidents. Media coverage, typically foreign media rather than the muzzled national media, is crucial. The lack of reaction in the absence of media coverage is instructive. The point was demonstrated when the Syrian regime massacred an estimated thirty thousand in Hama in 1982. The regime managed to keep the foreign media away from this secondary city, and barely a murmur of disapproval was heard. The South African regime instituted strict controls on foreign media in 1985, which had the effect intended. Live reports on the protest movement had broadened the international appeal of the anti-apartheid movement. Now, as South Africa disappeared from television screens, the breadth of the overseas movement shrank.

The austerity measures adopted by debt-ridden Third World countries, typically under pressure from the International Montary Fund, sparked an unprecedented wave of protests around the world. These 'IMF Riots' began in the mid-1970s and became widespread in the 1980s. Walton and Ragin (1989) report 77 separate incidents of strikes, riots, and demonstrations in 26 countries between 1976 and 1987. Most occurred in Latin America. Walton (1989) details 54 austerity protests in 13 Latin American countries from mid-1976 to mid-1987. In 1989 nearly 300 people were killed in riots triggered by an increase in the price of petrol in Venezuela.

The political impact of street demonstrations, riots, and insurrections varies. Walton (1989), analysing the results of protests, including strikes, against austerity measures in Latin America, notes that stunned governments frequently rescinded or ameliorated their austerity measures, or provided compensations. In addition, protests sometimes initiated a successful movement to depose a government, or added a push to a teetering regime. Particularly notable, the protests contributed to persuade external actors—foreign governments, the International Monetary Fund, private bankers—to retreat from austerity policies, at least for large-debt countries.

Wiseman (1986), on the other hand, concludes from his survey of urban riots in West Africa between 1977 and 1985, that they have been largely ineffective as a mechanism for changing policy. Only two out of the 46 riots he enumerated in the region during the 9-year period were successful in bringing about a change in government policy. And in these two cases the rioters had suffered grievous losses ranging from several dozen women killed in Conakry, Guinea, in 1977, to around 150 deaths in Monrovia, Liberia, in 1979. In the case of a few other riots, the regimes made some conciliatory moves, generally by releasing those who had been detained. Although some commissions of enquiry were set up, none resulted in findings that were hostile to the regime. Wiseman acknowledges, however, that in at least four cases riots, by indicating a decline in the regime's legitimacy, played some part in encouraging a subsequent military coup.

The power of the street was dramatically demonstrated in two revolutions: demonstrators took to the streets again and again to defy the Shah's regime in Iran, and urban insurgents challenged the Somoza regime in Nicaragua. Correspondents for *Newsweek* reported from Teheran in late November 1978:

> Hours after the 78-year-old leader of Iran's zealous religious opposition [the Ayatollah Khomeini] issued his cry for a *jihad*—a holy war—against Shah Mohammed Reza Pahlavi, it was heeded. In Teheran's central bazaar, demonstrators streamed out of mosques. Most were young. Many brandished chains—symbols of religious flagellation—or wore white mourning shrouds. All seemed willing to press their cause to the end. Shouting 'Death to the Shah' and 'God is great' they taunted the Shah's jittery troops, offering themselves as martyrs. Machine-gun and rifle fire rattled across the darkened, winter-chilled streets of Teheran. Yet another wave of violence swept Iran's capital the following day. By nightfall, officials put the death toll at 30; opposition leaders claimed that far more Iranians had been killed. (Butler, 1978: 44)

In Nicaragua, teenagers fought Somoza's National Guard. With rifles and pistols they confronted trained soldiers who brought in tanks and were supported by planes. A *Newsweek* correspondent reported in late August 1978:

Gangs of masked teenagers darted through the rubble-strewn streets of Matagalpa, Nicaragua's third-largest city. Ducking into doorways, the young gunmen paused to fire their pistols and rifles at government patrols and then regrouped behind barricades of bricks and burning refuse. Inside shuttered houses, families huddled together listening to the chatter of machine guns, the roar of homemade firebombs and the wail of ambulance sirens . . .

The battle for Matagalpa was, for the most part, a children's crusade; many of the snipers were teenagers, and some were as young as 12 . . . Finally, after six days and about 50 fatalities, resistance ceased. The National Guard swept through the town and announced that 'all is peaceful and tranquil'. The teenagers and a few Sandinist guerrillas melted into the surrounding hills, vowing to return and fight again. (Labich, 1978: 44)

Street demonstrations played a key role in bringing down the Shah's regime, and urban insurrections overwhelmed the Somoza regime eventually. These events provide support for the view that revolutions, in this day and age, are born in confrontations in the urban arena.

The Urban Bases of Contemporary Revolutions

Chinese Communists, led by Zhou Enlai, were driven from Shanghai in April 1927. A Communist commune established in Canton in December of that year was short-lived. The attempt to base a socialist revolution on the urban proletariat had failed. The Communist Party moved to build up peasant support through land and tax reforms in the areas it controlled. From guerrilla origins it raised peasant armies which were victorious in a civil war fought on conventional lines. Mao Zedong, an early advocate of the rural strategy, was to declare the establishment of the People's Republic of China in Beijing on 1 October 1949. The Chinese experience demonstrated, or so it seemed, that Third World revolutions would have to be based on the peasantry. The rural-based guerrilla movements in Indonesia, Vietnam, Algeria, Mozambique, Angola, and Guinea-Bissau reinforced this point of view (B. Moore, 1966; Skocpol, 1979). However, the struggle in these countries was directed foremost against a colonial regime. We need to distinguish such wars of national liberation from revolutionary movements which challenge a national government for control.[21] I will argue that contemporary revolutions are largely urban in character.[22]

National wars of liberation have invariably been fought in predominantly rural countries. More than 80 per cent of the country's population lived in rural areas when the Dutch abandoned Indonesia in 1949, as the French left Indo-China in 1954, and when the Portugese granted independence to Guinea-Bissau, Mozambique, and Angola in 1974–5. Algeria was a more urbanized country but still about two-thirds rural when the French finally made peace in 1962. In every case, the colonial power was challenged by a rural-based guerrilla movement that curtailed the production, processing, and transport of crops for export, thus striking at the very backbone of the colonial economy. If the colonial regime's capability to extract agricultural surplus was thus impaired, it invariably continued to control the cities effectively; this was the case even in Algeria, where the French destroyed the large terrorist organization in Algiers. Colonial regimes withdrew of their own accord, and at a time of their choosing, even if precipitately in the case of the French after their defeat at Dien Bien Phu. The benefits to be derived from maintaining the colonial presence had been curtailed, and the expeditionary force, the metropolitan government, and/or powerful sectors of public opinion were led to weigh the costs of protracted warfare. The colonial power eventually accepted that its interests would be better served by a harmonious relationship with a formally independent country. Indeed, most colonies were granted independence before any armed resistance emerged.

Wars of national liberation do not threaten the very existence of the metropolitan power but, rather, its circumscribed overseas interests. Revolutionary movements, in contrast, seek the annihilation of the regime they confront. They threaten its supporters with exile, imprisonment, and the firing squad. Essential characteristics of the revolutionary movement are that it employs extra-legal means in challenging a national government, and that elements outside the government and the security forces play the principal role: it is 'popular' in the sense that it bases itself outside the ruling élite. Revolutionary movements have seized power in four countries since the triumph of the Chinese Revolution in 1949. A brief look at the cases of Bolivia, Iran, and Nicaragua will show that these were essentially urban struggles. The case of Cuba is more ambiguous.[23]

If Ernesto 'Che' Guevara inspired revolutionaries throughout the Third World, he exalted rural guerrillas in his *Guerrilla Warfare*. The argument of the primacy of the rural guerrilla *foco* was further developed by Debray (1967). The austere, difficult, and dangerous life experienced by the Cuban *guerrilleros*, and their encounter with rural poverty, profoundly marked Fidel Castro, Che Guevara, and their fellow fighters, many of whom came similarly from the urban middle class. The fact that it was Castro, and a nucleus of his comrades from the Sierra Maestra, who seized control of the destiny of Cuba added weight to their experience. And, indeed, the Cuban Revolution is unique in the last four decades in that attacks by rural-based guerrillas led to the disintegration of the army they confronted.

Still, a closer look at the Cuban Revolution reveals a more complex picture. First, a large proportion of the *guerrilleros*, 60–80 per cent according to one source, were drawn from the urban population. The urban underground was instrumental in recruiting them, and in enabling them to join the fighters in the mountains. Second, the urban underground provided the lifeline for the *guerrilleros*: supplying them with arms, information, money, and even food, and establishing the contacts through which they gained national and international recognition.[24] Third, the urban underground carried out a wide range of violent actions and sustained most of the casualties. More than five thousand bombings were reported in 1957 and 1958. Havana's international airport was burned. The kidnapping of the Argentine racing driver Juan Fangio, who was kept prisoner for two days, attacted world-wide attention. The underground in Cienfuegos collaborated with the conspirators of the naval uprising there in 1957. The most spectacular action was carried out by the Directorio Revolucionario, the student organization, when it stormed the presidential palace on 13 March 1957 but failed in its attempt to kill the dictator.[25] In comparison, there were probably never more than 300 *guerrilleros* in the Sierra Maestra at one time. This was the case even during the largest army offensive, in the summer of 1958, when 40 of their number fell. Finally, when the regime collapsed, rebel troops were too few to seize power. It was the urban underground that policed the streets and took over the administrative machinery. And it was a general strike that signified mass support for the rebels and discouraged attempts to establish a conservative successor regime (Karol, 1970: 164–80; Thomas, 1977: 256–63).

The Bolivian Revolution may be seen as compressed into three days of intensive fighting when the government was overthrown in 1952. Or it may be traced back to its roots in organized labour, and disaffected elements of the middle class. The artisan-labour movement had reached a degree of national coherence at the beginning of the Great Depression. During the Second World War, organized labour had become politicized and radicalized as its leadership shifted to the tin miners. In an uneasy alliance with the labour movement, the Movimiento Nacionalista Revolucionario (MNR) from 1946 onwards increasingly became the rallying point of middle-class opposition committed to revolution. The first major joint effort of the MNR and the labour movement occurred in 1949 when a nation-wide revolt erupted. Rebel forces seized control of every provincial capital and mining camp but failed in La Paz. Loyal government troops used the capital as a base and succeeded in suppressing the rebels, province by province.

The 1952 insurrection was launched in La Paz by Los Grupos de Honor and the national police, whose commanding officer had agreed to join the MNR in a coup. The Groups of Honour were paramilitary cells, their composition mainly lower middle class: artisans, less well-organized workers of small factories, and elements of the *clase popular*. When the insurrection appeared to be doomed, the police general sought asylum in a foreign embassy. However, armed workers from the mines and factories joined the fight in several provincial centres, turned the tide in La Paz, and cut off possible reinforcements for the capital. What had started out as a coup with limited civil participation, ended with the leader of the MNR, Victor Paz Estenssoro, presiding over a government which included three labour ministers, i.e. official representatives of the labour Left. Whichever way the Bolivian Revolution is delineated, rural elements were not involved until after the overthrow of the government (Malloy, 1970).

The success of the revolution against the Shah is the more remarkable for the odds it faced. Iran had a well-established government and enjoyed an economic boom. The Shah boasted a large modern army that was extremely well equipped. SAVAK, the secret police, had established a reign of terror: tens of thousands had been arrested and savagely tortured, hundreds had been executed. Many dissidents had gone into exile. And great numbers of young people had refused to return after completing their studies abroad.

Some elements in the anti-Shah movement stand out. Most conspicuous were the street demonstrations. On 8 January 1978, theological students in the holy city of Qom staged a massive protest and sit-in. It was broken up by security forces. Their action provoked retaliation, and the security forces started shooting. In two days of disturbances dozens were killed, according to one estimate; at least 70 persons, by another. In February, the Tabriz demonstration of sympathy and solidarity for those killed in Qom rapidly turned into a vehement protest against the Shah. The local Azerbaijani police refused to intervene; troops were called in and used force. The demonstration turned into a riot spearheaded by poor recent immigrants and radical students. An estimated 100 persons were killed. A pattern evolved as leadership from the clergy and the bazaar helped to organize massive memorial demonstrations, at the traditional forty-day mourning intervals, for those killed in previous confrontations. Demonstrators took to the streets again and again to face troops who were shooting to kill. Over 3,000 are believed to have died in the first eleven months of 1978.

A second key element in the overthrow of the Shah was worker protest. Strikes did not gather momentum until mid-September. By the end of October, oil production had dropped by nearly three-quarters. Many factories were forced to go on short time or to close for lack of energy supplies. The drying up of oil exports threatened to create foreign-exchange problems. By early November all public services—transport, telecommunications, ports, and fuel supplies—were paralyzed or nearly so. Strikes at major banks affected import credits; strikes at customs halted industrial production by shutting off raw materials and spare parts.

A third factor was that large sectors of the middle class, and especially university students, had become increasingly disaffected over the years. Several guerrilla groups sprang from the student milieu abroad and at home. Their impact was effectively circumscribed by SAVAK, their significance primarily symbolic until after the Shah's departure, when they played a major role in the overthrow of his successor regime (Graham, 1979: 216–41; Keddie, 1981: 243–57; Bill, 1978). Neither a rural guerrilla movement nor the rural masses played any role in the Iranian revolution.

In Nicaragua the junction of an organized guerrilla force with an urban insurrection was crucial to the overthrow of the Somoza regime. The Cuban example had inspired the establishment of the Frente Sandinista de Liberación Nacional (FSLN). However, as elsewhere in Latin America, rural campaigns ended in defeat throughout the 1960s and into the 1970s. In 1975, splits developed within the FSLN over strategy: whether to concentrate on rural guerrilla warfare or on organizing among urban workers, and whether to give priority to military action or political activity. Then a number of spontaneous urban uprisings forced the FSLN's hand. They were spontaneous in two senses: triggered by specific events, and erupting independent of the FSLN. In February 1978, an insurrection in Monimbó held out for five days. In August youngsters armed only with pistols, rifles, and home-made contact bombs, their faces covered with black-and-red bandannas, forced the National Guard back to their barracks in Matagalpa. The FSLN, faced with the alternative to stop such insurrectionary tactics or to lead them, launched uprisings in five other cities on 9 and 10 September, but eventually had to retreat to the hills. In April 1979, Estelí rose against Somoza once more—accounts differ whether this uprising was initiated by the FSLN or a local initiative the FSLN decided to support. In any case, a comprehensive plan of action was already taking shape. A general strike called for 5 June paralysed the country; guerrilla units spearheaded insurrections in the major towns; and within six weeks the dictator was put to flight (Chavarría, 1982; Booth, 1985; Black, 1981).[26]

Humberto Ortega (1980: 4), leader of the Tercerista tendency within the FSLN, summarized the events in an interview:

> We always took the masses into account, but more in terms of their supporting the guerrillas, so that the guerrillas as such could defeat the National Guard. This isn't what actually happened. What happened was that it was the guerrillas who provided support for the masses so that they could defeat the enemy by means of insurrection. We all held that view, and it was practice that showed that in order to win we had to mobilize the masses and get them to actively participate in the armed struggle. The guerrillas alone weren't enough, because the armed movement of the vanguard would never have had the weapons

needed to defeat the enemy. Only in theory could we obtain the weapons and resources needed to defeat the National Guard. We realized that our chief source of strength lay in maintaining a state of total mobilization: social, economic and political mobilization that would disperse the technical and military resources of the enemy. Since production, the highways and the social order in general were affected, the enemy was unable to move his forces and other means about at will because he had to cope with mass mobilizations, neighbourhood demonstrations, barricades, acts of sabotage, etc. This enabled the vanguard, which was reorganizing its army, to confront the more numerous enemy forces on a better footing.

Each of these revolutionary movements was largely urban in character. Even where rural guerrillas played a prominent role, in Cuba and in Nicaragua, their leadership, a large proportion of their fellow combatants, and much of their material support were drawn from the urban milieu. It was in the cities that all the confrontations took place in Bolivia and Iran, that the decisive battles were fought in Nicaragua, and that the Cuban *guerrilleros* found crucial support to establish their regime. Urban workers determined the outcome in the battle for the control of La Paz, paralysed the economy in Iran and Nicaragua, and thwarted efforts to snatch the fruits of victory from the hands of the *guerrilleros* who had put the Cuban dictator to flight.

How to explain the urban character of contemporary revolutions? The four revolutionary struggles successful in the 1950s and 1970s were carried out in countries that varied greatly in their level of urbanization. Little more than a fifth of the population lived in cities in Bolivia at the time of the revolution, slightly less than half in Iran and in Nicaragua, somewhat more than half in Cuba. The importance of urban elements in the revolutions of the latter three countries may be seen as a function of their high level of urbanization. In such a perspective, the rapid urban growth experienced by most Third World countries can be taken to presage the age of urban revolutions. As Guillén (1973: 238), the intellectual mentor of urban guerrillas in Uruguay and beyond, put it:[27]

Strategically, in the case of a popular revolution in a country in which the highest percentage of the population is urban, the center of operations of the revolutionary war should be in the city. Operations should consist of scattered surprise attacks by quick and mobile units superior in arms and numbers at designated points, but avoiding barricades in order not to attract the enemy's attention at one place. The units will then attack with the greatest part of their strength the enemy's least fortified or weakest links in the city . . . The revolution's potential is where the population is.

The demographic observation reflects an economic reality: the surplus is increasingly produced in the urban sector. Even in largely rural Bolivia it was tin-mining rather than agriculture that constituted the core element in the national economy. The state's ability to extract agricultural surplus is no longer crucial to its very operation. Rather, it is the urban economy that finances the state apparatus. The extreme example is Imperial Iran, which filled its coffers with petro-dollars while importing food. Since the state is dependent on the urban economy, there are limits to repression in the urban context. Managers, professionals, skilled workers, and even semi-skilled workers cannot be replaced in great numbers at short notice. To imprison them for any length of time, to force them into exile, or to kill them, entails severe economic losses. This means not only a reduction in the resources available to the state, but also a deterioration in living conditions for the population at large that is likely to foster discontent.

Difficulties in extracting surplus provide powerful motives for colonial powers to reconsider the merits of their direct control over distant lands. For national governments and their supporters the situation is quite different. For one thing, they can appropriate surplus via the money-printing press, though inflation has its political costs. For another, many governments can count on foreign assistance to tide them over a crisis. If patron countries find it in their interest, they may provide substantial economic and/or military support for many years. Even a bankrupt government is unlikely to abdicate of its own accord. It may fall victim to shifts within its political base, or it may be toppled by sections of the army dissatisfied with the lack of resources. But confronted with a revolutionary movement, the ruling group will desperately cling to power. If a national war of liberation threatens the tentacles of empire, a revolutionary movement attacks the very existence of political and economic élites. And while members of the élite can make provision to live out their days in comfort in exile, much of the indigenous middle class, unlike colonial civil servants, has nowhere to go. The level of resistance of the élite and the middle class tends accordingly to be high.[28] The MNR, it is true, could effectively appeal to a large part of a middle class that had been deeply divided ever since Bolivia's humiliating defeat in the Chaco War. In Cuba, large sectors of the middle class had been hostile to Batista since 1952, when, after three popularly elected administrations, he seized power in a *coup d'état*. Elements of the middle class in Iran and in Nicaragua had been similarly alienated from autocratic rulers. Still, it was only when the revolutionary movement made life in the cities insecure and brought the economy to a halt that a full-scale withdrawal of support from the government took place in these three countries. And only when they had thus been quite isolated did Batista, the Shah, and Somoza flee.

Colonial governments withdrew in spite of their success in maintaining control of the cities. But loss of control over rural areas is not a sufficient condition to persuade a national government to relinquish power to a revolutionary movement. For such a movement to succeed, it must confront the government in its urban location. Control over the capital city is usually critical, as the abortive revolution of 1949 demonstrated in Bolivia.

The city seems to hold specific attractions for guerrilla activity. For the students and professionals who invariably predominate among the guerrillas in the early stages, and frequently longer, the city constitutes familiar terrain: they know it and their presence does not attract undue attention. Furthermore, while most guerrillas in rural areas are outsiders, even if they are of rural origin, the urban guerrilla can maintain the cover of a conventional life until he/she arouses the suspicions of the authorities. Finally, the urban crowd promises the guerrilla anonymity.

Urban guerrilla movements appeared to pose a real threat to several South American regimes in the late 1960s and early 1970s. The Tupamaros of Uruguay in particular had considerable success over a number of years, and in 1972 were widely believed ready to take power. However, when a military regime was installed, it managed to destroy the organization within a few months. Here and elsewhere urban guerrillas proved to be vulnerable to brutal repression. Systematic torture revealed the urban guerrillas' vulnerability to security leaks because of their dependency on safe houses which, once uncovered, could be rapidly attacked by vastly superior armed forces. A circle was established as torture—at the beginning often of people arrested quite haphazardly—provided information that led to the capture of participants or sympathizers who under torture provided further information. Torture served also to deter recruitment to the ranks of the guerrillas and to scare away their sympathizers.[29]

Rural areas can shield small mobile guerrilla units. But the transition from rural guerrilla activity to peasant army faces impossible odds. The circumstances in which the Chinese Communists were able to control entire provinces and to build up their armies were unique. There was no government exercising hegemonic power over China. And the country's vastness inhibited the establishment of such power. The Japanese invasion further undermined any effort at central government. It gained the Communist forces recognition as nationalists fighting the invader. At times it forced a truce on the Communists' opponents. Today governments everywhere can rapidly dispatch army units across an entire country. Any attempt to move beyond the guerrilla stage in rural areas has to reckon with the immense firepower, the high mobility, and the efficient communications system that are the hallmark of the modern army.[30] Governments have shown little hesitation in razing rural settlements and relocating peasants to counter rural-based challenges.

A national government will not surrender to a revolutionary movement unless it is confronted in its urban location by forces it can no longer contain. Such forces are most unlikely to be constituted of a peasant army. They have to be established in an urban mass movement rather than a guerrilla organization that can be detected and repressed. Substantial elements of the urban population must come to reject a regime to the point where they are prepared to confront its armed forces in the streets. The extent and depth of such disaffection will be affected by social and political mobilization. And the organization of the urban mass movement will bear on its success or failure.

Urban mass movements confront governments with difficult choices. Large-scale repression is likely to spread disaffection from the regime and to radicalize the opposition. No government has yet responded by relocating urban populations on any scale. And if heavy arms are used in re-establishing control over a city, or major parts of it, the destruction wrought is bound to be huge. Somoza, though, went so far as to order the destruction of entire small towns and of neighbourhoods in Managua held by the Sandinistas.

Against heavy odds guerrillas have maintained their presence in a number of countries for many years. They face the even more difficult task of mobilizing an urban mass movement that will confront the fire-power of a modern army.[31] Power and privilege are centred in the city; it is in the city that they are effectively challenged.

8

Urban and Regional Systems: A Suitable Case for Treatment?

One of the common features of policy statements throughout the Third World is the frequency with which politicians and planners have averred that something must be done about the growth of cities and increasing levels of regional inequality. The city is devouring national resources at the expense of the countryside and the farmer; centralization is destroying the dynamism of the provincial city; there is urban bias, metropolitan bias, rural underdevelopment; development is required in the poorest regions. Despite repeated statements of intent, however, there are few examples, outside certain socialist countries, of any real determination to develop the rural areas, to decentralize industry, to control the rapid growth of metropolitan areas, and still less to reduce poverty in the poorest regions.

This chapter asks whether and in what ways urban settlement systems are distorted. Do too many people live in cities or in urban areas that have grown too large? Has urban bias development policy distorted the pattern of national development? Is there a case for arguing that Third World cities are becoming unmanageable and that their growth must be slowed? In short, are the arguments relating to 'distorted' settlement systems valid, and, if they are valid, what can be done to modify current trends? I also consider the kinds of policies that have been introduced to treat urban and regional imbalances and comment upon their effectiveness. In general, I am unimpressed by the spatial policies of most Third World countries and question the usefulness of most kinds of spatial policy as a means of reducing poverty and narrowing personal income disparities. Can a spatial strategy, in fact, achieve anything if fundamental social and economic reforms are absent? Is it only political changes and modifications in developmental style which can cure those imbalances? Is spatial policy necessary if such reforms take place?

Although asking such broad questions, I am trying hard not to overgeneralize. General statements about the appropriate treatment of urban bias, over-urbanization, and regional imbalance are often highly misleading. Too often policies useful in one country at one specific time are turned into panacea for all countries at all times. Rarely do the authors of such panaceas caution that what is good for one society may well be irrelevant for another; what is good in terms of settlement policy for India is unlikely to be terribly relevant to the problems of Hong Kong; Brazil's national settlement strategy is likely to be more than an ocean apart from those of most West African states; China's policies may be of academic interest to decentralizers in India, but they can scarcely be contemplated in policy terms, given the different political systems. In short generalizations are of only limited use; 'instant solutions' taken from the latest vogue generalization have wrought havoc in the planning field. Let nothing that may be said here contribute to such instant planning; the generalizations made in this chapter are intended for thought, not to be introduced in evangelistic glee.

Forms of Regional, Spatial, and Urban Imbalance

Urban bias

Most of the social sciences have contributed to the extensive literature on the appropriate balance between urban and rural development. During the 1950s many economists were involved in a long debate about the correct allocation of investment between agriculture and industry. Authors such as Papanek (1954), and Johnston and Mellor (1961) favoured agricultural development, arguing that the agricultural sector must be made more productive in order to provide the necessary surplus for industrial and urban development. Without such a surplus inadequate funds would be available for industrial and urban growth and no market would develop for urban products and services. Counterposed to this argument was the idea that industrial and urban growth were prerequisities for a more modern and productive agricultural sector. Surplus rural labour needed to be absorbed into more productive urban activities, thereby permitting the introduction of more modern and capital-intensive agricultural practices and encouraging the diffusion of modern ideas and institutions into the traditional rural areas (Currie, 1965). Linked to this debate were various controversies over the balance of advantages associated with balanced growth (Nurkse, 1952), a 'big push' towards industrialization (Rosenstein-Rodan, 1943; Prebisch, 1950), or a programme of balanced imbalances (Hirschman, 1958).

Meanwhile, some demographers and sociologists were arguing that most Third World countries were suffering from 'over-urbanization' (Davis and Hertz, 1954; UNESCO, 1957). Statistical studies showed that a much larger proportion of the populations of Third World countries were employed in non-agricultural activities, and especially in services, than had been the case in the developed countries at similar levels of urbanization. This finding was interpreted as evidence that rural–urban migration had taken place despite the lack of adequate numbers of productive industrial jobs in the urban areas. Unlike European and North American experience, where industrialization had induced urbanization, in the Third World urban growth often preceded the creation of productive jobs. As a result, urbanization was much less efficient in the Third World and was merely the outcome of excessive migration from areas of surplus rural labour. In this sense, over-urbanization 'explained' why urban misery coexisted with rural poverty.

Such views were countered by charges that the definitions of over-urbanization were invalid. Sovani (1964: 117) called them 'chimerical and so unusable' and others pointed out that even if it were valid to compare the present experiences of the developing countries with those of the now developed countries in the past, no policy conclusion could be drawn from such comparisons (Brookfield, 1973; McGee, 1971). The over-urbanization concept was also attacked by the pro-urban lobby which argued that the city was the historical centre of innovation and change. Pirenne (1925) and Mumford (1938) had shown that historically the city had been the climax of civilization and the stimulus to further development. Friedmann (1968) argued that urbanization should be adopted as a deliberate strategy for change: without it meaningful development would be held back.

Parallel debates may well have taken place in socialist countries relating to the correct interpretation of Marx's ideas about the elimination of differences between town and country and about the revolutionary potential of the urban proletariat and the peasantry. In the Soviet Union the collectivization of the peasantry under Stalin effectively resolved the issue in favour of urban-industrial development. In China the different revolutionary experience, the fear of invasion, and the lower levels of urban development encouraged a much less favourable attitude to urban growth, arguably even an 'anti-urban' stance (Frolic, 1976; Kirkby, 1985: 249–50). In Kampuchea, Phnom Penh and the other major cities were evacuated in 1975 both in an effort to avert starvation and to increase the chances of establishing an equitable, self-sufficient, and socialist country.

Throughout the world the debate continues. The advocates of both the rural and the urban sectors remain convinced that most of the failures of capitalist or of communist society may be attributed to the excessive allocation of resources to the other sector. Thus Jakobson and Prakesh (1974: 279) complain that 'the prevailing world-wide attitude among politicians of all persuasions is anti-urban, and in particular anti-big city', while Lipton (1977) argues that the major developmental issue in the Third World today is the need to stem 'urban bias'.

Lipton's argument is sufficiently influential that it is worth examining in more detail. He argues that 'urban bias' leads to an excessive concentration of resources in urban areas. This not only worsens inequality in Third World societies, since a majority of the poor normally live in the countryside, but also slows the pace of national economic growth. Lipton's contribution lies in his attempt to show that the distortion leads both to greater inequality and to a slowing of economic growth. The basic cause of this distortion is that power in most Third World countries is held by urban groups who distort the allocation of resources in favour of the cities (Lipton, 1977: 1).

> The most important conflict in the poor countries of the world today is not between capital and labour. Nor is it between foreign and the national interests. It is between the rural classes and the urban classes. The rural sector contains most of the poverty, and most of the low cost sources of potential advance; but the urban sector contains most of the articulateness, organization and power. So the urban classes have been able to 'win' most of the rounds of the struggle with the countryside; but in doing they have made the development process slow and unfair.

In consequence, 'a shift of resources to the rural sector . . . is often, perhaps usually, the overriding developmental task.' He is not without support. For example, Gugler (1988*b*: 84) argues 'there would appear to be an approach that promises a more efficient allocation of labour between the rural and the urban sector as well as a reduction in the extreme inequalitites that characterize most Third World countries. It will aim at improving rural living standards by channelling productive resources to the rural areas and/or by directing a larger share of income to them.' Similarly Lefeber (1978: 22) notes that 'the required policies are known. The problem is to break through the deadlock caused by the dominance of congruent urban class interests.'

But is the urban/rural dichotomy too simple? Does it even exist in the terms described? Lipton's thesis covers most Third World countries, including Africa, Asia, and Latin America, even though most of the evidence he presents is drawn from South Asia. Clearly, it is unlikely that any simple generalization of such importance will apply equally well in all parts of the Third World. It is also interesting that several strands of evidence suggest that urban bias may be an inappropriate term even to describe the situation in India. For a start, it is possible that during the 1960s it was the urban poor who suffered deterioration in their standards of living rather than the rural poor. Dandekar and Rath (1971: 33) point out that between 1960–1 and 1967–8 private consumer expenditure of the bottom 20 per cent of rural Indians remained steady while that of the bottom quintile of the urban poor deteriorated. In addition, Byres (1974) has pointed out that the limited net marketed surplus flowing from the agricultural to the industrial sector in the 1950s and 1960s was hardly consistent with an acute case of urban bias, and

Mitra (1977) and Byres have shown that during much of the 1950s and 1960s the terms of trade moved against the cities. It is also pertinent to point out that the rate of urban growth in India has been very slow by Third World standards, increasing only from 17.6 per cent of the population in 1951 to 23.7 per cent in 1981 (Mohan and Pant, 1982).

If there are certain empirical objections to the applicability of the urban-bias thesis to India there are also alternative interpretations of the Indian scene. Thus Bose (1973: 20) argues that 'the lack of an urban lobby in the Indian Parliament and in the state legislatures is responsible for the continued neglect of problems of urban development.' And if this interpretation is difficult to reconcile with urban bias, Byres's (1974: 252) conclusion is still more critical 'the most important contradiction in the Indian socio-economic system in coming decades is that posed by an urban bourgeoisie intent upon industrialization but frustrated by a strong and increasingly powerful class of rich peasants and small and medium landlords, on their way to becoming capitalists and exercising growing political power.'

If urban bias is not universally accepted by students of India, where might it apply? It is most likely to be relevant to those countries in which urbanization has been very rapid, as in Africa or in Latin America. But even in countries in which urban growth has been rapid, would everyone agree that urban bias has been the explanation? In Uganda, for example, the proportion of the total population living in urban areas increased from 5 to 10 per cent between 1960 and 1975, a rate surpassed by few Third World nations (World Bank, 1979: 164). Yet, despite this finding, Muench (1978) argues firmly against the idea of urban bias. He points out that the urban poor have suffered considerable neglect in Kampala as a result of politically inspired regional policies which have favoured provincial élites in the name of greater regional equality.

Of course, there are many countries in the world where urban–rural disparities are too wide, having been brought about by policies that have clearly favoured the urban areas at the expense of the rural groups. But the generalization can be applied too broadly and, like all generalizations based on a dichotomy, it is likely to ignore other important processes that contribute to the same socio-economic pattern. A particular danger with the urban-bias thesis is that it portrays most power conflicts in Third World countries as arising 'primarily from where people live (rural versus urban areas) rather than, say, from the economic sectors in which they derive a livelihood (industry versus agriculture) or from their position in the class hierarchy' (Griffin, 1978: 108). In fact much of the developmental literature has pointed out that many of the difficulties Lipton portrays stem not from urban bias but from policies that favour some rural groups at the expense of others. Thus Lefeber (1978: 8) points to a common situation, in which rich agriculturalists support, or at least do not oppose, policies which keep agricultural prices down so as to reduce the cost of urban food. Such a policy is less biased against rural areas than against the small and medium-sized farmer who is unable to benefit from other government policies such as export credit, loans, and subsidies in the same way as the large farmer. The large farmer supports urban bias in one set of policies so as to benefit from rural bias in another set. In such circumstances, it is more appropriate to analyse the problems of the poor farmer in class terms than in terms of urban bias.[1] As Griffin has said 'Lipton has tried to explain too much, indeed virtually everything, in terms of urban bias. In the end it becomes a brilliant obsession.'

A more limited and less simplified generalization might find wider support. Bias in the allocation of resources in most Third World countries is typically directed against the poor. They lack access to critical resources such as land, water, and credit and receive little for their labour or their products. This bias applies to all poor people, but because most of the Third World poor live in the rural areas there appears to be a measure of urban bias. The discrimination is not against the rural areas, but against the poor of those areas. It is normal for the larger, commercial farmers to prosper even in areas such as Asia, where the evidence suggests that conditions for large numbers of the rural poor are deteriorating (Friedmann and Douglass, 1976; Griffin and Khan, 1978). Here lies the paradox of the green revolution; increasing agricultural production sometimes allied with increasing poverty. Those who have resources such as land and water increase their production while those who lack resources suffer as a consequence of the changes occurring about them (falling prices due to increased agricultural production by the big producers, higher fertilizer costs, rising costs of land, increasing mechanization, reduced access to water, etc.). But it is also important to remember that the urban poor are not well treated in most Third World countries. Low incomes, under-employment, and poor housing, infrastructure, and services are the lot of millions of families. If they generally fare better than their poor rural cousins, it is by little compared to the neglect they suffer compared to their richer urban neighbours.

As a consequence of these processes it may well be true that there is a net transfer of resources to urban areas.[2] Indeed, it would be most surprising in a rapidly urbanizing world if that were not the case. But are the evils of Third World countries due to the speed of urban growth, or is urban growth a symptom of other processes? In any case is the transfer of resources to urban areas at the expense of rural areas always undesirable? If higher productivity and greater welfare can be achieved in urban areas through a relative concentration of resources is that not development? Should such a situation be described as urban bias? It is surely better, and surely follows Lipton's meaning more closely, to apply the term only when an excessive allocation of resources to the urban areas reduces economic growth and distorts the distribution of income. Thus, if agriculture is neglected because funds are being invested in urban areas fuelling land speculation, government bureaucracy, and the import of luxury consumption goods, we might conclude legitimately that urban bias is the problem. Unfortunately, while such a situation may appear to obtain in many Third World nations, it is not easy to measure accurately. Obviously cost-benefit analysis and other forms of economic appraisal can approximate the returns on urban versus rural investment, but great care needs to be taken given the artificiality of prices and foreign exchange rates in so many Third World countries.

But even if we can measure the degree of urban bias by noting that funds are being directed into urban investment at rates of return far lower than those available in rural projects, even if we can show that urban programmes are less equitable than rural programmes, the fundamental issue comes down not to the desirability of urban versus rural development but to whether policy can be changed. In short, who makes the decisions, who holds power, and how do they use that power? Even if we agree with Lipton that rural areas should receive higher priority in decision-making, how can that change be achieved? And if it is achieved, would the outcome be an improvement? If the Chinese have improved welfare for the poor through policies which have sometimes starved the urban areas of infrastructure and services, has the much more anti-urban stance of the Kampucheans had a similar effect? The Brazilians poured money into the Amazon region, but did they improve welfare? Merely posing these issues demonstrates that urban bias is not unlike the jabberwocky. As Alice said, 'it seems very pretty but it's rather hard to understand!'

Closely linked to the idea of urban bias is the belief that rural development has been distorted by the pattern of urban growth. Specifically, it has been argued that there is an absence of towns at the base of the urban system in most Third World countries. As a result of a distorted development process, new central places are required to service the rural areas and to reinvigorate agricultural development. Johnson (1970) has noted that there is one central place for every 16 villages in Europe compared with one for every 157 villages in the Middle East. This gap in the urban-size distribution lowers agricultural productivity because there is nowhere to sell the produce and no opportunities for buying consumer goods which might stimulate farmers to increase production. Johnson also sees the small city as a means of providing services for the rural areas. At one level, Johnson is clearly right. Distant market centres do make the selling of agricultural produce more difficult and encourage the emergence of exploitative middlemen. But, the creation of new centres is at best a partial remedy, for marketing difficulties and exploitative middlemen are endemic in most poor countries simply because of the absence of good transportation, storage facilities, and credit. The poor farmer can neither move his produce to the city nor store it until prices improve. Similarly, the development of new urban centres may provide health and education services, but cannot guarantee their use. More often than not the cost of health provision is a greater barrier to attendance at clinics than is distance; new urban centres would only be effective if services and drugs were free. In my opinion, it is likely that Johnson's gap in the urban system is less a cause than a consequence of lower agricultural productivity; central places fail to emerge because farmers have low purchasing power. While the establishment of new central places would no doubt be of some assistance, it is difficult to see it as the major plank of a development programme or to agree with Johnson's principal conclusion that 'the raising of average incomes in under-developed countries will require town-building programmes' (p. 177).

Nevertheless, a closely parallel strategy is becoming popular with several international organizations and has been the subject of several recent books (Rondinelli, 1983; Hardoy and Satterthwaite, 1986; Kammeier and Swann, 1984; Cheema and Rondinelli, 1983). Rondinelli, in particular, has argued that secondary city development is essential in developing countries. While he is aware that some secondary cities perform an exploitative role in the rural areas, the productive regional centres should be strengthened. If national governments would invest more funds in such secondary centres they '. . . can play important roles in balancing the distribution of urban population and economic activities, in stimulating rural development, and in generating more socially and geographically equitable distributions of the benefits of urbanization' (Rondinelli, 1983: 197). Unfortunately, the diversity of secondary cities means that no universal set of policies can be devised favourable to their development. As most writers readily admit, more understanding of the nature of these centres and of their development potential is necessary. So, too, is a better understanding of the role and limitations of policy towards such centres. For as Hardoy and Satterthwaite (1986: 398) note, 'many Third World governments have adopted special programmes for small and intermediate urban centres . . . but those programmes which have not been designed to serve explicit social and economic goals and to suit each centre's local and regional context have rarely succeeded in achieving the hoped for development objectives.'

Other writers, of course, would argue that the stimulation of secondary cities, without radical modification of existing developmental approaches, is doomed to failure. This is broadly what Friedmann and

Douglass (1976: 372) suggest in their argument for agropolitan development. In Asia the countryside should be transformed by 'introducing and adapting elements of urbanism to a specific rural setting. This means: instead of encouraging the drift of rural people to cities by investing in cities, encouraging them to remain where they are by investing in rural districts and to transmute existing settlements into a hybrid form we call agropolis or city in the fields.' The spatial planning component in this strategy is the agropolis, designed '. . . to have a total population of between 15,000 and 60,000' and to permit the desired mix of local production, democracy, and cultural integration. The agropolitan district is self-reliant, self-financing, and self-governing; its political autonomy is limited only by the 'concurrent needs of all other districts and the combined needs of the larger community of which they form part' (Friedmann and Weaver, 1979: 203). The ultimate aim of this strategy is to reduce severely the power of transnational corporations over local areas. Equal access to the bases of social power allows the local community to express a rightful territorial interest. 'The territorial interest, then, becomes in every case controlling over subordinate, including corporate, decisions' (p. 204). In these circumstances, the territorial interest is the equivalent to the local societal interest. I shall return to this radical suggestion on pp. 259–62.

Distortions in the urban-regional system

If many observers believe that too many resources are devoted to urban areas to the detriment of the countryside, there is also no shortage of politicians and planners complaining of excessive spatial concentration within the urban-regional system. Indeed, the literature on regional development is full of statements about 'excessive' regional inequalities, 'over-developed' metropolises, 'over-centralized' population distributions, and 'unbalanced' settlement systems. Sometimes these distortions are blamed upon the inefficient or the inequitable model of development being pursued in a particular country, sometimes they themselves are seen to be a source of distortions. Not infrequently, such arguments are emphasized through ugly physical analogies; most commonly, the 'neglected periphery' and the 'blooming metropolis' are depicted in terms of a shrunken body with a swollen head. The swollen head has grown on the basis of resources that should have nourished the body. The head is variously described as over-large, the victim of an enormous influx of untutored rural migrants, the centre of crime, social deprivation, and inequalities. Similarly the shrunken body is the subject of emotional diagnosis; provincial politicians complain about centralization and its draining effect on regional dynamism; local services are inadequate because national governments provide insufficient resources; employment opportunities are limited because most of the highly productive industry is concentrated in the major city; most skilled manpower leaves the provincial cities for the metropolis.

In certain countries there is considerable truth in this analogy, but in general it is an over-simplified picture of reality; like all analogies, it must be used carefully. In anatomy a misshapen body can be diagnosed easily by comparison with the size and shape of the average human body; there may be marginal cases, but on the whole the norm is so prevalent that the dwarf or the giant stand out. But in the realm of human settlement systems what is the norm? There is no typical urban system against which the swollen head and shrunken body can be so easily compared. If there is no norm, then can we be sure that the exception is an exception? If there is no norm, can we be sure there is anything wrong with the presumed exception? The misshapen human body is exceptional because there are so many healthy bodies around with which to compare it; among human settlements it is probably more accurate to depict the misshapen as the norm. After all, more nations have primate cities than have non-primate centres, there are more countries with wide regional disparities than with equitable regional distributions of income. In short, we are comparing the urban system with an ideal; the body is misshapen only when compared to the ideal, but that ideal is not likely to be accepted by everyone with the same readiness. Few people agree on the best distribution of urban settlement; different ideologies and developmental goals lead to the recommendation of different ideal types. Equity and efficiency considerations may well indicate different urban forms, proponents of the free market are likely to recommend different urban distributions than socialists.

In addition to this normative problem the debates about human settlement systems frequently confuse a number of related but diagnostically distinct issues. Confusion over these issues often leads to policy recommendations that can do nothing to resolve more fundamental problems. Contributing to this confusion is the way we use the terms 'centralization' and 'urban primacy', the belief that there may be an optimum city size, and the failure to distinguish between the welfare of places and the welfare of people. I wish to examine these confusions briefly before going on to discuss the circumstances in which it may be justifiable to speak of distortions in the national settlement system.

Centralization is a much abused term. The essential problem is that it is used interchangeably both in a functional and in a geographical sense (Boisier, 1987). A centralized system may be one in which the president of the country makes all of the vital decisions, no one else in government or outside of it has real power. This might be described as *functional centralization*. Where, however, other agencies do wield real responsibility, there may still be centralization in the sense of geographical concentration since all of the agencies are concentrated in the national capital. Centralized decision making may be the outcome of either functional centralization or geographical concentration (Fig. 8.1).

But the centralization of power does not automatically imply a complete spatial concentration of resources. Power, for example, may be concentrated in the hands of a president who favours a policy of spreading investment throughout the country. Similarly, centralized decision making may characterize a nation even though its economic base is not concentrated in the capital city. Thus, in Venezuela, the government in Caracas makes most important decisions and most Venezuelans complain bitterly of the excessive level of functional and geographical centralization in decision making. On the other hand, the source of three-quarters of government wealth is the oil industry which is located in two regions in the far west and east of the country. Similarly, the government has long sought to deconcentrate employment away from the capital city.

In this chapter I shall use the terms *spatial concentration* or *spatial dispersal/deconcentration* to describe the geographical distribution of any variable, whether it be government revenues, power, industry, or population. If, for example, the distribution of population becomes

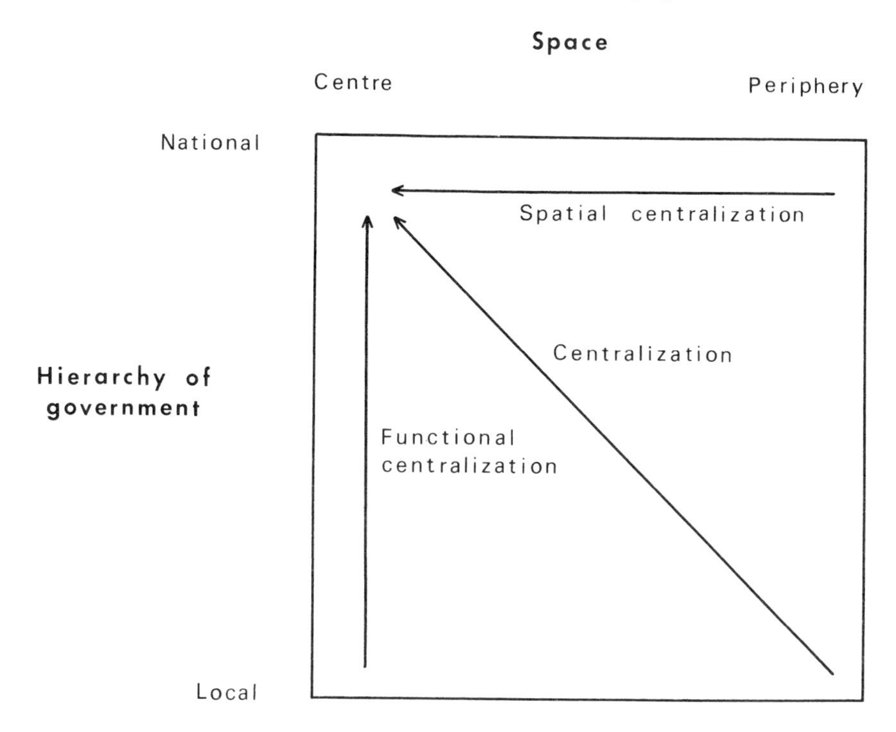

FIG. 8.1 Forms of the centralization of power

relatively more (less) concentrated in the major city, this would be called a process of *spatial concentration* (*deconcentration*). On the other hand, the greater concentration of power in the hands of a central government (as opposed to second-tier (regional) or third-tier (municipal) governments) or in the hands of a president (rather than the legislature or the people) will be described as *functional centralization*. Only when *functional centralization* and *spatial concentration* act in the same direction will I employ the unqualified term *centralization*. This terminology will prevent us falling into traps such as that of describing as decentralization the establishment of branch industries in the periphery (spatial deconcentration) when power clearly remains in the head offices of the capital city.

The meaning and the implications of the term 'urban primacy' are also much abused. As I observed in Chapter 2, the issue of urban size should be distinguished from that of a distortion in the national city size distribution. For while very large primate cities exist (Mexico City, Buenos Aires, and Bangkok), some primate cities are very small (Lomé, Asunción, Kathmandu, and Banjul) and certain large cities are not at all primate (São Paulo, Bogotá, and Calcutta). If we fail to distinguish between the size of the city and its position in the national city-size distribution we shall confuse discussion of different kinds of problem. Thus it is pointless to complain of urban diseconomies in Kathmandu or Bridgetown when neither city has more than 300,000 people. If the problem is that the city has grown larger than others in the nation because it has received most governmental revenues or controlled the benefits of foreign trade, we need to separate that problem from the issue of urban diseconomies.

The issue of an imbalance in the urban-size distribution raises a major problem which was alluded to earlier. To measure an imbalance we need to have a norm. But what is the norm against which to measure primacy when as Chapter 2 demonstrated, most nations have primate city distributions? Is the city-size distribution 'primate' when the largest city is three times or when it is twenty times the size of the second city? Should we measure primacy in economic and social terms or only in demographic terms?[3] The difficulty is that we lack an adequate general indicator of urban imbalance.[4] As a consequence, the desirable balance between cities remains in the minds of the politician, planner, or academic; value judgements and political commitments are as important in the recommendation of spatial deconcentration and decentralization as technical criteria.

Similar difficulties relate to the idea that major cities have become too large, or even that they have exceeded the maximum population at which all cities become inefficient and/or inequitable. Such a judgement normally rests on the assumption that urban diseconomies—for example, pollution or traffic congestion—have become so severe that the only answer is to deconcentrate both population and economic activity. Such an argument has several failings. First, it does not recognize that diseconomies are only one side of the argument; if urban agglomerations generate still greater urban economies, then the balance of economic advantage rests with spatial concentration. Second, even if diseconomies exceed economies, the best policy may be to reorganize the way the large city is run rather than to deconcentrate population and employment. Thus the best way to reduce an urban diseconomy such as traffic congestion may be to improve public transport, to cut the use of private cars, or to introduce parking meters. Air pollution can be reduced by the physical removal of polluting industries, but it can also be cut by fining errant companies. In short, urban problems are caused not only by size; more often than not urban diseconomies are an outcome of the rate of growth, the pattern of land use, the level of taxation or urban speculation, and other specific characteristics of individual cities rather than of size (Richardson, 1977; Townroe, 1979). Like the reputed finding that the monthly birthrate in Sweden rises during the seasonal in-migration of storks, the correlation between urban problems and size may be explained by other factors. Third, it assumes that the geographical dispersal of population and economic activity is both possible and without negative consequences. As pp. 257–9 show, such assumptions may prove dangerous.

Finally, the term 'regional balance' is often used in suspect ways. The principal problem is that there is only a limited correlation between the distribution of personal income and the pattern of regional income. Thus it is unreasonable to argue that a spatial deconcentration programme will automatically improve the distribution of personal income. Although equality of personal income guarantees regional income equality, the converse does not hold. Acute personal income inequality is consistent with total regional equality when the distributions of personal income and average per capita incomes are the same in every region but wide intraregional and interpersonal differences exist. Despite frequent exhortations to the contrary, many observers fail to distinguish between 'place' welfare and 'people' welfare. A region may be rich without its inhabitants participating equally in that wealth, another region may be poor but contain rich people. These points are easily illustrated by consideration of the case of India. In India measures of interregional disparity compare favourably with distributions in other nations, even though differences between urban and rural areas within regions and between income groups in those regions are very wide (Majumdar, 1977; Mathur, 1976).

Failure to distinguish between place and personal welfare encourages the adoption of regional programmes favouring particular regional groups. It also allows such a programme to pass as a measure to achieve greater equity rather than as a further source of interpersonal disparity. The convergence in regional incomes which occurred in the 1960s in Brazil, largely as a result of regional policies, failed to improve the situation of poorer groups in the poorest part of the country (Gilbert and Goodman, 1976). The upper-income groups in the north-east merely improved their position in comparison with the more prosperous groups in the richer south-east; an increase in regional prosperity brought an increase in personal welfare for some, but not for others. The term 'regional disparities' is dangerous if not supplemented by other measures of welfare for different income groups.

Having pointed to some of the principal errors which have appeared in discussions about national settlement systems and about regional income disparities, I wish to consider some of the more serious arguments put forward in support of, and against, particular kinds of spatial policy. Such arguments may be broadly categorized into those recommending (i) some kind of employment deconcentration programme; (ii) continued but better planned metropolitan expansion; (iii) neglect of the spatial problematic in favour of other kinds of policy change.

(i) *Employment deconcentration.* Employment deconcentration from the major cities is the most common kind of spatial policy. It may take the form of discouraging new economic activities from being located in the largest centres either through negative controls or through incentives to location in other parts of the country, it may seek to relocate existing economic activities to smaller cities, it may deconcentrate government activity, it may stimulate the expansion of smaller cities, it may even involve the construction of new cities. What all these approaches have in common is the general wish to control the growth of the largest cities, usually located in the most prosperous regions, and to stimulate the expansion of urban centres in peripheral areas. The transfer of economic activities from rich regions to poor is seen to serve several functions; it reduces the problems of the centre; it reactivates the provincial cities; it reduces regional disparity; and it helps lower-income groups in the provinces. Spatial deconcentration is seen to benefit both national and regional interests and is compatible with national economic growth.

Such economic deconcentration strategies often take the form of a 'growth centre' policy; a concept which holds a venerable place in the regional development literature (Friedmann, 1966; Darwent, 1969; Kuklinski, 1972; Moseley, 1975). At its simplest, the growth-centre notion conceives of an urban complex containing a series of industrial enterprises focused on a dynamic growth industry. The growth industry stimulates the emergence of ancillary companies the presence of which lowers its own operation costs. Specialized services and a skilled labour force emerge which help to maintain inter-regional competitiveness and generate new activities. As an agent of regional development, the growth centre serves a double function. It prevents the dissipation of agglomeration economies while achieving a more equitable distribution of economic activity: in Rodwin's (1961) words it permits 'concentrated decentralization'. In addition, it provides a focus for the development of the centre's hinterland. Industries which consume agricultural products will create a market for the region's farmers, as will the growing concentration of urban inhabitants. And in a traditional or conservative region the growth centre will perform an innovative role; it will encourage the spread of new ideas, techniques, machines, and products (Berry, 1972; Hansen, 1981).

The more intelligent regional analysts always accepted that the effects of a 'growth centre' strategy would be limited. Such an approach might reduce income inequalities and lighten the deepest pockets of poverty, but it would not remove them. Premissed on the assumption that, rightly or wrongly, industrialization was the archetypal national development policy, it was an incrementalist strategy which promised to improve the regional distribution of activity by raising urban efficiency and thereby increasing the opportunities for improving social services, employment, and productivity. It was also an opportunistic strategy in the sense that the less-developed countries had scarcely begun the transformations required by the development process (Friedmann, 1966). As Lewis (1962: 183) said of India, 'so much of the total locational outcome is yet to be settled that the question at issue virtually comes down to this: what will the territorial disposition of demand be in the Indian economy of 1970 and 1980?' Decisions about the future distribution of demand might recommend a policy of regional development and the stimulation of growth centres.

This argument has been strongly attacked by Friedmann and Weaver (1979: 187), who have blamed the growth-centre notion for accentuating unequal and inequitable patterns of development in less-developed countries. 'With the growth-centre doctrine as its principal tool, spatial development planning became the handmaiden of transnational capital.' The espousal of polarized development in the growth-centre doctrine allowed transnational corporations to dominate Third World nations. In the sense that the growth-centre notion is closely linked to the idea of industrialization and urban growth, this is a correct interpretation. But it is too simple to claim that there is an automatic convergence of interests between the practice of growth-centre planning and the interests of transnational corporations. Growth-centre planning may be compatible with industrial development but experience shows that transnational corporations are less than keen to move their plants from national capitals to growth centres in the periphery. Similarly, if growth centres are so well attuned to the ideology of transnational corporations, why do so many communist and socialist countries employ the same regional planning terminology and practice? Regional development in the USSR and Poland clearly differs in important respects from that in capitalist economies; nevertheless it is still clearly based on growth-centre strategy.[5] In the Third World Tanzania has clearly embraced the growth-centre strategy as a means of achieving greater regional balance and reducing urban bias (Sawers, 1989). Given such examples, is the form of regional development in most Third World countries a consequence of a belief in growth centres or the nature of the society in which those centres are introduced? To my mind, the trouble with the growth-centre strategy is not its underlying ideology but rather the blandness of the notion. Indeed, the idea is so flaccid that Marxist and right-wing regimes use the tool with equal facility.

The learned literature has described as 'growth centres' everything from villages of 5,000 people to the world's largest metropolis. Industrial areas, administrative centres, university cities, and sleepy market towns have all been labelled growth centres. As a result, the growth centre has been used as the intellectual rationale of every spatial strategy from Dodoma to the continued growth of Mexico City.[6] It has been used in one form or another in Tanzania, Brazil, Poland, Chile, China, and India. Its blandness has allowed it to be the handmaiden of whatever regime found it useful. It is not the regional approach that is the villain, but the development model which limits and dictates the manner of its use.

The problem with most growth-centre strategies is that they have failed to perform the multiple functions with which they were entrusted. They failed normally because such aims were too ambitious given the resources made available or because the spatial strategy was not supported by complementary sectoral policies. The blandness of the growth-centre notion has encouraged its rhetorical use in the least felicitous circumstances. But despite its inherent weaknesses and the consequent dangers, the growth-centre strategy should not be totally discarded. There are countries (Mexico and Venezuela perhaps) in which it is advisable to deconcentrate activities and major cities because topography, deficiencies in water supply, or pollution problems make continued expansion of the major cities undesirable; there are nations whose resources in poor or resource-rich areas may be best developed through some kind of urban-industrial programme; there are countries in which decentralization and employment deconcentration may be encouraged through a growth-centre strategy. Certainly no general case for employment deconcentration in Third World countries can be made, but as a limited contribution to certain kinds of economic, social, and political problems, it can sometimes be recommended. If the growth centre has been grossly misused by planners and politicians, it does not follow that the approach is useless. Where employment deconcentration and regional development based on industrial and urban expansion are to be recommended, the growth-centre approach may be appropriate. Most approaches have their uses if responsibly applied and if backed with adequate resources.

Employment deconcentration is not, however, a method of transforming society. It is a 'top-down' strategy of development intended only to mitigate certain malfunctions in society. It is not the answer to the problems of regional neglect or of limited public participation in decision-making. In this sense, the critique launched by Friedmann and Weaver (1979), and Stöhr and Taylor (1981) is well made. The latter have argued that 'development from below' is a requisite for true development. In order that the poor should benefit from economic growth, development strategies should be 'basic-needs oriented, labour-intensive, small-scale, regional-resource based, often rural-centred', and argue for the use of 'appropriate' rather than 'highest technology' (p. 1). Such an approach is far more radical than employment deconcentration *per se*.

(ii) *More efficient and equitable metropolitan expansion.* If a case can be made that employment deconcentration has a real, albeit limited, function, there can be little doubt that the need for such a strategy has been frequently exaggerated in many countries. Some of the more simplistic arguments have already been dismissed but in addition an argument that has gradually regained strength during the last few years is that policies to restructure the national settlement system are unnecessary. This argument is supported on the basis of three kinds of evidence, first, that regional polarization is a natural process in the path of national development and early state intervention will slow the pace of economic expansion; second, that metropolitan cities are rarely as inefficient or as inequitable as is often claimed, and, third, that solution of metropolitan problems is achieved more effectively through improved metropolitan planning than through employment deconcentration.

The argument that regional polarization is a natural corollary of the early stages of economic expansion finds intellectual and empirical support in the work of Kuznets (1966) and Williamson (1965). Both used historical data for developed countries and cross-section analysis to support the proposition that there is a long-term trend towards the equalization of personal (Kuznets) and regional (Williamson) incomes. Williamson found that in the ten countries for which temporal data were available, regional income disparities first increased and then declined as a more mature economic system evolved. More recently, Richardson (1977, 1989) has noted strong signs of what he has called 'polarization reversal' in the urban systems of certain countries.[7] In South Korea, Brazil, Mexico, and Colombia the growth rates of the largest cities have begun to fall as employment and population have begun to deconcentrate. While there is little sign that economic activity has moved far from the major cities, remaining well within the metropolitan zones, the slowing of growth is regarded as a positive sign (Townroe and Keen, 1984).

The implication of these findings is that government intervention and deconcentration are unnecessary because regional balance will occur naturally. Some economists have taken this point further to argue that government intervention in the distribution of economic activity is likely to waste scarce capital resources and thereby slow the rate of national economic growth. Thus in the longer term the country will be less able to redistribute income and remedy the problems of poverty. Regional balance and urban deconcentration should be left until a nation has achieved a higher level of development. The argument has been well put by a strong advocate of regional *laissez-faire* (Mera, 1976: 271): 'There is a fundamental conflict between high economic growth and decentralization of population. If a high rate of economic growth is to be achieved, further concentration of population into a few large metropolitan areas cannot be avoided.'

This view has often been supported by findings which attempt to show that large cities are more efficient and innovative than other urban centres. Such support includes evidence from Brazil, India, Sweden, and the United States which shows that industrial productivity is highest in the largest cities even when allowance is made for differences in capital per worker and size of enterprise (Richardson, 1973; Rocca, 1970); upon information from Japan, West Germany, Mexico, and the Soviet Union which shows that households are much better off in metropolitan areas; even when due allowance is made for higher living costs and for intervening variables such as age, colour, sex, and education (Hoch, 1972); on data which show that the per capita costs of social overhead capital tend to fall with increasing city size, or at least fail to rise (Richardson, 1973); and upon data from the United States which point to greater equality of incomes in metropolitan areas than elsewhere (Richardson, 1973). The sum of this evidence supports the view that urban economies exceed urban diseconomies even in today's largest cities. As Alonso (1969: 4) has put it: 'In brief, there is no basis for the belief that primacy or over-urbanization *per se* is detrimental to the efficiency goal of economic development. There are good grounds for believing in increasing urban size.'

Several writers go further and argue that even where there are signs of urban diseconomies, spatial deconcentration is not the answer; the fact that urban diseconomies such as crime, pollution, or traffic congestion tend to rise with city size can normally be explained by intervening variables (Alonso, 1971: 6; Richardson, 1973: 3). The fact that crime rates are so high in New York says more about the organization of, and the society in, that city than it does about urban centres of 10 million people. Improved urban organization and administration is a more appropriate approach to such diseconomies than employment or population deconcentration. Crime, for example, is better handled by a more efficient police force than by deconcentration; traffic congestion better controlled by higher car taxes or by parking meters than by migration control. The implicit policy recommendation stemming from these arguments is that the growth of large cities should rarely be discouraged.

Such a recommendation is not uncontroversial and the evidence on which it is based has been criticized on numerous grounds.

First, the tendency for regional income levels to converge is much weaker in contemporary less-developed countries than it was in the now developed nations. As Gilbert and Goodman (1976: 119–22) have argued, convergence is likely to be weak because: today's poor countries may never reach the levels of per capita income at which regional convergence begins; regional disparities in less-developed countries today are much greater than those characteristic of developed countries in the past; and convergence depends on effective government intervention and many governments show little sign of interest or ability to remedy regional inequalities. They conclude, 'There is no reason to assume that the processes which have led to convergence in the United States and other developed countries will function automatically and effectively in the less-developed nations.' If the process of development is different in the less-developed countries, then the pattern shown by cross-section studies is an unreliable guide to likely tendencies in the spatial distribution of resources and wealth.

Second, the efficiency of the larger cities has also been called into question (Johnston, 1976; Borukhov, 1975; Ternent, 1976). Gilbert (1976) argues that evidence of higher productivity in the largest cities should not be attributed only to agglomeration economies, for such 'economies' may derive from better urban infrastructure or higher-quality labour. In the latter case, it may be argued that higher productivity in the largest cities is to some degree achieved at the cost of lower productivity in smaller cities; if equivalent infrastructure or labour were available in medium-sized centres then the productivity of those centres might well rise. In addition, high productivity among private firms in large cities may be apparent because private companies are often subsidized indirectly by the state. If firms had to bear the full externality costs they impose, the higher productivity of large cities might well be less pronounced. If the firms were forced to pay some share of the diseconomies they create they might find the large city less attractive and many would transfer the higher productivity enterprises to intermediate cities. This process would reduce the apparent differential in industrial productivity between the smaller and the larger cities.

In fact, there was always some doubt about the supposed greater efficiency of the largest cities simply because so little work had been conducted in less-developed countries; most of the evidence in support of this case had been derived from studies in the developed world. And, while the few studies conducted in Latin America had provided conflicting results (Rocca, 1970; Boisier, 1973; Ternent, 1976), writers such as Richardson (1976) still argued that the advantages of increasing city size would be much greater in less-developed countries than in the developed world. Recent research, however, has not supported that case and Richardson (1989: 170) unashamedly admits that 'the earlier arguments about the superior productivity of very large cities may have been overstated. Highly productive cities may be found in all size classes, and the heavy concentration of economic activity in primate mega-cities may reflect the accidental consequences of government macro and sectoral policies as much as inherent efficiency advantages.'

Doubt has also been cast on the argument that all diseconomies in large cities are caused by intervening variables, viz. 'all large cities are locations for social problems, but not necessarily the same set of social problems, and their incidence in the big cities is higher merely because cities as a focus of civilization mirror society at large, perhaps in a magnified form' (Richardson, 1973: 3). While this is obviously correct, it is clear that major cities not only draw attention to societal problems but frequently accentuate them. Crime rates rise in large cities not only because of the more obvious concentrations of poor and wealthy, but also because it is more difficult to catch criminals. Traffic congestion is

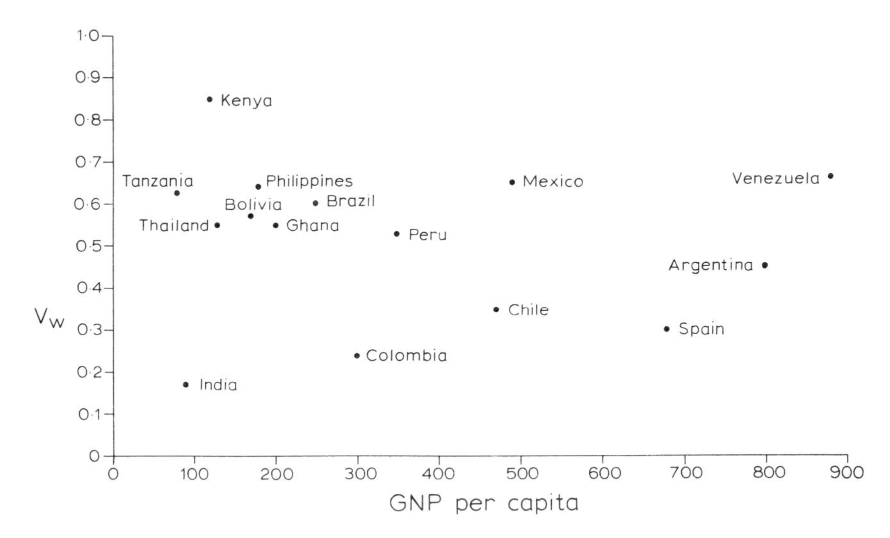

FIG. 8.2 Levels of regional income disparity in Third World nations

greater in metropolitan cities not only because there are more cars per capita, but also because transport problems are more complex. Solutions to such problems need to be larger, more expensive, and more radical than the solutions required in small cities. The solutions, however, may be very difficult to apply in most Third World metropolises.

Even if urban economies outweigh the diseconomies in large cities in less-developed countries, are the overall benefits equitably distributed? It can be argued that urban diseconomies most affect the lower-income groups, who are least able to escape them. Middle- and upper-income groups have the resources and knowledge to change their residential areas, command better public services, and influence political decisions, whereas the poor do not. Industrial zones are designated to keep pollution and lorries out of the high-income areas. The worst consequences of traffic congestion are inflicted on the middle- and lower-income groups living between the city centre and the more affluent housing areas. Urban renewal schemes rarely displace the rich but often dislocate low-income communities with inadequate consideration for the problems of the poor (Batley, 1982). Similarly, the distribution of the benefits in metropolitan areas rarely favours the poor. In most cities there is a clear demarcation line between the high-income and the low-income areas. Public roads, telephone, water, and electricity services in the high-income areas are superior, and where the public sector cannot provide adequate services, as in health and education, wealthy groups can resort to the private sector (Gilbert and Ward, 1985). The operation of the land and housing markets guarantee that the wealthy gain most from uncontrolled speculation.

Finally, it is obvious that the worst consequences of metropolitan expansion can be avoided by better urban planning. Critics and advocates of continued metropolitan expansion can agree that better planning is necessary. Nevertheless, it is by no means certain that better planning is possible. All around the less-developed world urban plans abound with little sign of effective implementation. Shibli (1974: 115) states that 'since 1947, the Karachi metropolitan area has not been planned systematically. The government did make an effort to prepare a plan, but did not provide the institutional, legal, and financial tools for the plan to be implemented.' In Jamaica 'as in so many countries, the ability to plan exceeds the administrative ability to deliver on announced proposals' (Trowbridge, 1973: 20). And, summarizing a study of thirteen nations, the Ford Foundation's International Urbanization Survey (1972: 21) notes that 'urban management in the developing countries is an *ad hoc* adaptation to circumstances always in danger of failure to adapt quickly enough to meet the pressure of those same circumstances.'

It can be argued that the absence of effective and equitable metropolitan planning is due to the fact that local politics are manipulated by important interest groups. Any attempt to raise taxes is immediately crushed, changes in zoning are made illicitly with the aid of bribes, major construction programmes are launched to cure problems that could be resolved in much cheaper ways. A glance at the attempts to give governments greater control over their urban areas supports the idea that real-estate interests are a powerful political lobby; such interests prosper because of the rising land values which bring chaos to other groups in the metropolitan areas. Effective planning is dependent on the political power balance and that balance has seldom favoured either equitable or even efficient metropolitan planning.

Even if major differences exist about the desirability of spatial deconcentration and the possibility of good metropolitan planning, most writers agree on certain issues. First, the concept of an optimum size of city is redundant (Richardson, 1989). Since city organization is at least as critical as size, and the nature of metropolitan economies varies considerably across the globe, no single size of city can be optimal. If a case is to be made in favour of deconcentration it has to be on the basis that certain goals can be achieved more cheaply, efficiently, or equitably in a more deconcentrated settlement system.

Second, whether or not deconcentration is favoured, more effective and fairer urban planning is a necessity in the large cities because there is no sign even in China or Cuba that metropolitan growth can be stemmed for very long. Such planning requires the implementation of policies unpopular with certain influential urban groups; a slowing in the growth of car ownership or at least severe constraints on car use within urban areas, the development of cheap forms of public transport, controls on pollution, and the introduction of effective land-use policies and land taxes. Urban politicians and vested-interested groups have to accept the modifications in the way large cities operate. Planning needs to be more effective both to maintain economic efficiency and to help the urban poor. Better planning is a necessity, not an unnecessary adornment to urban administration.

Third, each city requires different kinds of reforms. Some large cities have special problems which increase urban diseconomies; Caracas is located in a narrow valley and its region lacks water; Mexico City suffers from subsidence of the centre, from water shortages, and from atmospheric conditions which aggravate air pollution. Where it is more difficult to avoid urban diseconomies deconcentration may be a strong contender for political consideration on efficiency grounds alone. In other cases it may be quite unnecessary and may dissipate developmental potential.

Fourth, more should be done to compensate the sufferers of noise, pollution, and congestion. If large cities do indeed generate vital urban economies, companies will remain even if they are forced to pay the costs they impose on other city dwellers. Should firms decide to move because of the increased charges, then some of the arguments for decentralization will have been demonstrated to practical effect.

Finally, more should be done to encourage the seemingly spontaneous process, at least in the largest cities, towards policentric development. Many of the problems of traffic congestion stem from the concentration of development in a single city centre. If more encouragement were given to the emergence of secondary employment centres, either within the urban area or beyond it, then many of the advantages of city size might be obtained without the disadvantages. Sensible policentric development may be the answer to the major problems of the world's mega-cities (Richardson, 1989).

(iii) *The irrelevance of the spatial problematic.* Implicit in some of the previous arguments is the notion that the most severe social problems are handled rather poorly by spatial policies. Frequently, sectoral policies are far more effective than spatial programmes even in influencing the national settlement system. There can be little doubt that this view is correct. In addition, it is clear that little can be done to remedy spatial problems if substantial reforms are not made in society at large.

Some, indeed, have taken this argument further. They have argued that rather than serving to remedy some of the faults of society, urban deconcentration and regional development policies serve to mystify the issue of what is really wrong with society. As Geisse and Coraggio (1972: 58) put it 'the centralization–decentralization [*sic*] dichotomy tends to distract attention from the really important problem in Latin America. That is the necessity of permitting vigorous social restructuring and participation of all members of society in the benefits of economic growth, controlled and propelled by endogenous forces.' In similar vein, Singer (1975: 436) accuses those who wish to slow large city growth as attacking 'the consequences rather than the contradictions of capitalist development. Such misplaced criticism recommends solutions such as the control of urbanization, control of population, industrialization with less advanced (intermediate) techniques—[which] are utopian and tend to divert attention from the real problems of development.'

This argument has been developed most forcefully by Gore (1984) who emphasizes that the theory behind regional development planning is structured in such a way that it is biased against certain social and spatially defined groups. Rather than being 'scientifically sound' and therefore neutral, '. . . in reality, the propositions contained within the field are ideological' (Gore, 1984: 263). As such, the theory can be used to serve particular social interests while pretending to achieve some other kind of goal. The treatment of the regional problem can make the social situation worse and actually aggravate social inequalities.

This is not, of course, a new notion and is one that has been argued extensively by observers of the Latin American scene. Thus Jameson (1979: 69) has pointed out that in Peru, the ways in which regional programmes have been structured show that the consolidation of regional support for central government was the real motive behind plans for the substantial deconcentration of industry. Similarly, it is clear that regional policy is much less appropriate in some countries than in others. As Geisse and Coraggio (1972: 46) point out in some Latin American countries the difference in income between social groups in the metropolitan areas are as great as the differences between regions. In addition, since more poor people live in metropolitan areas in Argentina, Chile, or Uruguay than in the peripheral regions of those countries and are more socially aware than their provincial counterparts, dispersal policies are as likely to magnify as to reduce personal-income disparities. While the argument has much greater force in the southern cone of Latin America than in most other Third World countries, the point is well taken. Regional dispersal policies can serve either to reduce personal income inequality or to increase it.

The Practice of Spatial Deconcentration

If we take the number of deconcentration and regional development programmes as our criterion, there can be little doubt of the popularity of spatial planning in the Third World. During the 1960s and 1970s different forms of spatial planning swept Africa, Asia, and Latin America (Stöhr, 1975; UNCRD, 1976; UNRISD, 1971; Funnell, 1976). Despite widespread criticism during the late 1970s, its popularity with governments seems not to have diminished; for every government that has reduced its commitment to spatial development and planning, another has embraced it.

The forms that such a policy has taken, of course, has varied considerably both in terms of objectives and success. In general terms, however, spatial policy may be placed into one of three general categories: (1) policies to change the rural economy and thereby slow the rate of cityward migration; (2) policies to limit metropolitan growth through direct migration controls; and (3) policies to stimulate the growth of intermediate cities. Sometimes, of course, all three kinds of policy were used together.

Transforming the rural economy

Any policy that transforms the rural economy will affect the pace and form of urban development. Policies such as the redistribution of land to the poor may reduce urban/rural income disparities and slow urban growth by raising agricultural incomes. Other rural programmes, of course, may have the opposite effect; the green revolution and other efforts to raise agricultural productivity through incentives to commercial agriculture have accentuated landlessness and probably stimulated, rather than reduced, the flow of citybound migrants (Griffin and Ghose, 1979; Griffin and Khan, 1978; Kalmanovitz, 1977; Lipton, 1977).

It is clear that any radical approach to rural development is likely to affect substantially the pace and form of urban change. The precise nature of that change will obviously depend on the form of the rural policy. For that reason it is interesting to observe the distinctive approaches of two major countries to this problem. Both have poured resources into their rural areas but have done so in very different ways. Brazilian efforts to colonize the Amazon have embraced the capitalist ethic and used the transnational corporation as the main agent of change. The Chinese have used rural development to accelerate the introduction of socialism in an already populated countryside. The comparison is not intended to demonstrate the superiority of one mode of production over another, indeed such a comparison would be futile given the dangers of generalizing about two such vast and very different nations. It would also be artificial in the sense that policy in both countries, and especially in China, has changed profoundly through time; both sets of regimes have regularly modified their developmental objectives. What the comparison does demonstrate, however, is that regional development programmes cannot be separated from national development models. The programmes form an integral part of the process of national change and can be understood only in those terms. What the experiences also show is that the achievement of any regional goal is dependent on the resources and support given the programme. Whatever the differences between China and Brazil in their regional approaches, both have channelled substantial resources into certain non-urban areas.

The Brazilian frontier programme has gone through several phases and variations.[8] It has included, under the broadest definition, the building of Brasília, the construction of 10,000 miles of roads including the Belém–Brasíla highway and the *Transamazónica*, the setting up of regional agencies for this vast area, and the encouragement through huge tax incentives of private investment in the region. The programme has been effective in opening up a huge area for development; directly or indirectly, it has encouraged the migration of several million people into the states of Goîas, Mato Grosso, Pará, and Amazonas; large deposits of newly discovered minerals are now being extracted; huge timber and cattle ranches have been established; Manaus has been turning into a booming free-trade zone through tax reductions (Katzman,

1977; Kleinpenning, 1978; Mahar, 1989). As an example to the world of the speed with which a committed government in alliance with big business can act, there can be no doubt that the Brazilian experience has been impressive.

On the other hand, major questions must be asked about the success of the Brazilian exercise in terms of its effects on the poor, on the indigenous population, and on the environment. Major queries can also be posed about the benefits that the programme will produce for the majority of Brazilians. At one stage, between 1970 and 1974, a major aim of the Amazon programme was the resettlement of poor families from the north-east. The drought of 1970 had convinced the military government of the need to encourage families without land in the arid parts of the north-east to move west to open up the land along the sides of the new roads. In the absence of land reform in the north-east, land colonization was seen as the best possibility for alleviating the poverty of so many *nordestinos*. Unfortunately, the chance offered the poor was very much against the odds; the settlers received little advice or credit and were provided with insufficient infrastructural support to develop their land. By mid-1975 only a few thousand had been settled compared to the planned 100,000 (Katzman, 1977: 81). Apart from the construction of the roads, too little was spent on the Amazon programme. In addition the lack of effective governmental administration and the infertility of so much of the land meant that colonists were often victims rather than beneficiaries of the colonization process (Bunker, 1985; Moran, 1985). Their fate was all too often made worse by the danger of losing the land that they had cleared and occupied. Undeterred, and sometimes even abetted, by the government bureaucracy, large landowners established legal rights to the lands occupied by the poor (Foweraker, 1981; Moran, 1985). As a result, fewer colonists entered the region than had been expected. Having failed to take their 'chance', the small settlers were deemed to have little future in the region and emphasis shifted to the large company as colonizer. These companies were offered the chance of reducing their income-tax liabilities by an amount equal to the sum they invested in the Amazon region, up to half of their total tax bill. In addition, tax exemptions or deductions and the waiving of import duties on equipment were granted. These incentives led to the creation of vast cattle ranches, huge timber projects and the opening of major mining areas. The Amazon progamme had always contained a resource-development objective, but after 1975 it was dominant (Kleinpenning, 1982: 7). The opening of the Amazon now aimed to help resuscitate the slowing Brazilian economic miracle; the poor of the north-east, and indeed the indigenous Amazon population, were conveniently forgotten. If the Brazilian miracle was more notable for its economic growth rate than its ability to redistribute income, then the Amazon programme became part of that miracle.[9] If agriculture and mining have boomed, it has been to the benefit of a few large companies (Fearnside, 1986). It was also at the cost of further increasing Brazil's huge mortgage with the world's bankers: Brazil's foreign debt is the highest in the world and in 1985 its total long-term debt service was equivalent to some 35 per cent of its total exports of goods and services (World Bank, 1987: 237). In population terms, the area has grown dramatically, the population of *Amazonia Legal* having grown from 5.2 millions in 1960, to 7.2 millions in 1970, and 11.2 millions in 1980. However, this has not managed to transform the spatial structure of the country because 'the "surplus" agricultural population is not migrating *en masse* to the North' (Kleinpenning and Volbeda, 1985: 13). As a result, São Paulo and Rio de Janeiro have continued to expand, stimulated in part by the profits from the Amazon being reinvested in commercial and construction activities. As a spatial deconcentration policy it has been only a partial success. As to the environmental effects, comments vary from the critical to the cataclysmic. The latest estimates suggest that in 1988 12 per cent of the Amazonian forest had been cleared, some 600,000 square kilometres, and that the pace of deforestation accelerated markedly from the mid-1970s (Mahar, 1989: 7). Goodland and Irwin (1975) and Davis (1977) have pointed out that once cleared, the Amazon forest rapidly suffers from soil erosion and leaching; even cattle pasture is being degraded (Hecht, 1981). Most dramatically, Calder (1974) has argued that continued clearance of the *selva* will change the world's climate, raising the temperature of the globe with dire consequences. Similarly critical have been the comments on the treatment afforded to the Brazilian Indians who have been issued the command 'integrate or perish' (Hecht and Cockburn, 1990). Spectacular the Amazon programme may well be, but it demonstrates most of the faults of uncontrolled capitalist development; it generates economic growth, without improving the extreme maldistribution of wealth and income. As a spatial programme it clearly demonstrates the paramount importance of the style of national development on the formulation of subsidiary policies.

The last point can be drawn equally from the Chinese experience. The policies adopted under Mao say more about Chinese socialism than about rural development or spatial policies *per se*. The range of policies adopted after 1949 sought to modify the nature of the whole economic and social system. Whatever the effect of explicit spatial policies, implicit spatial policies would have had a dramatic effect. The combination of birth control, the redistribution of land from private landlords to the peasantry, state ownership of most economic activities, and limits on improving urban living conditions would have had a favourable effect on the rural population even without explicit spatial policies. The Chinese experience is undoubtedly interesting as an example of a spatial policy, but it would be absurd to classify it mainly as the outcome of such a policy. It would also be erroneous to overemphasize the extent to which Maoist policies favoured the rural sector. Clearly, the control of urban growth, increasing self-reliance in the countryside and developing the resources of the western region have been important features of those policies. But such an approach has been accompanied by other policies much less favourable to the rural sector.

As Leeming (1985: 18) points out the state has invested much more in industry than in agriculture. 'In spite of intermittent Maoist enthusiasm for mechanisation, Chinese farming has been systematically starved of investment in order to support the greedy heavy-industry sector.' Similarly, one of the bases of agricultural policy has been to keep the price of grain low, a policy bound to favour the urban sector. And, while it is true that the government did dramatically slow the pace of urban growth, the cities continued to expand. Between 1949 and 1979, Kirkby (1985: 10) estimates that the urban population expanded nearly three times.

None the less, Chinese development since 1949 is distinctive in so far as it has been much less pro-urban than policy in most other parts of the Third World. First, unlike the tendency in most less-developed countries, the terms of trade have moved in favour of agriculture and against industry especially since 1966 (Lardy, 1978; Leeming, 1985). Second, the Chinese have attempted to increase economic self-sufficiency among their rural communes. As far as possible communes have been encouraged to set up industrial plants to produce for their own needs—a deliberate attempt to limit the geographical areas of the markets of urban industries. Regions are encouraged to develop their own resources so as to reduce transport flows and to commit local people to the drive for national development. The mobilization of resources that would otherwise be wasted is one of the most interesting elements in the Chinese experience. Third, the Chinese have tried to change cultural values in favour of the rural areas. Urban bureaucrats and youths have been 'sent down' to the countryside either for short- or long-term stays. Most significantly, control over population flows have restricted the urban population to three-fifths of what it would have been without such controls (Kirkby, 1985: 130).

The outcome of these policies has been impressive in many respects. Perhaps the most notable achievement has been that all members of Chinese society have received an adequate minimum income protecting them from malnutrition and guaranteeing access to basic essentials. If the standard of living in China remains low, there can be little comparison with the lot of the average Chinese before the revolution nor any doubt that the distribution of income is much more equitable. This, together with the elimination of starvation and exploitation, is the main achievement of China, which compares well against the experiences of most other Third World countries. These achievements led to many calls during the 1970s to the effect that the world should learn from Chinese experience. During the 1980s, of course, the Chinese began to criticize their own system much more vociferously and with the events of Tiananmen Square in 1989, western enthusiasm for Chinese experience has waned considerably.

With respect to the spatial outcome of Chinese development, some of the results have been both unexpected and unwelcome. First, China has not managed to eliminate inter-regional or rural–urban disparities. In 1957, the last date for which reliable figures for agricultural and industrial incomes can be calculated, the difference in the provincial incomes of Shanghai and Honan was of the order of eight times. At that time regional inequalities compared badly to most developed countries but were not atypical for most Third World countries (table 2.1). Lardy's (1978) estimates suggest, however, that there has been a consistent, if slow, convergence in regional income levels since 1957: a trend less than typical for most Third World countries (Gilbert and Goodman, 1976). Urban–rural disparities remain, even if they have been softened, and the levels of wages, consumption, and services continue to be much higher in the cities (Murphey, 1976: 327). This contrast, together with the 'indefinable glitter of urban life', has attracted migrants to the urban areas. Frolic (1976: 154) argues that despite totalitarian controls, 'planners simply have been unable to limit the population of their largest cities, or to effectively control population movement.' Even if China will continue to be a very rural nation industrial and urban development must become more important. In the future, indeed, rapid urban growth appears to be inevitable. As Kirkby (1985: 243) argues the coming years will 'see an excess in the rural labour force running into hundreds of millions . . . the central problem for China's spatial planners is where to accommodate the vast population no longer needed on the land'.

The significant changes in the distributions of population and economic activity since 1949 have not been achieved cheaply; indeed, a great deal of production has been sacrificed. Such losses have been caused in large part by considerations for defence which have been a major factor in spatial policy. The uneasy relationship, first, with the United States and later with the Soviet Union encouraged a policy of economic deconcentration away from the coastal region. The spread of economic production and the building up of the interior was part of the defence against invasion. Nor should it be forgotten that the policy of rural bias and deconcentration has not been entirely successful in economic terms. Lardy (1978) notes that the Chinese have accepted that people should come before per capita income growth but queries whether the same goals might not have been achieved more effectively through modified policies.

Finally, it is important to emphasize that while China has achieved a considerable level of economic deconcentration, political centralization remains strongly entrenched. China is a highly centralized nation; all policy initiatives come from the top down and local needs are regularly sacrificed to national priorities. What has been achieved in China is local self-sufficiency rather than local independence. The country is therefore some way from the ideal balance that Friedman and Weaver (1979) advocate between territorial and functional needs, and it is indeed ironic that China has been the model for 'bottom-up' strategies of development as recommended by Stöhr and Taylor (1981). Within China there has long been an incipient totalitarianism which could prove disastrous if the political machine should fall into the wrong hands (Morawetz, 1979).

Since 1949 China has adopted ambitious and highly innovative spatial policies as part of a unique social and economic experiment. For this reason it is rightly recommended as an example to other Third World nations where equity and self-sufficiency rank low in the order of national priorities. But it would be erroneous to suggest that China's policy has been consistent or even anti-urban. Violent changes in policy have marked China's development approach; periods of sending people down to the countryside have been interspersed with a willingness to allow urban growth (Ma, 1989). Since Mao's death, of course, China's development policy and its attitude to town and country has changed again. As such the Chinese experience must be carefully considered before adopting any part of it. For while it is a marvellous laboratory for other countries to observe, Chinese policies have clearly not been an unconditional success. Nor, even where there are clear signs of success can the policies be easily transferred. Perhaps the most important lesson to be learned from China is that it is very much a special case due to its size, its history, and its revolution. Its recent experience is very different not only from other Third World countries but also from other communist nations. If China's experience is relevant and useful to countries in other parts of the world, it would be wrong to ignore both the problems that it has faced and the difficulties that transferring the experience might bring.

Migration control

Many city administrations in the Third World have felt the need to introduce policies to slow the rate of migration from the rural areas. Incapable of supplying the housing or infrastructure required to accommodate the growing population, they have frequently advocated the idea of influx control. In fact very few countries have introduced such a policy and in most places it has had limited success.

By far the best-known experiment in migration control has been in China. Beginning with the Great Leap Forward (1957–8) and accelerated during the Cultural Revolution (1966–8), millions of young and unemployed people were 'sent down' to the countryside for short or long stays (Murphey, 1976: 316). Between 1968 and 1978 some 8 million young people were exhorted to move to the rural areas both as a means of reducing pressure on the urban areas and as a means of lessening the cultural gap between urban and rural areas (Thompson, 1979: 305). While this policy had an important effect upon urban growth, it probably had less influence than the combination of improved rural conditions and the regular controls over migration available to the Chinese authorities. In a society where housing, work, and even food are allocated by the state, the movement of families can be easily controlled. Acquiring permission to travel in China has been likened to the process by which a soldier in a developed western country applies for a two-week pass. As a result, China has been one of the few countries in the world that has managed to slow the rate of urban expansion and, at times, even to reverse it (Kirkby, 1985).

In Kampuchea, of course, migration policy was still more drastic. Within hours of the arrival in Phnom Penh of a new government in April 1975, the city was being evacuated. By the end of the month the cities of Ream, Poipet, and Pailin had received similar treatment. Urban dwellers were simply directed to rural areas where they were to be absorbed. The rationale behind this policy was a mixture of the strategic and the Utopian. The fear of starvation in the refugee-swollen cities was clearly of importance, as was the possible threat of a counter-revolution. But there was also a longer-term aim of increasing the self-reliance of the country, the wish to control the bourgeoisification of middle-class groups and the goal of increasing agricultural production (Shawcross, 1979). Whatever the objectives, however it is difficult to find support for such a policy in Kampuchea or indeed anywhere else.

Outside the socialist world, the only nation in which controls on population movement have been effective has been the Republic of South Africa. Unless they have a job, Africans are not permitted to live in urban areas and even those with jobs have to work in the city for fifteen years before they are permitted to bring their families. The Republic's policy has long been one of permitting Africans residence in the cities only when their labour was required (Wilson, 1972). For an African the loss of his job means that he is forced to move to his respective 'Homeland' unless he manages to avoid the authorities. Clearly, there are many people who are living illegally in Soweto or Crossroads, but there can be little doubt that for many years the policy severely limited the movement of black Africans to the city. However unfair the policy may be, it limited the growth of the major cities:

Johannesburg's African population, for example, increased annually by only 2 per cent between 1960 and 1970 (Fair and Davies, 1976: 155).[10] Recently, however, economic and political pressures have forced some modification in the policy of influx control (Smit *et al.*, 1982). There are now greater possibilities for blacks to stay temporarily in white areas but the most profound change has been the expansion of the urban black population in the homelands. In 1960, the urban black population numbered only 33,000; by 1980, it had grown to almost 1.1 millions (Smit *et al.*, 1982: 95–7). South Africa is no longer so different from the rest of Africa in terms of its rate of urban growth.

Elsewhere controls on population movement have always been effective. In Indonesia, the governor of Jakarta limited migration by decree in 1970. All rural–urban migrants were required to obtain six-month permits by depositing a sum equivalent to twice the cost of their bus fare home. If after six months a job had been obtained, the deposit was returned. In addition to this policy, attempts were made to limit employment in the so-called formal sector: the number of tricycle taxis in the city was severely limited, an activity which employed a quarter of a million people in 1972 (Simmons, 1978: 15). The only real effects of this policy were to make life for the poor much harder and to increase the possibility of corruption. It did little to limit population expansion. Similarly, in Tanzania efforts to persuade the unemployed to leave Dar es Salaam and return to their rural homes met with little success.

In general, migration control does not recommend itself as a suitable policy in most Third World countries. It can be effectively applied only where there is an efficient and authoritarian government or where a genuine rural development programme reduces the differences between urban and rural living standards. In many cases its effect is likely to be harmful to the poor; in most countries it will have little effect beyond increasing the scope for corruption.

Employment deconcentration

By far the most common response to urban and regional imbalances is a programme to limit growth of the larger cities and to stimulate the economies of medium-sized centres. The prime candidate for inclusion in such a programme is invariably the industrial sector. Existing or new companies are persuaded to locate their plants in peripheral regions through a combination of tax incentives, the provision of infrastructure, the construction of industrial estates, and occasionally compulsion. In general, however, the stick has been used much less than the carrot; most governments have feared that too strong a programme of deconcentration would dissuade foreign companies from investing in the country, would lower efficiency in the industrial sector, or would offend important national business interests. Hence, few governments have compelled companies to move from the larger cities or even required that new companies should locate in peripheral regions. Even where some measure of compulsion has been used, as in Chile, Colombia, and Venezuela, the measure has been leavened by various incentives or confined to specific kinds of industries (UNCRD, 1976; Stöhr, 1975; Gilbert, 1974*a*; Townroe, 1979).

Various incentives have been used to attract industries to new locations. Since 1966, the Mexican government has encouraged North American companies to locate across the border by offering tax concessions on imported raw materials used in the production of 'exports' back to the United States (Baerresen, 1971). This policy has been gradually enlarged and, as a result of the plummeting value of the Mexican peso since 1982, a huge expansion has taken place in *maquiladora* employment. In 1989, it was estimated that more than 400,000 Mexicans were working in foreign owned *maquilas;* increasingly in plants owned by Japanese as well as by US companies (Mexico, 1991; Gibson and Corona, 1985). In Brazil major tax incentives attracted numerous companies to the principal cities of the poverty-stricken north-east, and even to the exotic manufacturing centre of Manaus some 1,200 kilometres from the mouth of the Amazon (Kleinpenning and Volbeda, 1985). Experience shows, therefore, that provided the incentives are sufficiently attractive, manufacturing companies will locate in a wide variety of 'undesirable' locations. But such incentives have to be gauged carefully, since industry tends to favour those areas that are closest to their preferred locations. Thus, in north-east Brazil new industry was located only in the three largest cities and in Peru new industrial estates attracted companies only to the most prosperous and attractive city, Arequipa (Gilbert, 1974*a*; Jameson, 1979; Goodman, 1972). In India the second and third development plans led to the establishment of 486 industrial estates throughout the country, but only estates close to the largest cities prospered; since 1963 most estates have been set up near the major cities. Manufacturers can be persuaded to locate in lower priority centres, but usually the incentives are inadequate, too widely spread, or constantly change as governments re-evaluate both their programmes and their commitment to industrial dispersal. As a consequence, few programmes have been wholly effective in shifting the location of industrial activity and many dispersal strategies have been declared redundant.

Where industry has been encouraged to move away from the largest cities, the issue is whether such deconcentration has had a positive effect on the poorer regions. In some cases, such as the banishment of the Chilean car industry to Arica in the far north of that elongated country, the results were little less than ludicrous. Cross-movements of parts and finished vehicles from the major industrial centre and major market in Santiago to the assembly plants in Arica were grossly inefficient (Johnson, 1967; Gwynne, 1986). More often industrial dispersal has increased regional output but created few jobs. Indeed, deconcentration programmes have tended to suffer the same difficulty as national industrialization programmes: capital-intensive technology has created few local jobs and little demand for local inputs. Thus the Mexican border programme and the Brazilian 34/18 tax incentive scheme have had limited effects on local poverty (Sklair, 1989; Goodman, 1972). The industry that has been established has failed to stimulate local enterprise or the surrounding regions in the way suggested by growth-centre theory. Indeed the effects on areas surrounding growth centres have been uniformly discouraging. Studies of the regions around Kuala Lumpur (Robinson and Salih, 1971), Medellín (Gilbert, 1975), and Ciudad Guayana (Travieso, 1972) have found that either as a result of weak 'spread' effects and/or substantial 'backwash' effects, the regions beyond the immediate vicinity of the growth centres receive little in the way of positive economic or social benefits. Large-scale industrialization within a growth centre is a poor means of developing a poor region in the absence of fundamental changes in the agricultural economy, the marketing system, and the pattern of land holding. Industrial deconcentration may serve to reduce pressure on the metropolitan areas, but unless accompanied by other, often more radical, programmes brings little benefit to poorer regions.

The building of new cities

The most direct, and often most dramatic, form of spatial-deconcentration policy has been the construction of new cities. New cities have performed a variety of spatial functions: as satellite centres to accommodate the growing population of a nearby metropolitan city; to house the population engaged in the development of a major new mineral resource; or to function as a new political capital.

Many satellite cities have been built along the lines of the British new towns. Efforts have been made to attract industry so as to provide jobs for the displaced populations and to prevent the emergence of 'dormitory suburbs'. Often such programmes have been a response to some special problem which has made metropolitan growth difficult to manage. Thus the partition of India led to a flood of migrants into Karachi which stimulated the construction of two satellite cities, Korangi and North Karachi (UNRISD, 1971). In other cases the aim has been to avoid perceived diseconomies associated with metropolitan expansion. In Venezuela new cities are being established in the Tuy Valley as a means of deconcentrating employment and population from Caracas's narrow valleys. In China Greater Shanghai now has over sixty satellites with populations rising to 60,000 people (Thompson, 1979: 305). Occasionally, too, new cities serve more debatable goals, as for instance in South Africa.[11]

New cities have often been associated with the development of mineral resource programmes. In Venezuela the widely publicized Ciudad Guayana project is part of an ambitious programme to develop the iron-ore, bauxite, and hydro-electric potential of the region (Friedmann, 1966; Rodwin, 1969). Today the city has a population of some 400,000, the Guri dam supplies over half of Venezuela's electricity, and its factories produce most of Venezuela's steel and all of its aluminium. Whether such a spectacular example of modern urban and industrial expansion has been an effective way of spending Venezuela's oil reserves is debatable given the social conditions in the city, the lack of stimulus afforded the surrounding region, and the limited numbers of jobs created (CEU, 1977; Travieso, 1972; Izaguirre, 1977). Nevertheless, whatever the criticisms, the plan has had a major impact on the spatial organization of the Venezuelan economy.

Similarly spectacular new cities have been built in several countries as new national or provincial capitals. Brasília, Islamabad, and Ankara were all built to serve as examples of the new future facing their nations. Their architecture and urban design reflected this goal; if Brasília was to serve as a national symbol nothing should be spared in its construction and nothing permitted to destroy the image created by Oscar Niemeyer (Epstein, 1973). Chandigarh, the new capital of the Indian Punjab, commissioned Le Corbusier to produce a city evocative of India's future; a future offering the poorest the means of living a dignified life (Sarin, 1979: 136–7). Such capitals represented a break with the past and with existing metropolitan centres which had dominated national life. They promised a new direction for development, located inland, away from the coast which had dominated previous political and economic history. More recently, several African nations have begun to construct new capitals. In Nigeria the government began construction of a new federal capital at Abuja, a sparsely populated area in the centre of the country. Not only would this city help reduce the problems of urban growth in Lagos, but it was hoped that the new capital would provide greater balance between the three major ethnic groups, the Hausa, the Ibo, and the Yoruba. Unfortunately, it seems as if the construction of the new capital has accentuated ethnic tensions more than resolving them (Moore, 1984). Elsewhere in Africa, new capitals have also been built in Tanzania (Dodoma) and Malawi (Lilongwe), and there are further plans to build new capitals in the Ivory Coast, in Mauritania and, possibly, in Liberia (Potts, 1985; Armstrong, 1985). In each case the transfer has been motivated by the wish to redress regional imbalances and to help establish a new national identity by moving administrative power away from the former colonial capital.

If there have been more cases of capital city transfer in Africa it is because most African nations have only recently achieved independence and because so many have had capitals that were so inconveniently located. Most existing capitals had been located to perform colonial functions not to serve as the administrative centre of a newly emergent country. Nevertheless, there are increasing signs that Latin American countries are beginning to think harder about the locations of their capitals. In 1987, Argentina approved the transfer of the Federal Capital to Viedma-Carmen de Patagones, twin towns located 1,000 kilometres south from Buenos Aires. The aim in this case is to help reduce the power, and thereby the growth, of the existing capital, to help stimulate the regional development of Argentina's frontier, particularly that of Patagonia, and, finally, to encourage a new national image by transforming and modernizing the state apparatus.[12] Argentina, therefore, follows the example of Belize, which moved its capital to Belmopan, and may one day be followed by Peru which is thinking of moving its capital from Lima to the eastern slopes of the Andes.

Dramatic though many of these new cities are, and despite the imaginative and optimistic philosophies underlying their construction, few have been totally effective. In few parts of the Third World has their growth radically altered the national settlement system. As the Ford Foundation (1972: 9) note, '. . . in a time span of some 20 years, they have a combined population of little more than one million—together, they comprise a city roughly eqivalent to Recife or Bangalore.' Nor has their development slowed the growth of metropolitan São Paulo, Caracas, or Karachi; all expand as rapidly and as chaotically as ever. The governments of Brazil and Pakistan may now be physically located in new capitals, but the old power centres still exert considerable political influence. Nor do the new cities live up to their ideals of creating new life-styles for their inhabitants. As Epstein (1973: 109) notes, 'the *barraco* or shack is at least as typical of Brasília as the colonnades of the Dawn Palace, the towers of the Congress, or the rectangular apartment buildings of the superblocks.' More often governments have created artificially divided twin cities. Ciudad Guayana contains the planned city and the next-door unplanned city of Puerto Ordaz where most of the poor live (Peattie, 1987; Macdonald and Macdonald, 1979). Brasília has experienced a series of conflicts between the planners, who have tried to exclude the spontaneous settlements from the urban area, and the poor who wish to live and work in the city (Epstein, 1973). Nor, of course, have most new cities been cheap to build. Vast sums have been spent on some of these cities which arguably could have been better spent on other projects. It is possible that these capitals have served

Table 8.1

New capital cities in Africa, Asia, and Latin America since 1950

Year[a]	Country	New capital	Former capital
1956	Brazil	Brasília	Rio de Janeiro
1957	Mauritania	Nouakchott	Saint Louis (Senegal)
1959	Pakistan	Islamabad	Karachi
1961	Botswana	Gaberone	Mafeking (South Africa)
1963	Libya	Beida[c]	Tripoli and Benghazi
1965	Malawi	Lilongwe	Zomba
1970[b]	Belize	Belmopan	Belize City
1973	Tanzania	Dodoma	Dar es Salaam
1975	Nigeria	Abuja	Lagos
1983[b]	Ivory Coast	Yamoussoukro	Abidjan
1987	Argentina	Viedma/Carmen de Patagones[d]	Buenos Aires

[a] Year decision announced and/or approved.
[b] Inauguration date.
[c] Moved back to Tripoli in 1969.
[d] Present cities—name of the new capital to be determined.

their nations in other ways; increasing national integration, evoking national pride, and breaking economic and sentimental attachments to the past. But set against these advantages is their cost and the limited effect they have had in helping the poor. As Renaud (1981: 116) cryptically recalls, 'there has yet to be a planning document for a new capital city that investigates the costs and benefits for the option of not building it.'

Models of Development and Spatial Policies

The failure of decentralization and national settlement strategies

If the standard of judgement is that an adequate spatial policy should help the poor in the poorer regions and organize the distribution of settlement so as to help lower-income groups, few explicit spatial policies have been effective. Metropolitan areas have been allowed to develop in ways which have favoured the higher-income groups and created problems suffered at least proportionately by the poor; regional programmes have either served to boost national economic growth rates or have been pushed less than wholeheartedly.

Regional policies have normally been embraced enthusiastially only when they have promised to accelerate national economic growth. Perhaps the example *par excellence* is the Guayana programme in Venezuela. Publicized widely as an example to the Third World of how to implement both an urban planning and a regional development policy in a poor country, the Guayana programme was first a national growth strategy and only second an urban and regional policy. Although a regional and social dimension was explicit in the adoption of the programme, its primary appeal to the national government lay in the economic potential of the region's rich iron, bauxite, and water resources. Had there been another explanation, then the regional agencies set up in other parts of Venezuela would have been given greater support than they were (Friedmann, 1966). The continued flow of resources into the Guayana and the rapid demise of regional development programmes in other parts of the country can be explained only in terms of the former's contribution to national growth and the lack of similar potential in the rest of the country. Similar experiences are common throughout the Third World. Major regional development projects have been adopted *only* when their principal contribution was likely to accrue to the nation as a whole.

The corollary of this argument is that when urban and regional policy goals have conflicted with the paradigm of national economic growth, the latter has been the winner. Government after government has established regional programmes espousing goals of greater equity and regional balance. Almost invariably these policies have been counteracted by national programmes which have tended to accentuate regional disparities and encourage the growth of major urban areas. Of course, the introduction of regional programme has sometimes reduced the level of spatial concentration that would have resulted from the national programme alone. But in general the weakness of regional and deconcentration programmes suggest that they were never priorities and never intended to achieve a real balance (Boisier, 1979; Jameson, 1979).

The growth-before-distribution ethos also pervades the formulation of regional-development programmes even when politics have dictated that a genuine policy of deconcentration be espoused. Thus, in the Brazilian north-east resources have been channelled for redistributive reasons away from the south-east. New industrial employment has been created in the main cities of the north-east and the programme has helped to raise the regional product of the area. Such a programme was not motivated originally by national growth objectives. The locational inefficiency involved in locating industry so far from the country's main markets and supply areas was not negligible and without considerable incentives private industry would not have responded to the programme. But while the aims of the policy clearly did not reflect the national growth ethos, the method of implementation did. Industrial expansion for large companies was certainly not slowed by the enormous tax reliefs offered under the 34/18 mechanisms. No change was invoked in the capital-intensive nature of Brazilian industry because of the failure to offer stronger incentives to labour-intensive companies (Goodman, 1972). Nothing was done to redistribute land or to introduce other social measures alien to the national growth strategy (Dickenson, 1980). The inevitable corollary has been that while the regional product increased, intra-regional disparities widened. Whether measured in terms of urban–rural differentials or in terms of the income shares of rich and poor, the latter have seen little improvement in their economic or social situation. There is also some evidence to suggest that real per capita incomes of the poor in several cities actually declined during the 1960s despite the programme (Gilbert and Goodman, 1976). In the north-east certain elements of the programme helped the poor, but the benefits were undermined by the lack of an equity component and by the regional effects of national policies concerned with the control of inflation and hence wage levels. Stabilization policies throughout Latin America have tended to hurt numerous groups, including the poor in peripheral regions.

If nothing else, regional policies have served to maintain, and sometimes even to increase, the share of poor regions in per capita national product. In this sense the arguments of the early Frank (1967) model which argued that rural area, national metropolis, and world 'metropole' were linked in an exploitative relationship which led inevitably to the impoverishment of the periphery was too simple. Most peripheral economies are growing either because of the forces of the market or through help from the regional policies of the national government. The problem is that while the poor regions may benefit from regional policies, the poor of those regions benefit insufficiently. Too often the benefits go to industrial and large-scale agricultural groups in the region. A reduction in inter-regional disparities is accompanied by growth in intra-regional disparities and personal inequalities.

What I am arguing is that major regional programmes and indeed programmes for planning the major cities are essential but in themselves are insufficient. As many writers are now suggesting, it is the nature of the development model or style that is critical (Dunham, 1979; Friedmann and Weaver, 1979; Stöhr and Tödtling, 1977; Richardson, 1981; Hinderink and Titus, 1988). If that model is consistently failing to create sufficient jobs or to redistribute income, then there is little that specific regional policies can do to resolve regional problems. As Richardson (1979: 144) correctly concludes 'spatial policy instruments often have a smaller impact than the implicit spatial effects of macro and sectoral policies.' In this one sense the experience of certain European countries is illustrative. Most have now introduced regional policies to remedy their spatial inequalities and to limit the growth of their major cities. In most European nations regional inequalities are tiny by comparison with those of Third World nations. But the connection between regional policy and the reduction of regional income disparities should be viewed with extreme care. It would seem most likely that social policies have been of greater importance than regional. In Britain regional disparities would have continued to be much larger had it not been for the introduction of free public education and health programmes, pensions for the old, grants for retraining, and unemployment benefits. In Europe regional policy was introduced into generally more equitable social systems than exist in the Third World. In the latter not only are existing inequalities extreme, but no real effort seems to have been made in most countries to reduce them (Cornelius, 1975). The continual search for methods to raise per capita income more rapidly is tending to perpetuate existing inequalities. In this context, regional development and decentralization policies are bound to be ineffective (Dewar *et al.*, 1986: 188). To bring about more balanced regional development and a fairer share of wealth requires much more than regional policy, it requires a national commitment to a much modified development programme in which equality occupies a much more dominant position.

The promise of alternative approaches

The degree to which alternative approaches favouring equity, or even the protecton of non-renewable resources can be introduced is clearly a matter of politics. While some developed capitalist countries such as Sweden or Britain have achieved a measure of personal equality, there are few Third World capitalist nations that are improving the distribution of income or raising the standard of living of the poor. Similarly, there are few signs of success in modifying the spatial structure or the national settlement system. Reacting to this unfavourable experience, much of the literature on regional development and national settlement systems is increasingly arguing that some kind of radical change is required in the capitalist-development model (Friedmann and Weaver, 1979; Stöhr and Taylor, 1981; Gore, 1984). It is not sufficient to introduce new kinds of spatial or urban policy because the operation of the capitalist system will undermine those policies. Greater equality is required to improve the living standards of the poor. Only under a more equitable system is it possible to modify successfully the urban structure, redress the imbalance between urban and rural areas, remove the worst regional disparities, and develop agricultural potential in a non-exploitative fashion. Only by removing some of the power of the transnational corporations and the 'comprador' bourgeoisie can true change take place.[13]

The example of China has clearly been foremost in many minds, although I have already examined some of the implications of that experience and suggested that socialism does not offer a prescribed policy and attitude towards urban development. This point is perhaps best illustrated in the different attitudes towards urban development taken by Marx, Lenin, Stalin, Mao, and Castro. Marx's view of the peasant attitudes was hardly flattering and his formula for revolution and future progress was linked integrally with urban and industrial development. Of course, as a European he was the child of an urban civilization. While his attitude did not change as the result of a brief study of India, greater experience in the Third World might have modified his views (Marx, 1853). By contrast, Mao Zedong and Castro, while both giving priority to industrialization, gave ample encouragement to the rural sector. In part their attitudes were influenced by the source of their revolutionary support. If Marx and Lenin felt that revolution could come only from the industrial proletariat who were subject to the full contradictions of capitalism, both Mao and Castro came to power as the result of rural support. Whether, of course, this is the most appropriate criterion on which to formulate national development policy as opposed to stimulating revolution is another issue. But these examples demonstrate that the theory and practice of socialism have few clear lessons with respect to spatial development.

Similarly, socialism in practice has not always resolved the problems that afflict poor nations. While some rightly question whether most communist countries are socialist, it is instructive to observe that few communist nations have managed to remove the worst forms of regional inequality. As Fuchs and Demko (1979: 304) have pointed out for Europe,

> judged from various perspectives—regional contrasts, urban–rural and urban–urban comparisons, and intra-urban distinctions—the socialist states studied exhibit marked spatial inequalities. The persistence of these inequalities can be explained in terms of the priority placed on efficiency or military security as opposed to equity in industrial location decisions, the favouring of investment in 'productive sectors' rather than social infrastructure, a desire to defer urbanization costs as reflected in constraints on urban growth, the growing scale requirements of service and human welfare facilities, and the continuation of substantial differences in income for various occupation groups.

As we have already seen, regional disparities are still evident in China; if regional inequality is a problem in peripheral capitalist nations it is likely to continue to be an issue under a socialist regime. If socialism is able to reduce personal income disparities, then regional disparities will diminish, but anything vaguely resembling complete personal and regional equality seems unlikely under many forms of socialism.

Another issue that a consideration of socialist and communist experience raises is whether it is possible to slow urban growth. We have seen that the Chinese under Mao sometimes banished people from the urban areas, whereas European communist states have permitted urban growth even if they have controlled metropolitan expansion. The decisive question is whether or not urban growth is inevitable. The Chinese have certainly accepted that Shanghai and its satellite cities serve an important function. If their growth has been limited, it has certainly not been stopped (Kirkby, 1985; Fujima, 1987). Similarly, Cuban experience suggests that having slowed the growth of Havana until rural conditions were improved, the national capital and other major cities have once again begun to expand (Susman, 1987). Now that China seems likely to accept a path of development more dependent on advanced technology and manufacturing, rapid urban growth is highly probable. Clearly, China will not suddenly accept the uncontrolled urban growth so typical of capitalist Third World nations, but as Frolic (1976: 159) notes, 'until she industrializes more rapidly, it is too early to say whether she has found a permanent solution to problems of urbanization and modernization'.

Finally, socialist experience is ambivalent on an increasingly important issue in the development debate; the appropriate attitude to the physical environment. The saving of energy and the conservation of renewable resources seem of increasing importance in the modern world even if decision makers are still scarcely recognizing the policy choices involved. In China policy seems to satisfy the seemingly sensible dictum of Brookfield (1979: 120) that 'to the several objective functions of "development" another must be added: to make the best use of natural and human resources where they are to be found.' The experience of the Soviet Union is almost the complete reverse; no improvement on the worst record of developed capitalist countries can be found there. The devastation of the forest reserves, the virgin land programme, and policies towards pollution hardly point to a viable socialist policy towards the environment. The Soviet Union has acted rather like Brazil; natural resources are something to be exploited to maximize economic growth. The distributive aims of the two countries may differ, but the goal of economic expansion has been predominant.

This last point is perhaps fundamental. Friedmann and Weaver (1979) have suggested that a more rural-based policy is essential for Africa and Asia. Implicitly, they have recommended socialism as the way of achieving such a programme. They may be right, but socialism is not of itself enough; there must also be a commitment to reducing the large-scale corporative element in decision making. In this their plea for the reassertion of the territorial criterion over the functional makes sense. It has been the imposition of developmental models on national and international space which has created so many of the problems of urban and rural development, inequality, and environmental waste. Such imposition has been integral to the capitalist model in the sense that transnational corporations have been encouraged to exploit the world's resources and permitted to dominate the markets and productive systems of the Third World. But socialist experience at least in Europe, does not offer a clear alternative (Bahro, 1978). Much of the current crisis in Eastern Europe can also be attributed to excessively centralized, corporative decision-making processes.

If the spatial policies in peripheral capitalist nations have failed because of the nature of the development models being applied, then alternatives must be sought. Unfortunately, at first sight there is no simple alternative available. Socialism offers differences and in terms of equity many improvements, but it clearly lacks certain fundamental ingredients which are essential if the problems of poverty in the world are to be resolved. The only undeniable lesson of socialism is that it demonstrates that changes in such key issues are not made by technicians on the basis of objective studies. Rather, they stem from political decisions made, sometimes, with the support of a majority of the population.

Notes

CHAPTER 1

1. Certain cities on the east coast of Africa, such as Dar es Salaam, were founded by Arabs. In addition, there was an indigenous urban culture among the Baganda who established Kampala.
2. Cuba did not gain its independence until 1902.
3. 'By primary urban generation I mean that essentially independent emergence of urban forms through the restructuring of a society that was previously at a folk level of integration and which was subject to no or negligible stimulus from already existing societies. . . . I would suggest that lower Mesopotamia, the Nile valley, the Indus valley, the North China Plain, Meso-America, and the Central Andes be treated as regions of primary urban generation. Regions where the diffusion of traits from already urbanized societies can be shown to have either initiated and/or significantly accelerated the transformation from folk to urban society . . . I shall designate as regions of secondary urban generation' Wheatley (1970: 395).
4. See Chapter 2 for an explanation of log-linear city-size systems.
5. For a discussion of the validity of these arguments see Chapter 8.

CHAPTER 2

1. According to Auerbach (1913), Zipf (1941), and Stewart (1958), all city-size distributions tend to a pattern described statistically by the log-normal or Pareto distributions. Zipf went further and argued that where the value of q in the equation $R_i = K P_i^{-q}$ equals unity, the city-size distribution has reached maturity (Rank of city i = constant times the population of city i to the power of $-q$). While there are many reasons for questioning the wisdom of Zipf's judgement, most national city-size distributions do record values ranging from 0.8 to 1.2 (Richardson, 1973). It also represents a useful rule of thumb according to which the largest city should be twice the size of the second, three times the size of the third, and four times the size of the fourth. Thus we can define normality on the four-city index as 0.48 (i.e. 1.0 divided by the sum of $1.0 = 0.5 + 0.33 + 0.25$). If we include cases in which the largest city is up to 25 per cent larger or smaller than in the mature case, we have upper and lower limits of non-primacy. Similarly, if we define high primacy as a condition where the largest city is more than twice the anticipated size we have the following limiting values:

	Four-city index
High primacy	0.65 to 1.00
Primacy	0.54 to 0.65
Non primacy	0.41 to 0.54
Low primacy	0.00 to 0.41

The flimsy basis for this definition is obvious. In addition, the four-city index is arbitrary in the sense that it excludes most of the rank-size distribution and the use of population as the sole indicator excludes other demographic and economic variables. Similarly, the basis of our standard, the rank-size rule of Zipf, is of doubtful theoretical value, once being described by Christaller (1966: 59) as 'a most incredible law'. Its only virtues are computational simplicity and wide usage.

2. High primacy is especially common in Latin America and the Caribbean where 15 out of 22 countries fall into that category compared to 6 out of 26 in Africa, 6 out of 27 in Asia, and 5 out of 24 in Europe. These data support earlier arguments by Browning (1972) and Morse (1971). High primacy is also more common among countries with small populations (Mehta, 1964; Davis, 1962; Linsky, 1965). Of the fifteen countries with less than 4 million inhabitants, in my calculations all but two are primate and nine are high primate.
3. This aspect was recognized by Friedmann in his later writing, most notably in Friedmann (1972–3).

CHAPTER 3

1. For analyses of the incorporation of rural populations, see Pearse (1970) on Latin America, and Gugler and Flanagan (1978: 180–3) on West Africa.
2. Lipton (1977: 146–53, 430–4) provides data on urban–rural differentials in wages, incomes, and expenditures for 19 developing countries and discusses the shortcomings of such data. Jamal and Weeks (1988) argue that the income gap between urban wage-earners and the rural population narrowed in most countries in sub-Saharan Africa, even disappeared in some, due to the dramatic fall in the real wages of urban workers since the 1970s.
3. For data from the World Fertility Survey and other sources on infant and early childhood mortality in the 1960s and 1970s, see the first edition of this book. The rural rate is reported higher than the urban in all but two of the 44 countries covered.
4. Two major distortions in the data have to be considered. Characterization as urban or rural depends on residence at the time of the survey. Thus some of the births and deaths reported as urban occurred in rural areas, before the migration of the mother, and to that extent high rural mortality rates increase the urban average. And the major source of error, under-reporting of children who have died, is more common in rural than in urban areas.
5. The magnitude of urban–rural differentials is a function of definitions of urban versus rural population, definitions which vary considerably among countries.

6. Kearney (1986) reviews a great deal of literature on migration while distinguishing modernization, dependency, and articulation perspectives.
7. Yap (1977) provides a critical review of the econometric studies that have dominated research since the late 1960s.
8. See for instance Songre *et al.* (1974) on migration from Burkina Faso.
9. To some extent, of course, the immigrants encountered in an urban sample constitute a selection. The least successful are most likely to have returned to their homes. This option is not always available, but it is certainly important in sub-Saharan Africa and India.
10. Once migrants are established in town they are likely to attract new migrants from their home area, to initiate 'chain migration'. However, the concept of 'migrant stock', i.e. the number of migrants of common origin living at a given destination, suffers from a severe disability as an explanatory variable for subsequent migration, since people may continue to migrate from the same origin to the same destination for the very reasons that induced the initial migration (Shaw, 1975: 83–5).

    Chain migration encourages direct moves, even over large distances, in contrast to 'stage migration', i.e. migrants moving from their place of birth to one or several intermediary destinations before reaching their final destination. There is intuitive appeal to the notion that rural emigrants go first to small towns and, after spending an adaptation period there, move on to metropolitan areas. Or the concept of stage migration may be stretched to span two generations: the first makes it to the local town, the second into the city. However, we lack the data to substantiate the view that stage migration is a significant mode of rural–urban migration in Africa, Asia, or Latin America (Simmons *et al.*, 1977: 29, 58, 95).
11. Several government programmes provide incentives for farmers to take cropland out of production in the United States.
12. The only sizeable Third World countries in which the rural population is actually declining are Argentina, Brazil, Chile, Colombia, Cuba, South Korea, Taiwan, Turkey, and Venezuela. United Nations (1991) shows a decline in the rural population of China as well, but this is based on assumptions of a drastic and sustained curtailment of fertility that are problematic.
13. Balán (1983) discusses migration in Latin American as a consequence of the expansion and contraction of economic opportunities in the countryside, emphasizing the social organization of agrarian production.
14. For a more detailed presentation and critique of this and other econometric models of rural–urban migration, see Rosenzweig (1988).
15. If there is any 'law' of migration that holds generally, it is that migration is selective in terms of age. Most types of migration recruit disproportionately among young adults. The effect is sufficiently strong to mask those types of migration that recruit primarily from other age groups, most notably retirement migration. Shaw (1975: 133) thus concludes his review with the proposition: 'The propensity to migrate varies inversely with age. Persons in their late teens, twenties and early thirties are more migratory than those in other age groups. The greatest propensity to migrate is observed for those in the age group 20–29 years.'

16. The sex ratios presented in table 3.3 are calculated for total national and total urban populations rather than from age-specific data. The reason is twofold: total population data are more generally available; and they avoid the problems arising from the widespread—and gender-specific—misreporting of age. Sex selectivity in rural–urban migration, as a rule, is more significant among adults than among children. The urban sex ratios presented thus doubly understate the sex ratios of adult migrants: because they include all ages, and because they include the urban-born.
17. I am indebted to Francine van de Walle for this observation.
18. Eviota and Smith's (1984) account of the migration of women in the Philippines provides historical depth. It suggests that women came to predominate in the urban population only after the Second World War.
19. Osterling (1988) reports that many families who had made a permanent move to the city returned to a Peruvian village after the mid-1970s: urban conditions had become difficult for blue-collar workers, and peach cultivation in the village had become increasingly lucrative.
20. An attendant serious problem is the spread of AIDS and venereal diseases in urban areas and their transmission to the countryside.
21. Goldstein (1978) reviews the literature on various forms of temporary migration in Asia, and in particular in South-East Asia. He emphasizes that temporary migration, and in particular circular migration, is much more common than the standard studies based on census data reveal. J. Clyde Mitchell (1985) reviews some of the theoretical debates over temporary migration.
22. B. Banerjee (1984) provides evidence from a survey in Delhi that cultural factors affect family separation as well: migrants from eastern and hill areas of Uttar Pradesh were more likely to leave their wives in the rural area than migrants from other regions.
23. One solution is for migrants to establish a farm within commuting distance from their place of work. Typically the urban agglomeration provides a ready market for garden produce, but it is frequently difficult for outsiders to gain access to land. Such constraints do not obtain around Kampala, where substantial numbers of those employed in the city have established farms (Gugler, 1975*b*).
24. Perrings's (1979: 130–1, 139, 236) detailed account of the labour policies pursued by the mining companies on both sides of the border suggests that even before 1940 they were not as divergent as is commonly assumed. Some of the most obvious dissimilarities in labour utilization can be accounted for by the fact that the wages of skilled workers recruited from Europe for Katanga were higher than those of their counterparts who came from the South African mines, while the wages of African workers were lower in Katanga.
25. Migrants can maintain a pattern of circular migration where they do not face competition from established urban workers. In a country such as India the construction industry is labour-intensive to an extreme degree. Large numbers of unskilled workers are recruited, through middlemen, from rural areas on a temporary basis. They work at wages so low, and in conditions so miserable, that they are unacceptable to urban workers (Bellwinkel, 1973).

Elsewhere, groups of migrants sometimes control a particular niche in the urban economy to such an extent that its members can maintain a pattern of circular migration as they replace one another on the job. Jellinek (1978) reports such a pattern among ice-cream vendors in Jakarta, Stretton (1985) for construction workers in Manila.

26. In some areas the visiting pattern is reversed as wives visit their spouses in town. Frequently they come on extended visits. Weisner (1972) provides a fine description and analysis of families thus operating two households, one in Nairobi, the other in the rural area. Also, children may stay with the father to take advantage of better educational facilities in the city.
27. This song, reported by Saxena (1977: 176), has been translated by C. Saskia Gugler.
28. In Vunamami on the island of New Britain, Papua New Guinea, the rural community was sufficiently affluent in the early 1960s for about half the wage-earners, including the better educated, to find local employment. By the age of thirty they returned home to such local employment in order to educate their children locally and to establish a copra and cocoa farm for the future (Salisbury, 1970: 164–71). Such a situation is exceptional, but it serves to highlight how rural opportunities in employment and in cash crops shape the economic context in which the decision whether to return or not is made.

CHAPTER 4

1. For the millenarian history of Cairo, Africa's largest metropolis, see Abu-Lughod (1971).
2. For an analysis of the ill-fated attempt to restructure the market for consumer goods in Chile when the Unidad Popular government came to power in 1970, see Griffith-Jones (1978).
3. In India the growth of cities with a population over 100,000 has been a function of the expansion of employment in services and in trade and commerce, rather than in industry, in every decade since 1911 (Asok Mitra *et al.*, 1980: 42–8).
4. Techniques designed for highly industrialized countries tend to be inappropriate for less-developed countries for a number of additional reasons. Not only do they require high levels of investment per worker, but they tend towards large-scale production and require sophisticated management techniques, are linked, through inputs and outputs, with an advanced technology system, presuppose a highly educated and skilled labour-force, and produce products in demand in affluent consumer societies (Frances Stewart, 1981).
5. Where corruption plays a role in the procurement of capital goods, it will bias the choice of technology towards a more capital-intensive pattern because the more expensive the purchase, the greater the scope for the pay-off. In addition, the selection of technically complex equipment reduces the risk of detection because of the greater ease of dissimulating the pay-off element in the price (Winston, 1979).
6. Bruton (1987), in an overview of the problems of technology choice and

factor proportions, emphasizes the importance of a continuous search for appropriate proportions, a search that has to be based on learning from production experience in a specific setting.

7. R. A. Berry (1975) offers a comprehensive discussion of voluntary unemployment and presents data from Colombia suggesting its importance there. His study, as well as those of Sabot (1979: 152–62) in Tanzania and Mazumdar (1981: 271) in Malaysia, indicated that the pool of unemployed comprised predominantly the young and married women. Presumably most of them were dependants.
8. Squire (1981: 71) summarizes data on rates of unemployment by level of education in 10 Third World countries. In 8 cases the rate is highest, frequently by a large margin, for those who have had secondary education. However, Kenya, the only African country represented, shows a regular decline in unemployment as level of education increases; this may reflect an educational system that was only beginning to catch up, in 1970, with manpower requirements. It is noteworthy that unemployment is lower among those with post-secondary than among those with only secondary education in every country, including countries such as the Philippines, Sri Lanka, and India, that are notorious for widespread unemployment among college graduates; presumably college students are in a better position to prepare for their transition into the labour-market.
9. Udall and Sinclair (1982) review the evidence for the 'luxury unemployment hypothesis'.
10. Most studies focus on earnings instead, either because the research's primary concern is with urban poverty or because earnings data are more easily available and are assumed to reflect productivity.
11. Such underemployment can be a quite stable feature among self-employed who have low overhead costs and are assured of a minimum number of customers because of personal ties and/or locational advantages.
12. A comparison with the Japanese permanent employment system springs to mind, but the latter precisely fails to provide for workers outside the major firms, and indeed the substantial proportion of casual workers in these firms. The analogy of the commitment to full employment in socialist countries is more accurate. In pre-revolutionary Cuba most workers in the sugar-fields and the sugar-mills were unemployed for a major part of the year. Since the Revolution they are offered employment throughout the year, and major efforts have been directed toward absorbing them in productive activities during the off-season. As in family enterprise, the problem becomes to what extent full employment, while desirable on equity grounds, only hides unemployment instead of employing workers to productive ends.
13. Bromley (1981) discusses the beggar–donor relationship and provides an account of begging in Cali, Colombia. On the basis of street counts and several case studies he establishes that the beggars of Cali are relatively few in number and very diverse in characteristics. They are generally severely deprived and disadvantaged individuals—even those whose begging involves deceiving the public are seriously disadvantaged. The mean income derived from a day's begging appears to be less than half the

minimum wage, and most beggars view begging as a shameful activity. It is very much a last resort. The press, in contrast, widely exaggerates the number of beggars as well as their earnings and portrays most beggars as shameless frauds. This image provides the ideological underpinning for repressive measures taken against beggars.

14. Illegal activities frequently come to be tolerated as necessary to the very survival of a major part of the urban population. The acceptance of urban squatting is the most salient example.
15. More than a million menials work as messengers and guards in the Indian Civil Service. According to one calculation they are on the average usefully employed for 12 minutes a day (Maddison, 1971: 95). If this estimate appears exaggerated, few would deny that there is a good deal of redundancy in this category as well as in other ranks of the bureaucracy in India—as in many other countries.
16. The conspicuous use of labour may be argued to be preferable to the conspicuous consumption of imported luxury goods precisely because it provides local employment; in addition it saves usually scarce foreign exchange. Such an argument takes for granted extreme income inequalities and, in the case of female domestics, sex discrimination in the labour-market. The image of the nursemaid who fulfills her charge's every whim while her own children suffer from neglect is all too disturbing.
17. The judgement that some labour is unproductive because its output is ethically undesirable, e.g. prostitution, can also be traced back to Adam Smith. A Scottish professor of moral philosophy, he was less willing than his neo-classical successors to accept the legitimacy of given preferences and was prepared to state that some preferences are better than others.
18. Fiala (1983) shows that inequality is a significant factor in expanding the service sector. His analysis of 23 less-developed countries found concentration of income in the top 20 per cent of the population in 1960 to be associated with an expanded service sector in 1970.
19. Indigenous labour is inevitably cheap relative to the incomes of foreign tourists. The internationalization of prostitution is a particularly revolting consequence, the Thai sex industry only the most blatant example.
20. An analysis of census data for the metropolitan areas of 26 major Third World cities found that in the 1960s net migration accounted on average for 37 per cent of the growth in their populations, but for 63 per cent of the growth in the population aged 15–29 (United Nations, 1985).
21. The amount of research, much of it sponsored by the International Labour Office, and writing on the informal sector is truly remarkable. It is the more extraordinary given the severe shortcomings of the concept. For a discussion of the reasons for the rapid diffusion and official adoption of the formal/informal distinction, see Bromley (1978) and Peattie (1987*a*). A comprehensive review of informal sector studies is provided by Moser (1984).
22. This probably explains why the Kenya Report arrived at the unusually low estimate that only 20 per cent of the income-earning opportunities in Nairobi in 1969 were provided by the informal sector, while suggesting that the share was much higher in other Kenyan cities and especially in small towns (ILO, 1972: 54).

23. A wide range of studies reviewed by Rempel (1981: 123–5) provides no consistent evidence that recent migrants are disproportionately represented in the informal sector. For barriers to entry into specific opportunities in Bogotá, see Peattie (1975).
24. The crime syndicates in Seoul were able fully to employ shoeshine boys because they controlled exclusive territories. Other investments in their territories included boarding-houses, tea-rooms, restaurants, and inns of prostitution (Kang and Kang, 1978).
25. Tokman (1978) reviews a number of such studies. Some of these assume that benign relationships between the sectors prevail and emphasize their complementarity; others focus on the subordination of informal-sector activities and argue exploitation. From the latter perspective, the analysis may be cast in terms of the articulation of a petty commodity mode of production with a capitalist mode (Moser, 1978).
26. Job inheritance constitutes an extreme case. In China, a policy that allowed retirees from the state sector to pass on their jobs to one of their children was revived in 1978 (Stacey, 1983: 239). In India, some collective-bargaining agreements provide for the preferential recruitment of offspring of retiring workers (Uma Ramaswamy, 1979).
27. A large-scale survey of small manufacturing enterprises in Cairo found that slightly more than half of the labour-force was constituted by the owners, their family members, and kin. There was, however, considerable variation according to the origin of owners. Immigrant owners were much more likely than those born in Cairo to employ members of their immediate family and other relatives (Meyer, 1989).
28. Domestic tasks extend beyond providing for the physical needs of household members to what Sharma (1986) calls 'household service work', in particular the cultivation and maintenance of ties with kin, neighbours, and friends who are sources of information and aid. Her study of households across a wide range of incomes in Shimla, India, shows the importance of such household service work—most of which is done by women.
29. Jones (1984) shows how problematic most estimates of the number of women engaged in prostitution are. He suggests that prostitution-related occupations constitute 7 to 9 per cent of female employment in Bangkok and Manila, cities that attract substantial sex tourism.
30. Domestic workers account for more than 20 per cent of all women in the paid work-force in Latin America and the Caribbean, according to census and labour-force surveys. In addition there are large numbers of women in domestic service who are not counted in the official statistics (Chaney and Garcia Castro, 1989).
31. A large number of studies have explored the relationship between women's work and child nutrition. Leslie (1989) provides a comprehensive review. She found few studies establishing a significant relationship between women's work status and extent of breast-feeding. On the other hand, a number of studies reported a negative association between maternal employment and child nutritional status. However, the causal relationship remains to be explored: concern about a child's nutritional status may cause

mothers to seek work. In fact, working mothers introduced breast milk substitutes earlier than non-working mothers in most cases.

32. For an overview of the different theoretical perspectives underlying various explanations of the disadvantaged position of women in the labour-market, see Anker and Hein (1986); for an evaluation of these perspectives on the basis of a study of small-scale enterprises in La Paz, Buechler (1986).
33. In the first edition of this book I lamented the dearth of serious research devoted to the incomes and working conditions of women. In the intervening ten years social scientists have carried out a good deal of research on women in the Third World, and there has been a flood of publications. The impetus of the Women's Movement was re-enforced, especially for the Third World, by the Women's Decade (1976–85) decreed by the United Nations. One of the most popular new areas of research on Third World women since the mid-1970s has been on women working in export industries, especially those women employed by multinational corporations. As Lim (1990) points out, the extensive literature on this subject is disproportionate to the relative importance of such employment for Third World women, the vast majority of whom are employed in agriculture, services, and non-export, non-multinational manufacturing activities. She suggests that the reason for this disproportionate concern was the historical coincidence of growing interest in women's changing roles world-wide with the expansion of export manufacturing in Third World countries. Many scholars and political activists embraced dependency theory and took an extremely critical view of multinational corporations. The irony is that these corporations are simultaneously under attack from powerful protectionist interests in the industrial countries as well as those in Third World countries who oppose either the accompanying trade liberalization or modern factory employment for women.
34. There are, however, serious health hazards for women working in export factories. 'Brown lung' is a common occurrence in the clothing industry. On the assembly lines for electronics some workers develop eye problems after only one year of employment, and many are exposed to dangerous chemicals (United Nations, 1989*b*: 149).
35. In some cases the contributions of men to the household are so meagre, and their demands such, that women prefer to raise their children on their own (Chant, 1985).
36. The Anti-Slavery Society has published reports on child labour in India (S. Banerjee, 1979); Jamaica (Ennew and Young, 1981); Morocco (Anti-Slavery Society, 1978); South Africa (Anti-Slavery Society, 1983); and Thailand (S. Banerjee, 1980). For an annotated bibliography on child labour, see International Labour Office (1986).
37. Elson (1982) explores the parallels between women and children with respect to their position in both the labour-market and the family.
38. Such abandoned youths in Mexico City are the subject of Louis Buñuel's classic film *Los Olvidados*; they have indeed been forgotten not only by their parents but also by social scientists.

CHAPTER 5

1. Cape Coast is relatively well provided by African standards, despite these low figures.
2. A judgement based on 360 interviews in four spontaneous housing areas in Bogotá conducted in 1978/9 (Gilbert and Ward, 1985).
3. Numerous examples exist of governments having demolished centrally located low-income housing to allow private enterprise to develop the area. All too often the grounds on which such decisions were taken have been problematic.
4. Statement attributed to Dr Mulder, then Minister of Rural Relations, by the *Guardian Weekly*, 8 Oct. 1978.
5. APRA is a party which had widespread support among low-income groups in both Lima and the north of Peru.
6. Because of the wealth of the Venezuelan state, considerable compensation is sometimes paid to private landowners whose land has been invaded. Another explanation is that landowners encourage the invasion of part of their land, then press the authorities to authorize and service it, which then raises the value of the rest of their land because it can now be serviced easily.

CHAPTER 6

1. 'Situational change' and 'biographic change' refer to the changes in the behaviour of individuals and may be distinguished from the changes a society undergoes over time, or 'historical change'. For a discussion of this conceptual distinction, an account of its intellectual history, and an application to West Africa see Gugler and Flanagan (1978: 97–117).
2. For an overview of research on kinship relations in contemporary Western Europe and North America, see Segalen (1986: 73–104).
3. Lest responses to a survey be discounted as unreliable, the contrast with what the respondents had to say about their relationships with co-workers may be noted: only a minority reported meeting co-workers off the job and most of these contacts were described as casual.
4. The close-knit network refers to a situation in which the people with whom an individual associates are known to one another. In the loose-knit network, in contrast, an individual relates to people who remain strangers to one another. Various means to enquire into, and describe, social relationships at the micro-level have been developed in network analysis. For an exposition of this conceptual methodology and a review of some key studies, see Hannerz (1980: 163–201), for an account of the subsequent movement toward structural analysis, Wellman (1988).
5. I am emphasizing integration in reaction to an academic tradition and persisting popular stereotypes proclaiming urban 'disintegration'. The next task is to investigate degrees of integration and sources of affinity. Claude Fischer's (1982) careful study of networks in northern California found significant differences between large cities and small towns. City residents were less involved with kin, neighbours, and fellow church members, more

involved with people drawn from the more voluntary contexts of work, secular associations, and just plain friendship. The structure of the networks differed as well: the networks of city dwellers were more loose-knit.

6. Findings from industrialized countries are quite consistent: rural and urban areas differ very little in terms of the prevalence of psychiatric impairment, though depression may be at a somewhat higher rate in cities (Webb, 1984). The most important study, based on data collected in the United States between 1960 and 1962, showed self-reported psychosomatic symptoms to be less prevalent in cities than in small towns or rural areas among both men and women (Srole, 1978). However the scores, while standardized for age, were not adjusted for differences in the socio-economic composition of the different types of communities.
7. Data for the US, Australia, New Zealand, and 7 European countries show that there the homicide rate is invariably higher in the major city than in the country at large. In the US a strong and perfectly monotonic relationship obtains between size of urban agglomeration and homicide rate: each category of city size has a higher homicide rate than all smaller city-size categories. However, the rural homicide rate is as high as the rate of cities that have a population of between 50,000 and 100,000, i.e. it is higher than the rate of smaller towns (Archer and Gartner, 1984: 104–7).
8. Jocano (1975: 100–22) gives an account of the activities of gangs of young men who have no steady jobs in Manila.
9. Such anonymity by no means characterizes all urban neighbourhoods. In a neighbourhood in the lower-class Tondo section of Manila, the outsider, especially the male intruding into a side street off the main thoroughfare with no clear purpose, runs the risk of being challenged, threatened, or actively molested by local guardians of the area (Hollnsteiner-Racelis, 1988). Similarly, Doshi (1974: 34) describes caste-based neighbourhoods in Ahmedabad, India, where everybody knows everybody else. Any stranger entering the neighbourhood is quickly surrounded by a group of boys and young children who enquire about his purpose.
10. The critique of the culture of poverty concept, fuelled by concern with its political implications, has been extensive and frequently heated: see Valentine (1968) and Leeds (1971).
11. Statements in several of Lewis's writings that his first visit to Cuba took place in 1947 are incorrect.
12. They consist of tapes, typed interviews, and copies of other materials that had been brought to the US before the Cuban authorities halted the research and confiscated manuscripts, interviews, tapes, photographs, and personal papers. In particular, all the completed questionnaires of the housing development study were taken (Butterworth, 1980: xxi).
13. The question to what extent the stigma of having come from a notorious slum was reinforced by racism is not addressed by Butterworth. Two-thirds of the residents were classified as black or mulatto (Butterworth, 1980: 24).
14. Another housing development, intended to accommodate poor families, including many squatters, is East Havana. Situated on a beautiful site overlooking Havana Bay, it was built to high standards in the very first years of the revolutionary regime and equipped with diverse services. Residents

interviewed informally in 1975 held that many of the families originally relocated there were unaccustomed to apartment living and had abused the buildings. Whatever the reasons, they had moved away and had been replaced by more educated and well-to-do families (Eckstein, 1977).

15. In the early days of their operation, the CDRs organized nightly vigilance in the housing development. However, the guards in the housing development—unlike the guards Lewis observed in Las Yaguas in 1961—were not permitted to carry firearms. Although this denial of arms to the guards was not unusual, it was resented and seen as another indication that government officials would never trust former slum dwellers (Butterworth, 1980: 112).
16. There is a striking parallel, in both theoretical orientation and political implications, between the hard-culture approach and the predominant focus on cultural barriers to change in the development literature of the 1950s. Resistance to a substantial redistribution of resources, and the disappointing results of such aid as was niggardly given, inspired approaches that were respectively ethno- and class-centric. In both cases the charge must be made that attention is diverted from the real issue, the exploitation of dominated groups, by blaming the victim (Ryan, 1971).

CHAPTER 7

1. Eckstein (1976) relates the concern with the urban poor in the 1960s to the Cuban Revolution. She notes that various branches of the US government, including the Defense Department, as well as private foundations, sponsored a large body of research on the urban poor in Latin America. There was little effort to define and describe urban poverty, but rather a preoccupation with the political implications of poverty. The research subsided in the 1970s, not because urban poverty was on the decline, but because the proposition that the urban poor are not vanguards of revolutionary movements was gaining wide acceptance.
2. In 1972 the World Bank broke with its policy of limiting funding to what were considered productive investments and began supporting urban low-cost settlements on a substantial scale. One is left to wonder whether this reflected a different appraisal of the prospects for urban unrest on the part of key decision-makers.
3. J. M. Nelson (1979) provides a comprehensive review of research on the political role of the urban poor in Third World countries.
4. Lloyd (1979: 76–81) distinguishes an ego-oriented cognitive map from the externalized analytical structure. Individuals relate their goals, and their image of routes by which these goals are to be achieved, to their own resources, central among them, especially for the poor, their personal network. However, society can also be observed from the outside as it were, and described as an externalized analytical structure.
5. For a recent attempt to explain why so many Third World states have only rather limited abilities to achieve social control and effective appropriation of resources, see Migdal (1988).
6. For a comprehensive discussion of patron–client relationships that also

provides a review of the literature on such relationships in a large number of Third World countries, see Eisenstadt and Roniger (1984).

7. The Mexican student movement of 1968 was co-ordinated by a National Strike Council drawn from the 128 participating schools and headed by a rotating committee whose composition changed weekly. There was thus a large reservoir of leaders rather than a handful of organizers who could be co-opted or repressed. The regime, unable to co-opt or repress individual leaders, had the demonstrators shot down indiscriminately at Tlatelolco, Mexico City, on 2 October. Probably as many as 50 people were killed outright in the Plaza of the Three Cultures. Many of the perhaps 500 wounded later died because doctors in the emergency wards of the city hospitals were not allowed to attend these victims until they had been placed under guard and interrogated. Some 1,500 others were arrested (Hellman, 1983: 175–83).
8. A comprehensive survey of such cleavages and their relation to national politics in the major Third World regions is provided by C. Young (1976). For a wide-ranging and penetrating discussion of ethnicity, ethnic conflict, and strategies of conflict reduction, see Horowitz (1985).
9. In addition to language, manifold clues allow the urban dweller to gauge the ethnic identity of a stranger: dress and ornaments; cultural modifications of physical appearance such as hair-style, beard, and facial scars; patterns of speech, even when a lingua franca is used; food and drink preferences; behaviour ranging from minor physical mannerisms to dance. Furthermore, particular occupations as well as residential areas are known to be the more or less exclusive preserve of one or another ethnic group. Singh (1976) provides an account of how such markers identify a person's region of origin, religion, and caste in India. Ease in categorizing encourages the rapid establishment of interaction with group members and the maintenance of distance toward outsiders. Easily discernible characteristics guide participants in communal riots in selecting their victims.
10. Squatting is not unknown in industrialized societies. During the Great Depression squatters set up shanty towns in many US cities, shanties which came to be called 'Hoovervilles', an ironic tribute to the then president. More recently, squatting in empty buildings has occurred in several Western European countries and the US.
11. The classic account by Mangin (1967) describes the pattern in Peru. A documentary, *Children of the Dust*, depicts the establishment of Ciudad de Dios, City of God, Lima, on Christmas Eve 1954.
12. Valladares (1978*a*) gives a detailed account of the relocation of squatters in Rio de Janeiro. She emphasizes how they managed to manipulate the system.
13. According to Castells (1983: 197) the Frente Popular Tierra y Libertad represented about 100,000 squatters in August 1976. The *ejido* organizations represent peasants collectively owning land.
14. These élite workers are variously described as white-collar employees, skilled workers, or well-paid workers in more modern industries (Collectif Chili, 1972; Handelman, 1975).
15. Nueva la Habana became the most famous of these militant *campamentos*. On

the outskirts of Santiago, it had a population of between 8,000 and 10,000 and was led by MIR. New Havana is the subject of a dissertation by Meunier (1976) who lived there from 1971 until the military coup; of a chronological account by Castelain (1975); of an extensive interview with a MIR organizer (Henfrey and Sorj, 1977: 130–48); and of a documentary filmed late in 1971, *Campamento*. It provided the prime focus for Petras's (1973) study of the squatter movement in Chile and one of the two locals of Spence's (1979) research on neighbourhood courts. According to the latter, the directorate went underground on the morning of the military coup, while some units of the settlement's militia went to Vicuña Mackenna and Puente Alto, areas which resisted for several days. On the night of 11 September the military came to New Havana, took the first sixteen men they found, and shot them immediately in the main square. Over the next days the army, the navy, and the police came in turns; sometimes there were four or five searches a day. Women were raped in front of their men, and children beaten in front of their parents. Residents were tortured in the settlement itself, one woman had both arms broken. Every morning fresh bodies were found at the entrance of the street between New Havana and the neighbouring *campamento*. And the military renamed the settlement: henceforth it was to be known as Nuevo Amanecer, i.e. New Dawn.

16. Bergquist (1986) provides accounts of close to a century of labour struggles in Chile, Argentina, Venezuela, and Colombia.
17. There has been considerable debate whether certain categories of workers have come to constitute a labour aristocracy in the process (Arrighi, 1970; Peace, 1975; Saul, 1975; Jeffries, 1978). The privileges enjoyed by part of the urban labour-force became an issue even in China during the Cultural Revolution (Blecher, 1988).
18. In some cases workers can sustain a prolonged strike by returning to their areas of origin to farm (Moorsom, 1979; Dandekar, 1986; 290).
19. The very privileges of protected workers demonstrate that the leverage they wield is little affected by the large numbers of unprotected workers and unemployed who very much want to join the protected labour-force. Instead, the industrial reserve army depresses the incomes and working conditions of the marginalized labour-force (Quijano, 1974; Breman, 1976).
20. Mesa-Lago (1978) gives a detailed account of the social security systems in Argentina, Chile, Mexico, Peru, and Uruguay and an analysis relating them to the strength of various pressure groups. On Brazil see Malloy (1979), who further discusses the contrast between Latin American countries where social-security schemes were introduced early, i.e. Chile, Uruguay, Argentina, Brazil, Peru, and Cuba, and such late adopters as Mexico, Colombia, Venezuela, and Central America.
21. Welch (1977) has emphasized the importance of the distinction between national wars of liberation and revolutions, and predicted that the African peasantry will not play a revolutionary role in the next several decades.
22. For a more detailed presentation of the argument, in juxtaposition to approaches that continue to stress the decisive role of rural elements, see Gugler (1988*c*).
23. The overthrow of the Lon Nol government in Cambodia in 1975 can be

argued to constitute a fifth case. The Khmer Rouge's transition from rural guerrilla to peasant army may then be seen to have been effected in circumstances not unlike those in China a generation earlier—which we will discuss shortly. The success of the rural guerrilla movement under Yoweri Museveni in Uganda in 1986 may also be taken to contradict the proposition advanced here, but there the government forces were disorganized to an extraordinary degree.

24. The dependence of rural guerrillas on urban support was highlighted when Che Guevara (1968) found himself without such support in Bolivia.
25. For a chronological listing of actions by the urban underground from 1952 to 1959, see Bonachea and San Martín (1974: 338–44). The composition of the first government after the victory of the revolution, in which the urban front was more strongly represented than the rural, may be taken as one indication of the importance of the urban movement (Thomas, 1977: 283–5).
26. The grim reality of street fighting is poignantly conveyed in Meiselas's (1981) photo essay.
27. Hodges (1973) provides a biographical account of Guillén, an introduction to his writings, and a discussion of his influence on the armed struggle in Latin America.
28. The position of élites in settler regimes is akin to that of national élites. Their followers tend to be even more committed to resistance than a national middle class: elements of the latter may anticipate accommodation with the new rulers, but for settlers defeat invariably means what they consider exile and destitution.
29. For an account of urban guerrilla warfare in Brazil, Uruguay, and Argentina, and excerpts from a large array of writings by and about the guerrillas, see Kohl and Litt (1974); for a comprehensive annotated bibliography, Russell *et al.* (1974); for an outstanding collection of documentary materials on the Tupamaros in Uruguay, drawn from a wide variety of sources, Mayans (1971). The Tupamaros, famous for their ingenuity and wit, are portrayed in Jan Lindquist's documentary *Tupamaros*.
30. Governments, rather than revolutionary movements, are as a rule the prime beneficiaries of external support. The United States in particular, quite apart from its open intervention in Vietnam and the Dominican Republic, has assisted many governments against their domestic opponents. Not only were foreign governments provided with equipment on a lavish scale but their troops were given training specifically geared to meet the guerrilla challenge. La Escuela de las Américas, located in the Panama Canal Zone, received its name in 1963 when a new curriculum emphasizing training in counter-insurgency and civic action was introduced. More than 20,000 officers and enlisted men, representing every Latin American country except Cuba, were trained there in the 1960s. In addition, small groups of Green Berets, stationed at Southern Command in the Panama Canal Zone, have worked with the troops of every Latin American nation except Mexico, Cuba, and Haiti. Fifty-two such missions, including parachute drops into guerrilla zones, were reported for 1965 alone. In 1966 and 1967, Green Berets assisted the Guatemalan army and suffered several losses at the hands of the guerrillas. In 1967 they set up a camp in Bolivia where they trained 600 raw

recruits of the Bolivian army who later that year tracked down Che Guevara and his comrades (Gott, 1971: 450–1, 488–9, Klare, 1972: 301, 306, 379–81; Lartéguy, 1970: 195–7). It is also the United States that, with its support for the Contras against the Sandinista government in Nicaragua from 1981 to 1990, provides a significant exception to the rule that governments are the prime beneficiaries of external support.

31. The failure of the urban guerrilla movements, while they were operating on a significant scale in Venezuela, Brazil, Uruguay, and Argentina, was precisely their inability to mobilize broad-based support and to gain control of the city streets. The Farabundo Martí Front for National Liberation in El Salvador demonstrated its strength in major urban offensives in 1981 and 1989, but each time had to retreat because of its inability to instigate an urban insurrection.

CHAPTER 8

1. For a development of this critique see Corbridge (1980).
2. I mean here both the transfer of surplus for investment in urban activities and the movement of skilled and educated personnel to the urban areas.
3. I have used only demographic measures here but clearly the degree of concentration of investment, industry, commerce, construction, etc. is at least as important. The only real justification for using demographic data alone is that in part they reflect the levels of concentration in economic activity. That is to say, people tend to migrate towards centres where jobs, opportunities and higher incomes are available.
4. See note 1 to Chapter 2.
5. Indeed from Poland several academic planners have helped diffuse the growth-centre strategy to capitalist nations in both the developed and the less-developed world.
6. Dodoma is the new capital of Tanzania.
7. Polarization reversal is not the same as regional per capita income convergence. The former applies only to population and economic concentration and not to per capita income and welfare indicators.
8. I am including broadly all those policies that have attempted to modify the spatial structure by opening up the frontier. The frontier programme includes, but is broader than, the Amazon region which covers the states of Amazonas, Pará, Acre, Rondônia, Roraima, Amapá, the western part of Maranhão, and the north of Goiás and Mato Grosso.
9. The term often used to describe Brazil's economic experience between 1968 and 1974 when the gross domestic product grew at an average annual rate of 10.1 per cent.
10. These are official figures and therefore exclude illegal immigrant. Nevertheless, the rate of expansion is still likely to be much lower than in the absence of such a policy.
11. See p. 137 for discussion of Cape Town's two new satellite cities.
12. After an initial burst of activity, however, little seems to have been done to

develop Viedma. Certainly, the Menem government seems reluctant to implement its predecessor's decision.

13. A 'comprador' bourgeoisie is one that imitates the bourgeois class of developed countries. Instead of investing in the local economy it expatriates profits; instead of developing industrial activity it imports manufactured products from abroad. The outcome is a distorted, dependent, and unequal pattern of social and economic development. 'Comprador' has been used to describe both colonial and neo-colonial élites.

Bibliography

ABRAMS, C. (1964), *Man's struggle for shelter in an urbanizing world*, MIT Press.

ABU-LUGHOD, J. L. (1971), *Cairo: 1001 years of the city victorious*, Princeton University Press.

—— (1976), 'Developments in North African urbanism: the process of decolonization', in Berry (ed.) (1976), 191–212.

—— (1980), *Rabat: urban apartheid in Morocco*, Princeton University Press.

—— and HAY, R. (eds.) (1977), *Third World urbanization*, Maaroufa Press/Methuen.

ACOSTA, M., and HARDOY, J. E. (1972), 'Urbanization policies in revolutionary Cuba', in F. F. Rabinovitz and F. M. Trueblood (eds.), *Regional and urban development policies: A Latin American perspective*, Latin American Urban Research, 2, Sage, 167–78.

AGARWAL, J. P. (1976), 'Factor proportions in foreign and domestic firms in Indian manufacturing', *Economic Journal*, 86. 589–94.

All-China Federation of Women (1987), 'Hebei Langfang diqu xiangzhen qiye zhaoshou tong-gong wenti dedao jiejue', *Funü gongzuo*, 7. 24–5. (English trans. (1989), 'The problem of hiring child labor by village-run and town-run enterprises in Langfang Prefecture, Hebei, has been resolved', *Chinese Education*, 22 (2). 65–8.)

ALONSO, W. (1968), *Industrial location and regional policy in economic development*, Center for Planning and Development Research, University of California.

—— (1969), 'Urban and regional imbalances in economic development', *Economic Development and Cultural Change*, 17. 1–14.

—— (1971), 'The economics of urban size', *Papers and Proceedings of the Regional Science Association*, 26. 67–83.

AMATO, P. W. (1970), 'Elitism and settlement patterns in the Latin American city', *Journal of the American Institute of Planners*, 36. 96–105.

AMIN, S. (1974), *Accumulation on a world scale: a critique of the theory of underdevelopment*, Monthly Review Press.

AMIS, P. (1984), 'Squatters or tenants: the commercialization of unauthorized housing in Nairobi', *World Development*, 12. 87–96.

ANDERSON, D. (1989), 'Infrastructure pricing policies and the public revenue in African countries', *World Development*, 17. 525–42.

ANGEL, S. (1983), 'Land tenure for the urban poor', in Angel *et al.* (eds.) (1983), 110–40.

— ARCHER R. W., TANPHIPHAT, S., and WEGELIN, E. A. (eds.) (1983), *Land for housing the poor*, Select Books.

ANKER, R., and HEIN, C. (1986), 'Introduction and overview' in Anker and Hein (eds.), *Sex inequalities in urban employment in the Third World*, St. Martin's Press, 1–56.

ANSTEY, V. (1936), *The economic development of India*, Longman, 3rd edn.

Anti-Slavery Society (1978), *Child labour in Morocco's carpet industry*. Anti-Slavery Society Report, Anti-Slavery Society, London.

——.(1983), *Child labour in South Africa*, Child Labour Series 7, Anti-Slavery Society, London.

APPALRAJU, J., and SAFIER, M. (1976), 'Growth centre strategies in less-developed countries', in Gilbert (ed.) (1976*a*), John Wiley, 143–68.

ARADEON, D. (1978), 'Regional assessment of human settlements policies in Nigeria', paper presented at Symposium on National Human Settlements Policies and Theory (University of Sussex, Feb. 1978).

ARCHER, D., and GARTNER, R. (1984), *Violence and crime in cross-national perspective*, Yale University Press.

ARIZPE, L. (1981), 'Relay migration and the survival of the peasant household', in J. Balán (ed.), *Why people move: Comparative perspectives on the dynamics of internal migration*, Unesco Press, 187–210.

ARMSTRONG, A. (1985), 'Ivory Coast: another new capital for Africa', *Geography*, 70. 72–4.

ARMSTRONG, W., and MCGEE, T. G. (1985), *Theatres of accumulation: studies in Asian and Latin American urbanization*, Methuen.

ARRIAGA, E. E. (1981), 'Direct estimates of infant mortality differentials from birth histories', in *World Fertility Survey Conference 1980: Record of Proceedings 2*, International Statistical Institute, 435–66.

ARRIGHI, G. (1970), 'International corporations, labour aristocracies, and economic development in Tropical Africa', in R. I. Rhodes (ed.), *Imperialism and underdevelopment: a reader*, Monthly Review Press, 220–67.

ARTLE, R. (1971), 'Urbanization and economic growth in Venezuela', *Papers and Proceedings of the Regional Science Association*, 27. 63–93.

ATHUKORALA, P., and JAYASURIYA, S. K. (1988), 'Parentage and factory proportions: a comparative study of Third-World multinationals in Sri Lankan manufacturing', *Oxford Bulletin of Economics and Statistics*, 50. 409–23.

AUERBACH, F. (1913), 'Das Gesetz der Bevölkerungskonzentration', *Petermann's Mitteilungen*, 59. 74–6.

AYENI, B. (1981), 'Lagos', in M. Pacione (ed.), *Problems and planning in Third World cities*, Croom Helm, 127–55.

BADCOCK, B. (1986), 'Land and housing policy in Chinese urban development, 1976–86', *Planning Perspectives*, 1. 147–70.

BAER, W. (1964), 'Regional inequality and economic growth in Brazil', *Economic Development and Cultural Change*, 12. 268–85.

BAERRESEN, D. W. (1971), *The border industrialization program of Mexico*, Heath Lexington.

BAHRO, R. (1978), *The alternative in Eastern Europe*, New Left Books.

BAKER, H. D. R. (1979), *Chinese family and kinship*, Columbia University Press.

BAKER, P. H. (1974), *Urbanization and political change: the politics of Lagos, 1917–1967*, University of California Press.

BALAN, J. (1976), 'Regional urbanization under primary-sector expansion in neo-colonial countries', in A. Portes and H. L. Browning (eds.), *Current perspectives in Latin American urban research*, University of Texas Press, 151–79.

—— (1983), 'Agrarian structures and internal migration in a historical perspective: Latin American case studies', in P. A. Morrison (ed.), *Population movements: their forms and functions in urbanization and development*, Ordina Edns., 151–85.

BALASSA, B. (1980), 'The process of industrial development and alternative development strategies', World Bank Staff Working Paper 438.

BANERJEE, B. (1984), 'Rural-to-urban migration and conjugal separation: an Indian case study', *Economic Development and Cultural Change*, 32. 767–80.

BANERJEE, S. (1979), *Child labour in India: A general review, with case studies of the brick-making and zari embroidery industries*, Child Labour Series 2, Anti-Slavery Society, London.

—— (1980), *Child labour in Thailand: a general review*, Child Labour Series 4, Anti-Slavery Society, London.

BANTON, M. (1957), *West African city: a study of tribal life in Freetown*, Oxford University Press.

—— (1965), 'Social alignment and identity in a West African city', in H. Kuper (ed.), *Urbanization and migration in West Africa*, University of California Press, 131–47.

—— (1973), 'Urbanization and role analysis', in A. Southall (ed.), *Urban anthropology: cross-cultural studies of urbanization*, Oxford University Press, 43–70.

BARAN, P. A. (1957), *The political economy of growth*, Monthly Review Press.

BARKIN, D. (1985), 'Global proletarianization', in Sanderson (ed.) (1985), 26–45.

BARNES, S. T. (1986), *Patrons and power: creating a political community in metropolitan Lagos*, International African Library, Manchester University Press/Indiana University Press.

BAROSS, P. (1983), 'The articulation of land supply for popular settlements in Third World cities', in Angel *et al.* (eds.) (1983), 180–209.

BARRATT-BROWN, M. (1974), *The economics of imperialism*, Penguin.

BATLEY, R. (1982), 'Urban renewal and expulsion in São Paulo', in Gilbert, Hardoy, and Ramírez (eds.) (1982), 231–62.

—— (1983), *Power and bureaucracy*, Gower.

BAUD, I. (1987), 'Industrial subcontracting: the effects of the putting out system on poor working women in India', in A. M. Singh and A. Kelles-Viitanen (eds.), *Invisible hands: women in home-based production*, Sage, 69–91.

BAUER, J. (1985), 'Demographic change, women and the family in a migrant neighbourhood of Teheran', in A. Fathi (ed.) (1985), *Women and the family in Iran*, E. J. Brill, 158–86.

BAUER, P. T. (1954), *West Africa trade*, Cambridge University Press.

BEDFORD, R. D. (1973), *New Hebridean mobility: a study of circular migration*, Department of Human Geography, Research School of Pacific Studies, Australian National University.

BELLWINKEL, M. (1973), 'Rajasthani contract labour in Delhi: a case study of the relationship between company, middleman and worker', *Sociological Bulletin*, 22. 78–97.

BENERIA, L. (1989), 'Subcontracting and employment dynamics in Mexico City', in A. Portes, M. Castells, and L. A. Benton (eds.), *The informal economy: studies in advanced and less developed countries*, Johns Hopkins University Press, 173–88.

—— and ROLDAN, M. (1987), *The crossroads of class and gender: industrial homework, subcontracting, and household dynamics in Mexico City*, Women in Culture and Society, University of Chicago Press.

BENTON, L. A. (1987), 'Reshaping the urban core: the politics of housing in authoritarian Uruguay', *Latin American Research Review*, 22. 33–52.

BEQUELE, A., and BOYDEN, J. (1988), 'Child labour: problems, policies and programmes', in A. Bequele and J. Boyden (eds.), *Combating child labour*, International Labour Office, 1–27.

BERG, E. J. (1961), 'Backward-sloping labour supply functions in dual economies: the Africa case', *Quarterly Journal of Economics* 75. 468–92.

BERGQUIST, C. (1986), *Labor in Latin America: comparative essays on Chile, Argentina, Venezuela, and Colombia*, Stanford University Press.

BERGSMAN, J. (1970), *Brazil: industrialization and trade policies*, Oxford University Press.

BERRY, B. J. L. (1961), 'City-size distributions and economic development', *Economic Development and Cultural Change*, 9. 573–87.

—— (1972), 'Hierarchical diffusion: the basis of development filtering and spread in a system of growth centres', in N. M. Hansen (ed.), *Growth centres in regional economic development*, Free Press, 108–38.

—— (ed.) (1976), *Urbanization and counter-urbanization*, Sage.

BERRY, R. A. (1975), 'Open unemployment as a social problem in urban Colombia: myth and reality', *Economic Development and Cultural Change*, 23. 276–91.

BHOOSHAN, B. S., and MISRA, R. P. (1980), *Habitat Asia. Issues and Responses 1: India*, Concept Publishing Company, New Delhi.

BILL, J. A. (1978), 'Iran and the crisis of '78', *Foreign Affairs*, 57. 323–42.

BIRKBECK, C. (1979), 'Garbage industry and the "vultures" of Cali, Colombia', in Bromley and Gerry (eds.) (1979), 161–83.

BLACK, G. (1981), *Triumph of the people: the Sandinista revolution in Nicaragua*, Zed Press.

BLAKE, G. H., and LAWLESS, R. I. (eds.) (1980), *The changing Middle Eastern city: processes and problems*, Croom Helm.

BLECHER, M. (1988), 'Rural contract labour in urban Chinese industry: migration control, urban–rural balance, and class relations', in Gugler (ed.) (1988(a)), 109–23.

BOISIER, S. (1973), 'Localización, tamaño urbano y productividad industrial: un caso de estudio de Brasil', *Revista Interamericana de Planificación*, 6. 87–112.

—— (1979), 'La planificación del desarrollo regional en América Latina', paper presented to the Seminario sobre Estategias Nacionales de Desarrollo Regional (Bogotá, Sept. 1979).

—— (1981), 'Chile: continuity and change—variations of centre-down strategies under different political regimes', in Stöhr and Taylor (eds.) (1981), 401–26.

—— (1987), 'Los procesos de decentralizaión y desarrollo urbano en el escenario actual de América Latina', *Revista de la CEPAL*, 31. 139–51.

BONACHEA, R. L., and SAN MARTIN, M. (1974), *The Cuban insurrection 1952–1959*, Transaction Books.

BOOTH, J. A. (1985), *The end and the beginning: the Nicaraguan revolution*, Westview Press, 2nd edn. (1st edn. 1982).

BORRERO-MUTIS, S. (1986), *Mapa económico de Colombia*, Separata de la revista Colombia Año 2000, FONADE, Bogotá.

BORUKHOV, E. (1975), 'On the urban agglomeration and economic efficiency: comment', *Economic Development and Cultural Change*, 24. 199–205.

BOSE, A. (1973), *Studies in India's urbanization 1901–1971*, Tata McGraw Hill.

BOSERUP, E. (1970), *Woman's role in economic development*, St. Martin's Press.

BRAND, R. R. (1972), 'Migration and residential site selection in five low-income communities in Kumasi (Ghana)', *African Urban Notes*, 7. 73–94.

BRANDT, W. (1980), *North–South: a programme for survival. Report of the Independent Commission on International Development*, Pan Books.

BRAZIL (1976), *Amazonia*, Brazilian Embassy, London.

BREESE, G. (ed.) (1969), *The city in newly developing countries: readings on urbanism and urbanization*, Prentice Hall.

BREMAN, J. (1976), 'A dualistic labour system? A critique of the "informal sector" concept', *Economic and Political Weekly*, 11. 1870–6, 1905–8, and 1939–44.

BRETT, S. (1974), 'Low income settlements in Latin America: the Turner model', in de Kadt and Williams (eds.) (1974), 171–96.

BROMLEY, R. (1978), 'Introduction—The urban informal sector: why is it worth discussing?', *World Development*, 6. 1033–9.

—— (ed.) (1978), 'The urban informal sector: critical perspectives', *World Development*, 6. 1031–198.

—— (1981), 'Begging in Cali: image, reality and policy', *International Social Work*, 24. 22–40.

—— (1988), 'Working in the streets: survival strategy, necessity, or unavoidable evil?', in Gugler (ed.) (1988*a*), 161–82. (Earlier version in Gilbert, Hardoy, and Ramírez (eds.) (1982), 59–78).

—— and C. GERRY (eds.) (1979), *Casual work and poverty in Third World cities*, John Wiley.

BROOKFIELD, H. C. (1973), 'On one geography and a third world', *Transactions of Institute of British Geographers*, 58. 1–20.

—— (1975), *Interdependent development*, Methuen.

—— (1978), 'Third World development', *Progress in Human Geography*, 2. 121–32.

—— (1979), 'Urban bias, rural bias, and the regional dimension: to the house of Tweedledee', in Rothko Chapel Symposium, *Towards a new strategy of development*, Pergamon, 97–121.

BROWNING, H. L. (1972), 'Primacy variation in Latin America during the twentieth century', in Instituto de Estudios Peruanos (ed.), *Integración y proceso social en América*, Lima.

BRUNDENIUS, C. (1981), 'Growth with equity: the Cuban experience (1959–80)', *World Development*, 9. 1083–96.

BRUNER, E. M. (1972), 'Batak ethnic associations in three Indonesian cities', *Southwestern Journal of Anthropology*, 28. 207–29.

—— (1973), 'Kin and non-kin', in A. Southall (ed.), *Urban anthropology: cross-cultural studies of urbanization*, Oxford University Press, 373–92.

BRUSH, J. E. (1962), 'The morphology of Indian cities', in Turner (ed.) (1962), 57–70.

BRUTON, H. (1987), 'Technology choice and factor proportions problems in LDCs', in N. Gemmell (ed.), *Surveys in development economics*, Blackwell, 236–65.

BRYANT, C. (1980), 'Squatters, collective action, and participation: learning from Lusaka', *World Development*, 8. 73–86.

BRYANT, J. (1969), *Health and the developing world*, Cornell University Press.

BUECHLER, J.-M. (1986), 'Women in Petty Commodity Production in La Paz, Bolivia', in J. Nash and H. Safa (eds.), *Women and change in Latin America*, Bergin & Garvey, 165–88.

BUNKER, S. G. (1985), 'Misdirected expertise in an unknown environment: standard bureaucratic procedures as inadequate technology on the Brazilian "planned frontier" ', in J. Hemming (ed.) (1985), 103–18.

BURGESS, R. (1978), 'Petty commodity housing or dweller control? A critique of John Turner's views on housing policy', *World Development*, 6. 1105–34.

—— (1982), 'The politics of urban residence in Latin America', *International Journal of Urban and Regional Research*, 6. 465–80.

—— (1985), 'The limits of state self-help housing programmes', *Development and Change*, 16. 271–312.

BURLAND, C. A. (1967), *Peru under the Incas*, Evans.

BURTON, R. F. (1856), *First footsteps in East Africa*, Everyman.

BUTLER, D., JENKINS, L., and CAME, B. (1978), 'Iran erupts again', *Newsweek*, 11 Dec. 1978. 44.

BUTTERWORTH, D. (1972), 'Two small groups: a comparison of migrants and non-migrants in Mexico City', *Urban Anthropology*, 1. 29–50.

—— (1980), *The people of Buena Ventura: relocation of slum dwellers in post-revolutionary Cuba*, University of Illinois Press.

BYRES, T. J. (1974), 'Land reform, industrialization and the marketed surplus in India: an essay on the power of rural bias', in D. Lehmann (ed.), *Agrarian reform and agrarian reformism: studies of Peru, Chile, China and India*, Faber, 221–61.

CALDER, N. (1974), *The weather machine*, BBC Publications.

CALDWELL, J. C. (1969), *African rural–urban migration: the movement to Ghana's towns*, Columbia University Press.

CALLAWAY, B. J. (1987), *Muslim Hausa women in Nigeria: tradition and change*, Syracuse University Press.

CARDOSO, F. H. (1972), 'Dependency and development in Latin America', *New Left Review*, 84. 83–95.

—— (1977), 'Current theses on Latin American development and dependency: a critique', *Boletín de Estudios Latinoamericanos y del Caribe*, 22. 53–64.

—— and FALETTO, E. (1969), *Dependencia y desarrollo en América Latina*, Siglo XXI.

CARROLL, A. (1980), *Pirate subdivisions and the market for residental lots in Bogotá*, City Study, World Bank.

CASTELAIN, C. (1975), 'Histoire du "Campamento Nueva Habana" (Chili)', *Espaces et Sociétés*, 15. 117–31.

CASTELLS, M. (ed.) (1973), *Imperialismo y urbanización en América Latina*, Editorial Gustavo Gili.

—— (1977), *The urban question: a Marxist approach*, Edward Arnold (orig. French edn. 1972).

—— (1983), *The city and the grassroots: a cross-cultural theory of urban social movements*, University of California Press.

—— and PORTES, A. (1989), 'World underneath: the origins, dynamics, and effects of the informal economy', in A. Portes, M. Castells, and L. A. Benton (eds.), *The informal economy: studies in advanced and less developed countries*, Johns Hopkins University Press, 11–37.

CEU (Centro de Estudios Urbanos) (1977), *La intervención del estado y el problema de la vivienda: Ciudad Guayana*, CEU, Caracas.

CHAKRABARTI, P. (1985), 'Some aspects of "complex" families in Calcutta', *Journal of Comparative Family Studies*, 16. 377–86.

CHAMBERLAIN, M. E. (1974), *Britain and India: the interaction of two peoples*, David & Charles.

CHANDLER, T., and FOX, G. (1974), *3000 Years of Urban Growth*, Academic Press.

CHANDRA, S. (1977), *Social participation in urban neighbourhoods*, National, New Delhi.

CHANEY, E. M., and GARCIA CASTRO, M. (1989), 'A new field for research and action', in E. M. Chaney and M. García Castro (eds.), *Muchachas no more: household workers in Latin America and the Caribbean*, Temple University Press.

CHANT, S. (1985), 'Single-parent families: choice or constraint? The formation of female-headed households in Mexican shanty towns', *Development and Change*, 16. 635–56.

CHAPMAN, M., and PROTHERO, R. M. (eds.) (1985), *Circulation in Third World countries*, Routledge and Kegan Paul.

CHASE-DUNN, C. (1989), *Global formation: structures of the world-economy*, Blackwell.

CHATTERJI, R. (1980), *Unions, politics and the state: a study of Indian labour politics*, South Asian Publishers.

CHAUDHURI, S. (1962), 'Centralization and the alternative forms of decentralization: a key issue', in Turner (ed.) (1962), 213–39.

CHAVARRIA, R. E. (1982), 'The Nicaraguan insurrection: an appraisal of its originality', in T. W. Walker (ed.), *Nicaragua in revolution*, Praeger, 25–40.

CHEEMA, G. S., and RONDINELLI, D. A. (eds.) (1983), *Decentralization and development: policy implementation in developing countries*, Sage.

CHRISTALLER, W. (1966), *Central places in Southern Germany*, Prentice Hall.

CLARKE, C. G. (1974), 'Urbanization in the Caribbean', *Geography*, 59. 223–32.

CLEAVES, P. S. (1974), *Bureaucratic politics and administration in Chile*, California University Press.

CLINARD, M. B., and ABBOTT, D. J. (1973), *Crime in developing countries: a comparative perspective*, John Wiley.

COHEN, A. (1969), *Custom and politics in urban Africa: a study of Hausa migrants in Yoruba towns*, University of California Press.

COHEN, M. (1978), 'Regional development or regional location. A comparison of growth and equity approaches', Working Paper 6, Development Planning Unit, University College London.

COLLECTIF CHILI (1972), 'Revendication urbaine, stratégie politique et mouvement social des "pobladores" au Chili', *Espaces et sociétés*, 6–7. 37–57.

COLLIER, D. (1976), *Squatters and oligarchs*, Johns Hopkins University Press.

CONNOLLY, P. (1982), 'Uncontrolled settlements and self-build: what kind of solution? The Mexico City case', in P. M. Ward (ed.), *Self-help housing: a critique*, Mansell, 141–74.

COOKE, P. (1987), 'Britain's new spatial paradigm: technology, locality and society in transition', *Environment and Planning A*, 19. 1289–301.
CORBRIDGE, S. (1980), 'Urban bias, rural bias, and industrialization: an appraisal of the work of Michael Lipton and Terry Byres', Department of Geography, University of Cambridge, MS.
—— (1986), *Capitalist world development: a critique of radical development geography*, Macmillan.
CORNELIUS, W. A. (1973), 'The impact of governmental performance on political attitudes and behaviour: the case of the urban poor in Mexico', in F. F. Rabinovitz and F. M. Trueblood (eds.), *National–local linkages: the interrelationship of urban and national polities in Latin America*, Latin American Urban Research, 3, Sage, 217–55.
—— (1975), 'Introduction', in W. A. Cornelius and F. M. Trueblood (eds.), *Urbanization and inequality: the political economy of urban and rural development in Latin America*, Latin American Urban Research, 5, Sage, 9–25.
COSTELLO, V. F. (1977), *Urbanization in the Middle East*, Cambridge University Press.
COULOMB, R. (1985), 'La vivienda de alquiler en las áreas de reciente urbanización', *Revista de ciencias sociales y humanidades*, 6. 43–70.
CROSS, M. (1979), *Urbanization and urban growth in the Caribbean*, Cambridge University Press.
CRUSH, J., and JAMES, W. (1991), 'Depopulating the compounds: migrant labor and mine housing in South Africa', *World Development*, 19. 301–16.
CUENYA, B. (1986), 'El submercado de alquiler de piezas en Buenos Aires', *Boletın de medio ambiente y urbanización*, 17. Suplemento Especial, 3–8.
CURRIE, L. L. (1965), *Una política urbana para los paises en desarrollo*, Tercer Mundo, Bogotá.
—— (1971), 'The exchange constraint of development: a partial solution to the problem', *Economic Journal*, 81. 886–903.
DANDEKAR, H. C. (1986), *Men to Bombay, women at home: urban influence on Sugao village, Deccan Maharashtra, India, 1942–1982*, Michigan Papers on South and Southeast Asia, Center for South and Southeast Asian Studies, University of Michigan.
DANDEKAR, V. M., and RATH, N. (1971), *Poverty in India*, Ford Foundation.
DANE (Departamento Administrativo Nacional de Estadística) (1977), *La vivienda en Colombia*, Bogotá.
DARWENT, D. F. (1969), 'Growth poles and growth centres in regional planning—a review', *Environment and Planning*, 1. 5–32.
DAVENPORT, T. R. H. (1977), *South Africa: a modern history*, Macmillan.
DAVIS, K. (1962), 'Los causas y efectos del fenómeno de primacia urbana con referencia especial a América Latina', XIII Congreso Nacional de Sociología, Mexico.
—— (1969), *World urbanization 1950–1970*, 2 vols., University of California.
—— and HERTZ, H. (1954), 'Urbanization and the development of preindustrial areas', *Economic Development and Cultural Change*, 4. 6–26.
DAVIS, S. H. (1977), *Victims of the miracle*, Cambridge University Press.
DAZA ROA, A. (1967), 'La repartición de los ingresos', *Revista del Banco de la República*, 40. 880–9.

DEBRAY, R. (1967), *Révolution dans la révolution?* Cahiers Libres 98, Paris: Maspero. (English trans. (1967), *Revolution in the revolution? Armed struggle and political struggle in Latin America*, Monthly Review Press.)

DE KADT, and WILLIAMS, E. (eds.) (1974), *Sociology and development*, Tavistock Publications.

DE TRAY, D. (1983), 'Children's work activities in Malaysia', *Population and Development Review*, 9. 437–55.

DEWAR, D. (1976), 'Metropolitan planning and income redistribution in Cape Town', Occasional Paper 1, Department of Urban and Regional Planning, University of Cape Town.

—— and ELLIS, G. (1979), *Low income housing policy in South Africa*, Citadel Press.

—– TODES, A., and WATSON, V. (1986), *Regional development and settlement policy: premises and prospects*, Allen and Unwin.

DICKENSON, J. P. (1980), 'Innovations for regional development in North-east Brazil, a century of failures', *Third World Planning Review*, 2. 57–74.

DIETZ, H. A. (1978), 'Metropolitan Lima: urban problem solving under military rule', in W. A. Cornelius and R. V. Kemper (eds.), *Metropolitan Latin America: the challenge and the response*, Latin American Urban Research, 6, Sage, 205–26.

DIXON, W. J. (1987), 'Progress in the provision of basic human needs: Latin America, 1960–1980', *Journal of Developing Areas*, 21. 129–40.

DOBBIN, C. (1972), *Urban leadership in Western India: politics and communities in Bombay City 1840–1885*, Oxford University Press.

DOEBELE, W. A. (1975), 'The private market and low income urbanization in developing countries: the "pirate" subdivision of Bogotá', Discussion Paper D75–11, Department of City and Regional Planning, Harvard University.

—— (1978), 'Selected issues in urban land tenure', in World Bank (1978), 99–207.

—— and PEATTIE, L. R. (1976), 'Some second thoughts on sites and services', mimeo.

DOEPPERS, D. F. (1976), 'The development of Philippine cities before 1900', in Yeung and Lo (eds.) (1976), 28–44.

DOMINGUEZ, J. I. (1978), *Cuba: order and revolution*, Harvard University Press.

DOSHI, H. (1974), *Traditional neighbourhood in a modern city*, Abhinav Publications.

DOS SANTOS, T. (1970), 'The structure of dependence', *American Economic Review*, 60. 231–6.

DRAKAKIS-SMITH, D. W. (1976*a*), 'Some perspectives on slum and squatter settlements in Ankara', paper presented to Institute of British Geographers, Annual Conference, Jan. 1976, Lanchester Polytechnic.

—— (1976*b*), 'Urban renewal in an Asian context: a case study in Hong Kong', *Urban Studies*, 13. 295–305.

—— (1981), *Urbanization, housing and the development process*, Croom Helm.

DRAKE, P. W. (1988), 'Urban labour movements under authoritarian capitalism in the Southern Cone and Brazil, 1964–83', in Gugler (ed.) (1988*a*), 367–98.

DUNHAM, D. (1979), 'What do regional theorists do after midnight?', Institute of Social Studies, The Hague, mimeo.

DUTT, R. C. (1968), *The economic history of India*, vol. 1, Trubner's Oriental Editions.

DWYER, D. J. (ed.) (1972), *The city as a centre of change in Asia*, University of Hong Kong Press.

DWYER, D. J. (ed.) (1975), *People and housing in Third World cities*, Longman.
—— (1987), 'Urban housing and planning in China,' *Transactions of Institute of British Geographers*, 11. 479–89.
EAMES, E. (1967), 'Urban migration and the joint family in a North Indian village', *Journal of Developing Areas*, 1. 163–78.
ECKSTEIN, S. (1976), 'The rise and demise of research on Latin American urban poverty', *Studies in Comparative International Development*, 11. 107–26.
—— (1977), 'The debourgeoisement of Cuban cities', in I. L. Horowitz (ed.), *Cuba communism*, Transaction Books, 3rd edn., 443–74.
—— (1988*a*), *The poverty of revolution: the state and the urban poor in Mexico*, Princeton University Press. (Augmented paperback edn., 1st edn. 1977).
—— (1988*b*), 'The politics of conformity in Mexico City', in Gugler (ed.) (1988*a*), 294–307. (Earlier version (1976), *British Journal of Sociology*, 27. 150–64.)
—— (1989), 'Power and political protest in Latin America', in S. Eckstein (ed.), *Power and popular protest: Latin American social movements*, University of California Press, 1–60.
—— (1990), 'Poor people versus the state and capital: anatomy of a successful community mobilization for housing in Mexico City', *International Journal of Urban and Regional Research*, 14. 274–96.
EDWARDS, M. (1981), 'Cities of tenants: renting as a housing alternative among the Colombian urban poor', doctoral dissertation, University of London.
—— (1982), 'Cities of tenants: renting among the urban poor in Latin America', in Gilbert, Hardoy, and Ramírez (eds.) (1982), 129–58.
EISENSTADT, S. N., and RONIGER, L. (1984), *Patrons, clients and friends: interpersonal relations and the structure of trust in society*, Themes in the Social Sciences, Cambridge University Press.
ELLIS, G., HENDRIE, D., KODY, A., and MAREE, J. (1977), *The squatter problem in the Western Cape: some causes and remedies*, South African Institute of Race Relations.
EL-SHAKS, S. (1965), 'Development, primacy and the structure of cities', Ph.D. dissertation, Harvard University.
—— (1972), 'Development, primacy and systems of cities', *Journal of Developing Areas*, 7. 11–36.
ELSON, D. (1982), 'The differentiation of children's labour in the capitalist labour market', *Development and Change*, 13. 479–97.
EMERSON, J. P. (1983), 'Urban school-leavers and unemployment in China', *China Quarterly*, 93. 1–16.
EMMANUEL, A. (1972), *Unequal exchange: a study of imperialism of trade*, New Left Books.
ENNEW, J., and YOUNG, P. (1981), *Child labour in Jamaica: a general review*, Child Labour Series 6, Anti-Slavery Society, London.
EPSTEIN, A. L. (1958), *Politics in an urban African community*, Manchester University Press.
—— (1981), *Urbanization and kinship: the domestic domain on the Copperbelt of Zambia 1950–1956*, Academic Press.
EPSTEIN, D. G. (1973), *Brasília—plan and reality: a study of planned and spontaneous settlement*, University of California Press.

EVERS, H.-D. (1976), 'Urbanization and urban conflict in Southeast Asia', *Asian Survey*, 15. 775–85.

—— (1976), 'Urban expansion and land ownership in underdeveloped societies', in Walton and Masotti (eds.) (1976), 67–79.

EVIOTA, E., and SMITH, P. C. (1984), 'The migration of women in the Philippines', in J. T. Fawcett, S.-E. Khoo, and P. C. Smith (eds.), *Women in the cities of Asia: migration and urban adaptation*, Westview Press, 165–90.

EYRE, L. A. (1972), 'The shanty towns of Montego Bay, Jamaica', *Geographical Review*, 62. 394–413.

FAIR, T. J. D., and DAVIES, R. J. (1976), 'Constrained urbanization: white South Africa and Black Africa compared', in Berry (ed.) (1976), 145–68.

FEARNSIDE, P. M. (1986), 'Agricultural plans for Brazil's Grande Carajás program: lost opportunity for sustainable local development?', *World Development*, 14. 385–410.

FEDER, E. (1977), *Strawberry imperialism: an enquiry into the mechanisms of dependency in Mexican agriculture*, Editorial Campesina.

FELDMAN, K. (1975), 'Squatter migration dynamics in Davao City, Philippines', *Urban Anthropology*, 4. 123–44.

FELIX, D. (1983), 'Income distribution and the quality of life in Latin America: patterns, trends and policy implications', *Latin American Research Review*, 18. 3–34.

FERNANDEZ-KELLY, P. (1983), *For we are sold, I and my people*, State University of New York Press.

FERREE, M. MARX, and GUGLER, J. (1983), 'The participation of women in the urban labor force and in rural–urban migration in India', *Demography India*, 12. 194–213.

FIALA, R. (1983), 'Inequality and the service sector in less developed countries: a reanalysis and respecification', *American Sociological Review*, 48. 421–8.

FIELD, B. (1987), 'Public housing in Singapore', *Land Use Policy*, 4. 147–56.

FINDLAY, A. M., and PADDISON, R. (1986), 'Planning the Arab city: the cases of Tunis and Rabat', *Progress in Planning*, 26. 1–82.

FISCHER, C. A. (1976), 'The ecology of modernization with special reference to Asia', in T. G. Lim and V. Lowe (eds.), *Towards modern Asia: aims, resources and strategies*, Heinemann, 108–21.

FISCHER, C. S. (1982), *To dwell among friends: personal networks in town and city*, University of Chicago Press.

—— (1984), *The urban experience*, Harcourt Brace Jovanovich, 2nd edn. (1st edn. 1976).

FISHER, W. B. (1983), 'Urban evolution in Islamic areas', in J. Patten (ed.), *The expanding city: essays in honour of Professor Jean Gottmann*, Academic Press, 77–102.

FITZGERALD, E. V. K. (1976), *The state and economic development: Peru since 1968*, Cambridge University Press.

FLANAGAN, W. G. (1977), 'The extended family as an agent in urbanization: a survey of men and women working in Dar es Salaam, Tanzania', Ph.D. dissertation, University of Connecticut.

Ford Foundation (1972), *International urbanization survey: findings and recommendations*, Ford Foundation.

FOSTER-CARTER, A. (1976), 'From Rostow to Gunder Frank: conflicting paradigms in the analysis of underdevelopment', *World Development*, 4. 167–80.
—— (1978), 'The modes of production controversy', *New Left Review*, 107. 47–77.
FOWERAKER, J. (1981), *The struggle for land: a political economy of the pioneer frontier in Brazil from 1930 to the present day*, Cambridge University Press.
FOX, R. (1975), *Urban population growth trends in Latin America*, Inter-American Development Bank.
FRANK, A. G. (1967), *Capitalism and underdevelopment in Latin America: historical studies of Chile and Brazil*, Monthly Review Press.
—— (1969), *Latin America: underdevelopment or revolution*, Monthly Review Press.
FRENCH, R. A., and HAMILTON, F. E. I. (eds.) (1979), *The socialist city: spatial structure and urban policy*, John Wiley.
FRIEDMAN, S. (1987), *Building tomorrow today: African workers in trade unions, 1970–1984*, Ravan Press.
FRIEDMANN, J. P. (1961), 'Cities in social transformation', *Comparative Studies in Society and History*, 4. 86–103.
—— (1966), *Regional development policy: a case study of Venezuela*, MIT Press.
—— (1968), 'The strategy of deliberate urbanization', *Journal of the American Institute of Planners*, 34. 364–73.
—— (1972–3), 'The spatial organization of power in the development of urban systems', *Development and Change*, 4. 12–50.
—— and DOUGLASS, M. (1976), 'Agropolitan development: towards a new strategy for regional planning in Asia', in UNCRD (1976), 333–87.
—— and WEAVER, C. (1979), *Territory and function: the evolution of regional planning*, Edward Arnold.
—— and WULFF, R. (1976), *The urban tradition: comparative studies of newly industrializing societies*, Edward Arnold.
FRÖBEL, F., HEINRICHS, J., and KREYE, O. (1980), *The new international division of labour*, Cambridge University Press.
FROLIC, B. M. (1976), 'Noncomparative communism: Chinese and Soviet urbanization', in M. Field (ed.), *Social consequences of modernization in communist societies*, Johns Hopkins University Press, 149–61.
FUCHS, R. J., and DEMKO, G. J. (1979), 'Geographic inequality under socialism', *Annals of the American Association of Geographers*, 69. 304–18.
FUJIMA, R. (1987), *Urbanization and urban problems in China*, Institute of Developing Economies, Tokyo.
FUNNELL, D. (1976), 'The role of small service centres in regional and rural development: with special reference to Eastern Africa', in Gilbert (ed.) (1976*a*), 77–112.
FURTADO, C. (1971), *Economic development of Latin America: a survey from colonial times to the Cuban Revolution*, Cambridge University Press.
GANS, H. J. (1962), 'Urbanism and suburbanism as ways of life: a re-evaluation of definitions', in A. H. Rose (ed.), *Human behaviour and social processes*, Houghton Mifflin, 625–48.
GARRETON, M., M. A. (1989), 'Popular mobilization and the military regime in Chile: the complexities of the invisible transition', in S. Eckstein (ed.), *Power and popular protest: Latin American social movements*, University of California Press, 259–77.

GEISSE, G., and CORAGGIO, J. L. (1972), 'Metropolitan areas and national development', in F. F. Rabinovitz and F. M. Trueblood (eds.), *Regional and urban development policies: a Latin American perspective*, Latin American Urban Research, 2, Sage, 45–60.
GELLNER, E. (1977), 'Patrons and clients', in E. Gellner and J. Waterbury (eds.), *Patrons and clients in Mediterranean societies*, Duckworth, 1–6.
GIBSON, L. J., and CORONA, A. (eds.) (1985), *The U.S. and Mexico: borderland development and national economies*, Westview Press.
GIDDENS, A. (1984), *The constitution of society*, Polity Press.
GILBERT, A. G. (1974*a*), *Latin American development: a geographical perspective*, Penguin.
—— (1974*b*), 'Industrial location theory: its relevance to an industrializing nation', in B. S. Hoyle (ed.), *Spatial aspects of development*, John Wiley, 271–90.
—— (1975), 'A note on the incidence of development in the vicinity of a growth centre', *Regional Studies*, 9. 325–33.
—— (ed.) (1976*a*), *Development planning and spatial structure*, John Wiley.
—— (1976*b*), 'The arguments for very large cities reconsidered', *Urban Studies*, 13. 27–34.
—— (1981*a*), 'Bogotá: an analysis of power in an urban setting', in M. Pacione (ed.), *Problems and planning in Third World cities*, Croom Helm, 65–93.
—— (1981*b*), 'Pirates and invaders: land acquisition in urban Colombia and Venezuela', *World Development*, 9. 657–78.
—— (1983), 'The tenants of self-help housing: choice and constraint in the housing markets of less developed countries', *Development and Change*, 14. 449–77.
—— (1986), 'Self-help housing and state intervention: illustrative reflections on the petty commodity production debate', in D. W. Drakakis-Smith (ed.), *Urbanisation in the Developing World*, Croom Helm, 175–94.
—— (1987*a*), 'Latin America's urban poor: shanty dwellers or renters of rooms?', *Cities*, 4. 43–51.
—— (1987*b*), 'Latin America, the world recession and the New International Division of Labour', *Tijdschrift voor economische en sociale geographie*, 77. 368–77.
—— and GOODMAN, D. E. (1976), 'Regional income disparities and economic development: a critique', in Gilbert (ed.) (1976*a*), 113–42.
—— HARDOY, J. E., and RAMÍREZ, R. (eds.) (1982), *Urbanization in contemporary Latin America*, John Wiley.
—— and VARLEY, A. (1991), *Landlord and tenant: housing the poor in urban Mexico*, Routledge.
—— and WARD, P. M. (1978), 'Housing in Latin American cities', in D. T. Herbert and R. J. Johnston (eds.), *Geography and the urban environment*, vol. 1, John Wiley, 285–318.
—— and WARD, P. M. (1985), *Housing, the state and the poor: policy and practice in three Latin American cities*, Cambridge University Press.
GINSBURG, N. (1973), 'From colonialism to national development: geographical perspectives on patterns and policies', *Annals of the Association of American Geographers*, 63. 1–21.
GLUCKMAN, M. (1960), 'Tribalism in modern British Central Africa', *Cahiers d'études africaines*, 1. 55–70.

GOLDSTEIN, S. (1978), *Circulation in the context of total mobility in Southeast Asia*, Papers of the East–West Population Institute 53, East–West Center, Honolulu.

—— (1990), 'Urbanization in China, 1982–87: effects of migration and reclassification', *Population and Development Review*, 16. 673–701.

GOODLAND, R. J. A., and IRWIN, H. S. (1975), *Amazonian jungle: green hell to red desert*, Elsevier.

GOODMAN, D. E., and REDCLIFT, M. (1981), *From peasant to proletarian: capitalist development and agrarian transition*, Blackwell.

GOODMAN, D. L. (1972), 'Industrial development in the Brazilian northeast: an interim assessment of the tax credit scheme of article 34/18', in R. J. A. Roett (ed.), *Brazil in the sixties*, Vanderbilt University, Press, 231–74.

GORE, C. (1984), *Regions in question: space, development theory and regional planning*, Methuen.

GORE, M. S. (1971), *Immigrants and neighbourhoods: two aspects of life in a metropolitan city*, Tata Institute of Social Sciences, Bombay.

GOTT, R. (1971), *Guerrilla movements in Latin America*, Doubleday.

GRAHAM, R. (1979), *Iran: the illusion of power*, Croom Helm, rev. edn. (1st edn. 1978).

GREENWOOD, W. (1933), *Love on the dole*, Jonathan Cape.

GREGORY, D., and URRY, J. (eds.) (1986), *Social relations and spatial structures*, Macmillan.

GRENNEL, P. (1972), 'Planning for invisible people: some consequences of bureaucratic values and practices', in Turner and Fichter (eds.) (1972), 95–121.

GRIFFIN, K. (1978), review of 'Why poor people stay poor', *Journal of Development Studies*, 15. 108–9.

—— and GHOSE, A. K. (1979), 'Growth and impoverishment in the rural areas of Asia', *World Development*, 7. 361–84.

—— and KHAN, A. R. (1978), 'Poverty in the Third World: ugly facts and fancy models', *World Development*, 6. 295–304.

GRIFFITH-JONES, S. (1978), 'A critical evaluation of Popular Unity's short-term and financial policy', *World Development*, 6. 1019–29.

GRIMES, O. F. (1976), *Housing for low income urban families*, Johns Hopkins University Press.

GRINDAL, B. T. (1973), 'Islamic affiliations and urban adaptations: the Sisala migrant in Accra, Ghana', *Africa*, 43. 333–46.

GUEVARA, E. (1960), *La guerra de guerrillas*, Departamento del Minfar, Havana. (English trans. (1961), *Guerrilla warfare*, Monthly Review Press.)

—— (1968), *The diary of Che Guevara; Bolivia: November 7, 1966–October 7, 1967. The authorized text in English and Spanish*, Bantam Books.

GUGLER, J. (1969), 'On the theory of rural–urban migration: the case of Subsaharan Africa', in J. A. Jackson (ed.), *Migration*, Cambridge University Press, 134–55.

—— (1971), 'Life in a dual system: Eastern Nigerians in town, 1961' *Cahiers d'études africaines*, 11. 400–21.

—— (1975*a*), 'Particularism in Subsaharan Africa: "tribalism" in town', *Canadian Review of Sociology and Anthropology*, 12. 303–15.

—— (1975*b*), 'Part-time farmers: the peri-urban holdings of urban workers in Kampala', *Journal of Eastern African Research and Development*, 5. 219–24.

—— (1976), 'Migrating to urban centres of unemployment in tropical Africa', in A. H. Richmond and D. Kubat (eds.), *Internal migration: the New and the Third World*, Sage, 184–204.

—— (1980), ' "A minimum of urbanism and a maximum of ruralism": the Cuban experience', *International Journal of Urban and Regional Research*, 4. 516–34.

—— (ed.) (1988*a*), *The urbanization of the Third World*, Oxford University Press.

—— (1988*b*), 'Overurbanization reconsidered', in Gugler (ed.) (1988*a*), 74–92. (Earlier version (1982), *Economic Development and Cultural Change*, 31. 173–89.)

—— (1988*c*), 'The urban character of contemporary revolutions', in Gugler (ed.) (1988*a*), 399–412. (Earlier version (1982), *Studies in Comparative International Development* 17(2). 60–73.)

—— (1991), 'Life in a dual system revisited: urban–rural ties in Enugu, Nigeria, 1961–1987', *World Development*, 19. 399–409.

—— and FLANAGAN, W. G. (1978), *Urbanization and social change in West Africa*, Cambridge University Press.

—— and LUDWAR-ENE, G. (1990), 'Many roads lead women to town in Subsaharan Africa', paper presented at World Congress of Sociology, Madrid, July 1990.

GUILLEN, A. (1973), *Philosophy of the urban guerrilla: the revolutionary writings of Abraham Guillén*, ed. D. C. Hodges, William Morrow. (First pub. (1966) as *Estrategia de la guerrilla urbana: principios básicos de guerra revolucionaria*, Ediciones Liberación, Montevideo.)

GUILLET, D. (1976), 'Migration, agrarian reform, and structural change in rural Peru', *Human Organization*, 35. 295–302.

GUTIERREZ, J., CAMAROS, J., COBAS, J., and HERTENBERG, R. (1984), 'The recent worldwide economic crisis and the welfare of children: the case of Cuba', *World Development*, 12. 247–60.

GWYNNE, R. N. (1978), 'Government planning and the location of the motor-vehicle industry in Chile', *Tijdschrift voor economische en sociale geografie*, 69. 130–40.

—— (1985), *Industrialisation and urbanisation in Latin America*, Croom Helm.

HALEY, A. (1977), *Roots*, Pan Books.

HALL, P. (1980), 'New trends in European urbanization', *Annals of the American Academy of Political and Social Science*, 451. 45–51.

HAMER, A. M. (1985), 'Decentralized urban development and industrial location behaviour in São Paulo, Brazil: a synthesis of research issues and conclusions', World Bank Occasional Paper 732.

—— STEER, A., and WILLIAMS, D. (1986), 'Indonesia: the challenge of urbanization', World Bank Staff Working Paper 787.

HANDELMAN, H. (1975), 'The political mobilization of urban squatter settlements: Santiago's recent experience and its implications for urban research', *Latin American Research Review*, 10 (2). 35–72.

HANF, T. (1988), 'Homo oeconomicus—homo communitaris. Crosscutting loyalties in a deeply divided society: the case of trade unions in Lebanon', in M. J. Esman and I. Rabinovich (eds.), *Ethnicity, pluralism, and the state in the Middle East*, Cornell University Press, 173–84.

HANNERZ, U. (1969), *Soulside: inquiries into ghetto culture and community*, Columbia University Press.

HANNERZ, U. (1980), *Exploring the city: inquiries toward an urban anthropology*, Columbia University Press.

HANSEN, N. (1981), 'Development from above: the centre-down development paradigm', in W. B. Stöhr and D. R. F. Taylor (eds.) (1981), 39–72.

HARDOY, J. E. (1975), 'Two thousand years of Latin American urbanization', in Hardoy (ed.) (1975), 3–56.

—— (ed.) (1975), *Urbanization in Latin America: approaches and issues*, Anchor/Doubleday.

—— and SATTERTHWAITE, D. (eds.) (1986), *Small and intermediate urban centres: their role in national and regional development in the Third World*, Hodder and Stoughton.

HARMS, H. (1982*a*), 'Historical perspectives on the practice and purpose of self-help housing', in Ward (ed.) (1982), 15–55.

—— (1982*b*), 'Production of housing in Cuba', *Proceedings of the Bartlett Summer School, 1981: the production of the built environment*, London.

HARRIS, J. R., and TODARO, M. P. (1968), 'Urban unemployment in East Africa: an economic analysis of policy alternatives', *East African Economic Review*, 4. 17–36.

—— (1970), 'Migration, unemployment and development: a two-sector analysis', *American Economic Review*, 60. 126–42.

HARRIS, N. (1983), *The end of the Third World*, Penguin.

HART, K. (1973), 'Informal income opportunities and urban employment in Ghana', *Journal of Modern African Studies*, 11. 61–89.

HARVEY, D. (1973), *Social justice and the city*, Arnold.

HECHT, S. B. (1981), 'Deforestation in the Amazon basin: practice, theory and soil resource effects', *Studies in Third World Societies*, 13. 61–108.

—— and COCKBURN, A. (1990), *The fate of the forest*, Verso.

HELLMAN, J. A. (1983), *Mexico in crisis*, Holmes & Meier, 2nd edn (1st edn. 1978).

HEMMING, J. (ed.) (1985), *Change in the Amazon Basin*, 2: *The frontier after a decade of colonisation*, Manchester University Press.

HENFREY, C., and SORJ, B. (1977), *Chilean voices: activists describe their experiences of the Popular Unity period*, Harvester Press.

HERRICK, B. H. (1965), *Urban migration and economic development in Chile*, MIT Press.

HINDERINK, J., and STERKENBURG, J. (1975), *Anatomy of an African town*, State University of Utrecht.

—— and TITUS, M. J. (1988), 'Paradigms of regional development and the role of small centres', *Development and Change*, 19. 401–23.

HINDSON, D. (1987), *Pass controls and the urban African proletariat in South Africa*, Ravan Press.

HIRSCHMAN, A. O. (1958), *The strategy of economic development*, Yale University Press.

HOCH, I. (1972), 'Income and city size', *Urban Studies*, 9. 299–328.

HODGES, D. C. (1973), 'Introduction: the social and political philosophy of Abraham Guillén', in D. C. Hodges (ed.), *Philosophy of the urban guerrilla: the revolutionary writings of Abraham Guillén*, William Morrow, 1–55.

HOLLAND, S. (1976), *Capital and the regions*, Macmillan.

HOLLNSTEINER-RACELIS, M. (1977), 'The case of "the people versus Mr. Urbano Planner y Administrador" ', in Abu-Lughod and Hay (eds.) (1977), 307–20.
—— (1988), 'Becoming an urbanite: the neighbourhood as a learning environment', in Gugler (ed.) (1988*a*), 230–41. (Earlier version in Dwyer (ed.) (1972), 29–40.
HOLMSTRÖM, M. (1984), *Industry and inequality: the social anthropology of Indian labour*, Cambridge University Press.
HOPKINS, A. G. (1973), *An economic history of West Africa*, Longman.
HOPKINS, N. S. (1972), *Popular government in an African town: Kita, Mali*, University of Chicago Press.
HOROWITZ, D. L. (1985), *Ethnic groups in conflict*, University of California Press.
HORVATH, R. J. (1972), 'A definition of colonialism', *Current Anthropology*, 13. 45–51.
HOSELITZ, B. F. (1953), 'The role of cities in the economic growth of underdeveloped countries', *Journal of Political Economy*, 61. 195–208.
—— (1957), 'Generative and parasitic cities', *Economic Development and Cultural Change*, 3. 278–94.
HUGO, G. J. (1983), 'New conceptual approaches to migration in the context of urbanization: a discussion based on the Indonesian experience', in P. A. Morrison (ed.) (1983), *Population movements: their forms and functions in urbanization and development*, Ordina Edns., 69–113.
—— (1985), 'Circulation in West Java, Indonesia' in R. M. Prothero and M. Chapman (eds.), *Circulation in Third World countries*, Routledge & Kegan Paul, 75–99.
HUMPHREY, J. (1982), *Capitalist control and workers' struggle in the Brazilian auto industry*, Princeton University Press.
—— (1987), *Gender and work in the Third World: sexual divisions in Brazilian industry*, Tavistock.
HUTTON, C. (1973), *Reluctant farmers? A study of unemployment and planned rural development in Uganda*, East African Publishing House.
IBGE (Instituto Brasileiro de Estatística) (1970), *Anuario Estatístico do Brasil, 1970*.
India, NIUA (National Institute of Urban Affairs) (1989), 'Rental housing in India: an overview', NIUA Research Study 31.
INGRAM, G. K., and CARROLL, A. (1981), 'Spatial structure of Latin American cities', *Journal of Urban Economics*, 9. 257–73.
INKELES, A., and SMITH, D. H. (1970), 'The fate of personal adjustment in the process of modernization', *International Journal of Comparative Sociology*, 11. 81–114.
Inter-American Development Bank (IADB) (1985), *Economic and social progress of Latin America: Annual report 1985*, Washington DC.
International Labour Office (1972), *Employment, incomes and equality: a strategy for increasing productive employment in Kenya*, ILO.
—— (1986), *Annotated bibliography on child labour*, ILO.
—— (1988), 1987–88 *Yearbook of Labour Statistics*, ILO.
—— (1990), *1989–90 Yearbook of Labour Statistics*, ILO.
ISARD, W. (1960), *Methods of regional analysis*, MIT Press.
ISHOC (International Symposium on Housing Organizing Committee) (1987), *Country profiles: housing and settlement conditions*, Yokohama.

IZAGUIRRE, M. (1977), *Ciudad Guayana; la estrategia del desarrollo polarizado*, Ediciones SIAP-Planteos, Buenos Aires.

JAKOBSON, L., and PRAKASH, V. (1974), 'Urban planning in the context of a "new urbanization" ', Jakobson and Prakash (eds.) (1974), 259–86.

—— (eds.) (1974), *Metropolitan growth: public policy for South and Southeast Asia*, Halsted Press.

JAMAL, V., and WEEKS, J. (1988), 'The vanishing rural–urban gap in sub-Saharan Africa', *International Labour Review*, 127. 271–92.

JAMESON, K. P. (1979), 'Designed to fail: twenty-five years of industrial decentralization policy in Peru', *Journal of Developing Areas*, 14. 55–70.

JEFFERSEN, M. (1939), 'The law of the primate city', *Geographical Review*, 29. 226–82.

JEFFRIES, R. D. (1978), *Class, power and ideology in Ghana: the railwaymen of Sekondi*, African Studies Series 22, Cambridge University Press.

JELIN, E. (1982), 'Women and the urban labour market' in R. Anker, M. Buvinic, and N. H. Youssef (eds.), *Women's roles and population trends in the Third World*, Croom Helm, 239–67.

JELLINEK, L. (1978), 'Circular migration and the *pondok* dwelling system: a case study of ice-cream traders in Jakarta', in P. J. Rimmer, D. W. Drakakis-Smith, and T. G. McGee (eds.), *Food, shelter and transport in Southeast Asia and the Pacific*, Department of Human Geography, Research School of Pacific Studies, Australian National University, 135–54.

—— (1988), The changing fortunes of a Jakarta street trader', in Gugler (ed.) (1988*a*), 204–23.

JENKINS, R. (1984), 'Divisions over the international division of labour', *Capital and Class*, 22. 28–57.

JOCANO, F. L. (1975), *Slum as a way of life: a study of coping behaviour in an urban environment*, University of the Philippines Press.

JOEKES, S. P. (1982), *Female-led industrialization. Women's jobs in Third World export manufacturing: the case of the Moroccan clothing industry*, Institute of Development Studies Research Report 15, Institute of Development Studies at the University of Sussex.

—— (1987), *Women in the world economy: an INSTRAW study*, Oxford University Press.

JOHNSON, E. A. J. (1970), *The organization of space in developing countries*, Harvard University Pres.

JOHNSON, L. L. (1967), 'Problems of import substitution: the Chilean automobile industry', *Economic Development and Cultural Change*, 15. 202–16.

JOHNSON, M. (1986), *Class and client in Beirut: the Sunni Muslim community and the Lebanese state, 1840–1985*, Ithaca Press.

—— (1988), 'Political bosses and strong-arm retainers in the Sunni Muslim quarters of Beirut', in J. Gugler (ed.) (1988*a*), 308–27.

JOHNSTON, B. F., and MELLOR, J. W. (1961), 'The role of agriculture in economic development', *American Economic Review*, 51. 566–92.

JOHNSTON, R. J. (1976), 'Observations on accounting procedures and urban-size policies', *Environment and Planning A*, 8. 327–39.

JONES, G. W. (1984), 'Economic growth and changing female employment structure in the cities of Southeast and East Asia', in G. W. Jones (ed.), *Women in the urban and industrial workforce: Southeast and East Asia*, Monograph 33, Development Studies Centre, Australian National University, Canberra, 17–61.

JOSHI, H., and JOSHI, V. (1976), *Surplus labour and the city: a study of Bombay*, Oxford University Press.

KALMANOVITZ, S. (1977), *Ensayos sobre el desarrollo del capitalismo dependiente*, Editorial Pluma, Bogotá.

KAMMEIER, H. D., and SWAN, P. (eds.) (1984), *Equity with growth? Planning perspectives for small towns in developing countries*, Asian Institute of Technology.

KANG, G. E., and KANG, T. S. (1978), 'The Korean urban shoeshine gang: a minority community', *Urban Anthropology*, 7. 171–83.

KAROL, K. S. (1970), *Les guérilleros au pouvoir: l'itinéraire politique de la révolution cubaine*, L'histoire que nous vivons, Robert Laffont. (English trans. (1970), *Guerrillas in power: the course of the Cuban revolution*, Hill & Wang.)

KATZENSTEIN, M. F. (1979), *Ethnicity and equality: the Shiv Sena party and preferential policies in Bombay*, Cornell University Press.

KATZMAN, M. T. (1977), *Cities and frontiers in Brazil: regional dimensions of economic development*, Harvard University Press.

KAY, C. (1989), *Latin American theories of development and underdevelopment*, Routledge.

KEARE, D. A., and PARRIS, S. (1982), 'Evaluation of shelter programs for the urban poor: principal findings', World Bank Staff Working Paper 547.

KEARNEY, M. (1986), 'From the invisible hand to visible feet: anthropological studies of migration and development', *Annual Review of Anthropology*, 15. 331–61.

KEDDIE, N. R. (1981), *Roots of revolution: an interpretive history of modern Iran*, Yale University Press.

KELES, R. (1973), *Urbanization in Turkey*, Ford Foundation.

—— and KANO, H. (1987), *Housing and the urban poor in the Middle East—Turkey, Egypt, Morocco and Jordan*, Institute of Developing Economies, Tokyo.

Kenya (1986), *Kenya Statistical Abstract*, Nairobi.

KHALAF, S. (1977), 'Changing forms of political patronage in Lebanon', in E. Gellner and J. Waterbury (eds.), *Patrons and clients in Mediterranean societies*, Duckworth, 185–205.

KIERNAN, V. G. (1972), *The lords of human kind: European attitudes to the outside world in the Imperial age*, Penguin.

KING, A. D. (1976), *Colonial urban development: culture, social power and environment*, Routledge & Kegan Paul.

KING, T. (1970), *Mexico: industrialization and trade policies since 1940*, Oxford University Press.

KIRKBY, R. J. R. (1985), *Urbanisation in China: town and country in a developing economy 1949–2000 AD*, Croom Helm/University of Columbia Press.

KLARE, M. T. (1972), *War without end: American planning for the next Vietnams*, Alfred A. Knopf.

KLEINPENNING, J. M. G. (1978), 'A further evaluation of the policy for the integration of the Amazon region (1974–1976)', *Tijdschrift voor economische en sociale geografie*, 69. 78–85.

KLEINPENNING, J. M. G. and VOLBEDA, S. (1985), 'Recent changes in population size and distribution in the Amazon region of Brazil', in Hemming (ed.) (1985), 6–36.

KOHL, J., and LITT, J. (1974), *Urban guerrilla warfare in Latin America*, MIT Press.

KOTHARI, S. (1983), 'There's blood on those matchsticks: child labour in Sivakasi', *Economic and Political Weekly*, 18. 1191–202.

KOWARICK, L., and BRANT, V. C. (ed.) (1978), *São Paulo 1975: Growth and poverty*, Bowerdean Press.

—— and CAMPANARIO, M. (1986), 'São Paulo: the price of world city status', *Development and Change*, 17. 159–74.

KUKLINSKI, A. (ed.) (1972), *Growth poles and growth centres in regional planning*, Mouton.

KUSNETZOFF, F. (1975), 'Housing policies or housing politics: an evaluation of the Chilean experience', *Journal of Interamerican Studies and World Affairs*, 17. 281–310.

KUZNETS, S. (1966), *Modern economic growth: rate, structure and spread*, Yale University Press.

LABICH, K., and MCGUIRE, S. (1978), 'Mutiny in Matagalpa', *Newsweek*, 11 Sept. 1978. 44, 46, and 49.

LACLAU, E. (1971), 'Feudalism and capitalism in Latin America', *New Left Review*, 67. 19–38.

LAHIRI, T. B. (1978), 'Calcutta: a million city with a million problems', in Misra (ed.) (1978), 43–72.

LAITE, J. (1981), *Industrial development and migrant labour in Latin America*, Manchester University Press/University of Texas Press.

—— (1988), 'The migrant response in Central Peru', in Gugler (ed.) (1988*a*), 61–73.

LANDE, C. H. (1977), 'Introduction: the dyadic basis of clientelism', in S. W. Schmidt, L. Guasti, C. H. Landé, and J. C. Scott (eds.) (1977), *Friends, followers, and factions: a reader in political clientelism*, University of California Press, xiii–xxxvii.

LANDER, L., and FUNES, J. C. (1965), 'Urbanization and Development', in *Urbanizacíon en Venezuela*, Documento URVEN 8, CENDES, Caracas. (Repr. in Hardoy (ed.) (1975), 287–337.)

LAQUIAN, A. A. (ed.) (1971), *Rural–urban migrants and metropolitan development*, Intermet, Toronto.

—— (1977), 'Whither site and services?' *Habitat*, 2. 291–301.

LARDY, N. R. (1978), *Economic growth and distribution in China*, Cambridge University Press.

LARTEGUY, J. (1970), *The guerrillas*, New American Library. (First published (1967) as *Les guérilleros*, Raoul Solar.)

LAUN, J. I. (1976), 'El estado y la vivienda en Colombia: análisis de urbanizaciones del Instituto de Crédito Territorial en Bogotá, in C. Castillo (ed.), *Vida urbana y urbanismo*, Instituto Colombiano de Cultura, Bogotá.

LAVELL, A. M. (1971), 'Industrial development and the regional problem: a case study of Central Mexico', doctoral dissertation, University of London.

LEAN, L. L. (1984), 'Towards meeting the needs of urban female factory workers in Peninsular Malaysia', in G. W. Jones (ed.), *Women in the urban and industrial workforce: Southeast and East Asia*, Monograph 33, Development Studies Centre, Australian National University, Canberra, 129–48.

LEEDS, A. (1969), 'The significant variables determining the character of squatter settlements', *América Latina*, 12. 44–86.

—— (1971), 'The concept of the "culture of poverty": conceptual, logical, and empirical problems, with perspectives from Brazil and Peru', in E. B. Leacock (ed.), *The culture of poverty: a critique*, Simon & Schuster, 226–84.

—— (1973), 'Locality power in relation to supralocal power institutions', in A. Southall (ed.) (1973), *Urban anthropology: cross-cultural studies of urbanization*, Oxford University Press, 15–41.

—— and LEEDS, E. (1970), 'Brazil and the myth of urban rurality: urban experience, work, and values in the "squatments" of Rio de Janeiro and Lima', in A. Field (ed.), *City and country in the Third World: issues in the modernization of Latin America*, Schenkman Publishing, 229–85.

—— (1976), 'Accounting for behavioral differences: three political systems and the responses of squatters in Brazil, Peru, and Chile', in Walton and Masotti (eds.) (1976), 193–248.

LEEMING, F. (1985), *Rural China today*, Longman.

LEFEBER, L. (1978), 'Spatial population distribution: urban and rural development', paper presented to the ECLA/CELADE Seminar on Population Redistribution, Aug. 1978.

LEFEBVRE, H. (1970), *La révolution urbaine*, Paris.

LEISERSON, A. (1979), 'Child labour in developing countries', *Internationale Entwicklung*, 4. 3–13.

LESLIE, J. (1989), 'Women's work and child nutrition in the Third World', in J. Leslie and M. Paolisso (eds.), *Women, work, and child welfare in the Third World*, AAAS Selected Symposium 110, Westview Press, 19–58.

LEWIS, J. P. (1962), *Quiet crisis in India: economic development and American policy*, Brookings Institution, Washington DC.

LEWIS, O. (1959), *Five families: Mexican case studies in the culture of poverty*, Basic Books.

—— (1970*a*), 'The culture of poverty', in O. Lewis, *Anthropological essays*, Random House, 67–80. (First pub. (1966), *Scientific American*, 215 (4). 19–25.)

—— (1970*b*), 'Urbanization without breakdown: a case study', in O. Lewis, *Anthropological essays*, Random House, 413–26. (First pub. (1952), *Scientific Monthly*, 75 (1). 31–41.)

—— LEWIS, R. M., and RIGDON, S. M. (1978), *Neighbors: living the Revolution. An oral history of contemporary Cuba*, University of Illinois Press.

LIEUWEN, E. (1961), *Venezuela*, Oxford University Press.

LIM, L. Y. C. (1990), 'Women's work in export factories: the politics of a cause', in I. Tinker (ed.), *Persistent inequalities: women and world development*, Oxford University Press, 101–19.

LIN, V. (1987), 'Women electronics workers in Southeast Asia: the emergence of a working class', in J. Henderson, and M. Castells (eds.), *Global restructuring and territorial development*, Sage, 112–35.

LINN, J. (1983), *Cities in the developing world: policies for their equitable and efficient growth*, Oxford University Press.

LINSKY, A. S. (1965), 'Some generalizations concerning primate cities', *Annals of the American Association of Geographers*, 55. 506–13.

LIPTON, M. (1977), *Why poor people stay poor: a study of urban bias in world development*, Temple Smith/Harvard University Press.

LITTLE, K. (1974), *Urbanization as a social process*, Routledge & Kegan Paul.

LLOYD, P. (1979), *Slums of hope? Shanty towns of the Third World*, Penguin.

LOMNITZ, L. (1977), *Networks and marginality: life in a Mexican shanty town*, Academic Press. (First pub. (1975) as *Cómo sobreviven los marginados*, Siglo Veintiuno Editores, Mexico City.)

—— (1978), 'Mechanisms of articulation between shantytown settlers and the urban system', *Urban Anthropology*, 7. 185–205.

—— (1988), 'The social and economic organization of a Mexican shanty town', in Gugler (ed.) (1988*a*), 242–63. (Earlier version in W. A. Cornelius and F. M. Trueblood (eds.) (1974), *Anthropological perspectives on Latin American urbanization*, Latin American Urban Research, 4, Sage, 135–55.

LONG, N., and ROBERTS, B. (1984), *Miners, peasants and entrepreneurs: regional development in the Central Highlands of Peru*, Cambridge University Press.

LOSADA, R., and GOMEZ, H. (1976), *La tierra en el mercado pirata de Bogotá*, Fedesarrollo.

LOWDER, S. (1987), *Inside the Third World city*, Croom Helm.

LUBECK, P. M. (1986), *Islam and urban labour in Northern Nigeria: the making of a Muslim working class*, African Studies Series 51, Cambridge University Press.

MA, X. (1989), 'Contemporary rural–urban migration of the Chinese population', *Chinese Journal of Population Science*, 1. 51–66.

MABOGUNJE, A. L. (1962), *Yoruba towns*, Ibadan University Press.

—— HARDOY, J. E., and MISRA, R. P. (1978), *Shelter provision in developing countries*, John Wiley.

MACDONALD, J. S., and MACDONALD, L. (1979), 'Planning, implementation and social policy: an evaluation of Ciudad Guayana 1965 and 1975', *Progress in Planning*, 11, parts I and II. 1–206.

MCGEE, T. G. (1967), *The Southeast Asian city*, Bell.

—— (1971), *The urbanization process in the Third World: explorations in search of a theory*, Bell.

—— (1976), 'The persistence of the proto-proletariat: occupational structures and planning for the future of Third World cities', *Progress in Geography*, 9. 3–38.

MCGREEVEY, W. P. (1971*a*), 'A statistical analysis of primacy and normality in the size distribution of Latin American cities, 1750–1960', in R. M. Morse (ed.), *The urban development of Latin America 1750–1920*, Stanford University, 116–29.

—— (1971*b*), *An economic history of Colombia, 1845–1930*, Cambridge University Press.

MADDISON, A. (1971), *Class structure and economic growth: India and Pakistan since the Moghuls*, W. W. Norton.

MAHAR, D. J. (1989), *Government policies and deforestation in Brazil's Amazon region*, World Bank.

MAJUMDAR, M. (1977), 'Regional income disparities, regional income change and federal policy in India, 1950–51 to 1967–68: an empirical evaluation', Occasional Paper 7, Department of Economics, Dundee University.

MAJUMDAR, T. K. (1978), 'The urban poor and social change: a study of squatter settlements in Delhi', in A. de Souza (ed.), *The Indian city: poverty, ecology and urban development*, Manohar, New Delhi, 29–60.

MALLOY, J. M. (1970), *Bolivia: the uncompleted revolution*, University of Pittsburgh Press.

—— (1979), *The politics of social security in Brazil*, Pitt Latin American Series, University of Pittsburgh Press.

MALPEZZI, S. (1984), 'Rent controls: an international comparison', paper presented to American Real Estate and Urban Economics Association.

—— (1989), 'Rental housing in developing countries: issues and constraints', Urban Development Division, World Bank.

—— and MAYO, S. (1987), 'User cost and housing tenure in developing countries', *Journal of Development Economics*, 25. 197–220.

MANGIN, W. (1967), 'Latin American squatter settlements: a problem and a solution', *Latin American Research Review*, 2. 65–98.

MARRIS, P. (1979), 'The meaning of slums and patterns of change', *International Journal of Urban and Regional Research*, 3. 419–41.

MARX, K. (1853), 'The results of British rule in India', *New York Daily Tribune*, 11 July 1853. (Repr. in K. Marx (1950), *On colonialism*, Lawrence & Wishart.)

—— (1967), *Capital*, 3 vols., International Publishers Edition.

MASON, E. *et al.* (1980), *The economic and social modernization of the Republic of Korea*, MIT Press.

MASON, M. (1978), 'Contemporary colonization processes in the northeast Mato Grosso', doctoral dissertation, University of London.

MASSEY, D. (1984), *Spatial division of labour*, Macmillan.

MATHEY, K. (1989), 'Recent trends in Cuban housing policies and the revival of the Microbrigade movement', *Bulletin of Latin American Research*, 8. 67–82.

MATHUR, O. P. (1977), 'The problem of regional disparities: an analysis of Indian policies and programmes', *Habitat International*, 2. 427–53.

MATOS-MAR, J., and MEJIA, J. M. (1982), 'Casual work, seasonal migration and agrarian reform in Peru', in P. Peek and G. Standing (eds.), *State policies and migration: studies in Latin America and the Caribbean*, Croom Helm, 81–119.

MAXWELL, N. (ed.) (1979), *China's road to development*, Pergamon, 2nd edn.

MAYANS, E. (ed.) (1971), *Tupamaros: antología documental*, CIDOC Cuaderno 60, Centro Intercultural de Documentación, Cuernavaca.

MAYER, P. (1971), *Townsmen or tribesmen: conservatism and the process of urbanization in a South African city*, Oxford University Press, 2nd edn (1st edn. 1962.)

MAZUMDAR, D. (1976), 'The urban informal sector', *World Development*, 4. 665–79.

—— (1981), *The urban labor market and income distribution: a study of Malaysia*, Oxford University Press.

MEHTA, S. K. (1964), 'Some demographic and economic correlates of primate cities: a case for re-evaluation', *Demography*, 1. 136–47.

MEISELAS, S. and ROSENBERG, C. (1981), *Nicaragua: June 1978–July 1979*, Pantheon Books.

MERA, K. (1973), 'On the urban agglomeration and economic efficiency', *Economic Development and Cultural Change*, 22. 309–24.

—— (1976), 'The changing pattern of population distribution in Japan and its implications for developing countries', in UNCRD (1976), 247–77.

MERRICK, T. W. (1976), 'Employment earnings in the informal sector in Brazil: the case of Belo Horizonte', *Journal of Developing Areas*, 10. 337–53.

MERTON, R. K. (1938), 'Social structure and anomie', *American Sociological Review*, 3. 672–82.

MESA-LAGO, C. (1974), *Cuba in the 1970s: pragmatism and institutionalization*, University of Arizona Press.

—— (1978), *Social security in Latin America: pressure groups, stratification, and inequality*, Pitt Latin American Series, University of Pittsburgh Press.

MEUNIER, C. (1976), 'Revendications urbaines, stratégie politique et transformation idéologique: le campamento "Nueva Habana" (Santiago) 1970–73', doctoral dissertation, École des Hautes Études en Sciences Sociales, Paris.

Mexico, Instituto Nacional de Estadística, Geografía e Informática (INEGI) (1984), *X Censo General de Población y Vivienda 1980*, Mexico City.

—— (1991), *Agenda Estadística 1990*, Mexico City.

—— (1989), *Agenda Estadística 1988*, Mexico City.

MEYER, D. R. (1986), 'System of cities dynamics in newly industrializing nations', *Studies in Comparative International Development*, 21. 3–22.

MEYER, G. (1989), 'Small businesses and socio-economic development in the Middle East: the example of manufacturing enterprises in Cairo', Research Paper, Berliner Institut für Vergleichende Sozialforschung.

MICHAELSON, K. L. (1976), 'Patronage, mediators, and the historical context of social organization in Bombay', *American Ethnologist*, 3. 281–95.

—— (1979), 'Power, patrons, and political economy: Bombay', in M. B. Leons and F. Rothstein (eds.), *New directions in political economy: an approach from anthropology*, Contributions in Economics and Economic History 22, Greenwood Press, 235–48.

MIGDAL, J. S. (1988), *Strong societies and weak states: state–society relations and state capabilities in the Third World*, Princeton University Press.

MILIBAND, R. (1969), *The state in capitalist society*, Quartet Books. (1st edn. 1967).

—— (1977), *Marxism and politics*, Oxford University Press.

MILLER, J. C. (1974), *Regional development: a review of the state of the art*, USAID, Washington DC.

MILLS, E. S. (1972), 'Welfare aspects of national policy towards city sizes', *Urban Studies*, 9. 117–24.

—— and TAN, J. P. (1980), 'A comparison of urban population density functions in developed and developing countries', *Urban Studies*, 17. 313–21.

MISRA, R. P. (ed.) (1978), *Million cities of India*, Vikas Publishing House.

MITCHELL, J. C. (1985), 'Towards a situational sociology of wage-labour circulation', in R. M. Prothero and M. Chapman (eds.), *Circulation in Third World countries*, Routledge & Kegan Paul, 30–53.

MITCHELL, M., and RUSSELL, D. (1987), 'South Africa in crisis: the role of the black trade unions', in W. Brierley (ed.), *Trade unions and the economic crisis of the 1980s*, Gower, 190–210.

MITRA, A. (1977), *Terms of trade and class relations*, Frank Cass.

——, MUKHERJI, S., and BOSE, R. (1980), *Indian cities: their industrial structure, immigration and capital investment 1961–1971*, Abhinav Publications.

MOHAN, R., and PANT, C. (1982), 'Morphology of urbanisation in India', *Economic and Political Weekly*, 17. 1–16.

MOORE, B., Jr. (1966), *Social origins of dictatorship and democracy: lord and peasant in the making of the modern world*, Beacon Press.

MOORE, J. (1984), 'The political history of Nigeria's new capital', *Journal of Modern African Studies*, 22. 167–75.

MOORSOM, R. (1979), 'Labour consciousness and the 1971–72 contract workers strike in Namibia', *Development and Change*, 10. 205–31.

MORAN, E. F. (1985), 'An assessment of a decade of colonisation in the Amazon Basin', in Hemming (ed.) (1985), 91–102.

MORAWETZ, D. (1979), 'Walking on two legs: reflections on a China visit', *World Development*, 7. 877–92.

MORSE, R. M. (1971), 'Trends and issues in Latin American urban research, 1965–1970', *Latin American Research Review*, 7. 3–52 and 19–76.

—— (1975), 'The development of urban systems in the Americas in the nineteenth century', *Journal of Inter-American Studies and World Affairs*, 17. 4–26.

MOSELEY, M. J. (1975), *Growth centres in spatial planning*, Pergamon.

MOSER, C. O. N. (1978), 'Informal sector or petty commodity production: dualism or dependence in urban development?', *World Development*, 6. 1041–64.

—— (1982), 'A home of one's own: squatter housing strategies in Guayaquil, Ecuador', in Gilbert, Hardoy, and Ramírez (eds.) (1982), 159–90.

—— (1984), 'The informal sector reworked: viability and vulnerability in urban development', *Regional Development Dialogue*, 5 (2). 135–78.

—— (1989), 'Gender planning in the Third World: meeting practical and strategic gender needs', *World Development*, 17. 1799–825.

MUENCH, L. H. (1978), 'The private burden of urban social overhead: a study of the informal housing market of Kampala, Uganda', doctoral dissertation, University of Pennsylvania.

MUMFORD, L. (1938), *The culture of cities*, Harcourt Brace.

—— (1975), *The city in history*, Penguin. (1st edn. 1961.)

MURPHEY, R. (1969), 'Traditionalism and colonialism: changing urban roles in Asia', *Journal of Asian Studies*, 29. 67–84.

—— (1972), 'City and countryside as ideological issues: India and China', *Comparative Studies in History and Society*, 14. 250–67.

—— (1976), 'Chinese urbanization under Mao', in Berry (ed.) (1976), 311–30.

MYERS, D. (1978), 'Caracas: the politics of intensifying primacy', in W. A. Cornelius and R. V. Kemper (eds.), *Metropolitan Latin America: the challenge and the response*, Latin American Urban Research, 6, Sage, 227–58.

MYRDAL, G. (1957), *Economic theory and underdeveloped regions*, Duckworth.

—— (1970), *The challenge of world poverty: a world anti-poverty program in outline*, Pantheon.

NAIPAUL, V. S. (1969), *The mimic men*, Penguin. (1st edn. 1967.)

—— (1977), *India: a wounded civilization*, André Deutsch.

NAVARRO, M. (1989), 'The personal is political: Las Madres de Plaza de Mayo', in S. Eckstein (ed.), *Power and popular protest: Latin American social movements*, University of California Press, 241–58.

NELSON, J. M. (1979), *Access to power: politics and the urban poor in developing nations*, Princeton University Press.

NELSON, N. (1988), 'How women and men get by: the sexual division of labour in

the informal sector of a Nairobi squatter settlement', in Gugler (ed.) (1988*a*), 183–203. (Earlier version in Bromley and Gerry (eds.) (1979), 283–302.)
NURKSE, R. (1952), 'Some international aspects of the problem of economic development', *American Economic Review*, 42. 571–83.
O'CONNOR, A. M. (1971), *A geography of tropical African development*, Pergamon.
—— (1976), ' "Third World" or one world?' *Area*, 8. 269–71.
—— (1983), *The African city*, Hutchinson.
ODELL, P. R., and PRESTON, D. A. (1973), *Economies and societies in Latin America: a geographical interpretation*, John Wiley.
ODONGO, J., and LEA, J. P. (1977), 'Home ownership and rural–urban links in Uganda', *Journal of Modern African Studies*, 15. 59–73.
OJO, G. J. A. (1966), *Yoruba palaces: a study of afins of Yorubaland*, University of London Press.
ONG, A. (1990), 'Japanese factories, Malay workers: class and sexual metaphors in West Malaysia', in J. M. Atkinson and S. Errington (eds.), *Power and difference: gender in island Southeast Asia*, Stanford University Press.
ORAM, N. D. (1976), *Colonial town to Melanesian city: Port Moresby 1884–1974*, Australian National University.
ORTEGA, H. (1980), 'Nicaragua: the strategy of victory (interview with Humberto Ortega, Commander-in-Chief of the Sandinista People's Army)', English lang. edn. of *Granma*, 27 Jan. 1980. 3–8. (First pub. (1979) as 'Nicaragua: La estrategia de la victoria (entrevista a Humberto Ortega, commandante en jefe del Ejército Popular Sandinista)', *Bohemia*, 71 (52). 4–19.)
ORWELL, G. (1937), *The road to Wigan Pier*, Victor Gollancz.
OSTERLING, J. P. (1988), 'Rural schools, community development and the issue of population retention in a Peruvian peasant community: the Huayopampa case study', in J. Havet (ed.), *Staying on: retention and migration in peasant societies*, University of Ottawa Press, 159–73.
OXAAL, I., BARNETT, T., and BOOTH, D. (eds.) (1975), *Beyond the sociology of development: economy and society in Latin America and Africa*, Routledge & Kegan Paul.
PADEN, J. N. (1973), *Religion and political culture in Kano*, University of California Press.
PALLOIX, C. (1975), 'The internationalisation of capital and the circuit of social capital', in H. Radice (ed.), *International firms and modern imperialism*, Penguin.
PALMA, G. (1978), 'Dependency: a formal theory of underdevelopment or a methodology for the analysis of concrete situations of underdevelopment?' *World Development*, 6. 881–924.
PAPANEK, G. F. (1954), 'Development problems relevant to agricultural tax policy', Papers and Proceedings of the Conference on Agricultural Taxation and Economic Development, Harvard.
—— (1975), 'The poor of Jakarta', *Economic Development and Cultural Change*, 24. 1–27.
PARISH, W. L. (1987), 'Urban policy in centralized economies: China', in G. S. Tolley and V. Thomas (eds.), *The economics of urbanization and urban policy in developing countries*, World Bank, 73–84.
PARR, J. B. (1980), 'Models of city size in an urban system', *Papers and Proceedings of the Regional Science Association*, 25. 221–53.

—— (1974), 'Regional differences within a nation: a comment', *Papers and Proceedings of the Regional Science Association*, 32. 83–91.

PARRY, J. P. (1979), *Caste and kinship in Kangra*, Routledge & Kegan Paul.

PAYNE, G. K. (1977), *Urban housing in the Third World*, Leonard Hill.

—— (1982), 'Self-help housing: a critique: the gecekondus of Ankara', in Ward (ed.) (1982), 117–40.

—— (1984), 'Ankara', *Cities*, 1. 210–15.

PEACE, A. (1975), 'The Lagos proletariat: labour aristocrats or populist militants?', in R. Sandbrook and R. Cohen (eds.), *The development of an African working class: studies in class formation and action*, Longman, 281–302.

PEARSE, A. (1970), 'Urbanization and the incorporation of the peasant', in A. J. Field (ed.), *City and country in the Third World: issues in the modernization of Latin America*, Schenkman, 201–12.

—— (1975), *The Latin American peasant*, Frank Cass.

PEARSON, L. B. (ed.) (1969), *Partners in development: report of the Commission on International Development*, Pall Mall Press.

PEATTIE, L. R. (1975), ' "Tertiarization" and urban poverty in Latin America', in W. A. Cornelius and F. M. Trueblood (eds.) (1975), *Urbanization and inequality: the political economy of urban and rural development in Latin America*, Latin American Urban Research, 5, Sage, 109–23.

—— (1979), 'Housing policy in developing countries: two puzzles', *World Development*, 7. 1017–22.

—— (1987*a*), *Planning: rethinking Ciudad Guayana*, University of Michigan Press.

—— (1987*b*), 'An idea in good currency and how it grew: the informal sector', *World Development*, 15. 851–60.

PEEK, P. (1982), 'Agrarian change and labour migration in the Sierra of Ecuador', in Peek and Standing (eds.) (1982), 121–45.

—— and STANDING, G. (eds.) (1982), *State policies and migration: studies in Latin America and the Caribbean*, Croom Helm.

PEIL, M. (1976), 'African squatter settlements: a comparative study', *Urban Studies*, 131. 155–66.

—— (1981), *Cities and suburbs: urban life in West Africa*, Africana Publishing.

—— with SADA, P. O. (1984), *African urban society*, John Wiley.

PEREZ-PERDOMO, R., and NIKKEN, P. (1979), *Derecho y propiedad de la vivienda en los barrios de Caracas*, Universidad Central de Venezuela/Fondo de Cultura Económica. (English trans. in Gilbert, Hardoy, and Ramírez (eds.) (1982), 205–30.)

PERLMAN, J. (1976), *The myth of marginality: urban poverty and politics in Rio de Janeiro*, University of California Press.

PERLO, M. (1979), 'Política y vivienda en México 1910–1952', *Revista Mexicana de Sociología*, 41. 769–835.

PERRINGS, C. (1979), *Black mineworkers in Central Africa: industrial strategies and the evolution of an African proletariat in the Copperbelt 1911–41*, Heinemann/Africana Publishing.

PETRAS, E. M. (1973), *Social organization of the urban housing movement in Chile*, Council on International Studies, Special Studies 39, State University of New York.

PFEFFERMAN, G. P., and WEBB, R. (1979), 'The distribution of income in Brazil', World Bank Staff Working Paper 356.

Philippines, National Economic and Development Authority (1984), *Philippine Statistical Yearbook 1985*, Manila.

PHILLIPS, D. R., and YEH, A. G. O. (1983), 'China's experiments with modernisation: the Shenzhen special economic zone', *Geography*, 68. 289–300.

PIRENNE, H. (1925), *Medieval cities*, Princeton University Press.

POLANYI, K. (1945), *Origins of our time: the great transformation*, Victor Gollancz.

PORTES, A. (1972), 'Rationality in the slum: an essay on interpretative sociology', *Comparative Studies on History and Society*, 14. 268–86.

—— (1979), 'Housing policy, urban poverty, and the state: the favelas of Rio de Janeiro, 1972–76', *Latin American Research Review*, 14 (2). 3–24.

—— (1985), 'Latin American class structures: their composition and change during the last decades', *Latin American Research Review*, 20. 7–39.

—— BLITZER, S., and CURTIS, J. (1986), 'The urban informal sector in Uruguay: its internal structure, characteristics, and effects', *World Development*, 14. 727–41.

—— and WALTON, J. (1976), *Urban Latin America: the political condition from above and below*, University of Texas Press.

—— (1981), *Labor, class, and the international system*, Academic Press.

POTTS, D. (1985), 'Capital relocation in Africa: the case of Lilongwe in Malawi', *Geographical Journal*, 151. 182–96.

POULANTZAS, N. (1969), 'The problem of the capitalist state', *New Left Review*, 58. 67–78.

—— (1973), *Political power and social classes*, New Left Books.

—— (1976), 'The capitalist state: a reply to Miliband and Laclau', *New Left Review*, 95. 63–83.

POWELL, J. D. (1980), 'Electoral behaviour among peasants', in I. Volgyes, R. E. Lonsdale, and W. P. Avery (eds.), *The process of rural transformation: Eastern Europe, Latin America and Australia*, Comparative Rural Transformation Series, Pergamon, 193–241.

PRADILLA, E. (1976), 'Notas acerca del "problema de la vivienda" ', *Ideología y Sociedad*, 16. 70–107.

PREBISCH, R. (1950), *The economic development of Latin America and its principal problems*, United Nations.

PRESTON, S. H. (1988), 'Urban growth in developing countries: a demographic reappraisal', in Gugler (ed.) (1988*a*), 11–31. (Earlier version (1979), *Population and Development Review*, 5. 195–215.)

PRYER, J., and CROOK, N. (1988), *Cities of hunger: Urban malnutrition in developing countries*, Oxfam.

QADEER, M. A. (1974), 'Do cities modernize the developing countries? An examination of South Asian experience', *Comparative Studies in Society and History*, 16. 266–83.

QUIJANO OBREGON, A. (1974), 'The marginal pole of the economy and the marginalised labour force', *Economy and Society*, 3. 393–428.

RAMACHANDRAN, P. (1974), *Pavement dwellers in Bombay City*, Tata Institute of Social Sciences, Bombay.

RAMASWAMY, E. A. (1973), 'Politics and organized labour in India', *Asian Survey*, 13. 914–28.

—— (1984), *Power and justice: the state in industrial relations*, Oxford University Press.

—— (1988), *Worker consciousness and trade union response*, Oxford University Press.

RAMASWAMY, U. (1979), 'Tradition and change among industrial workers', *Economic and Political Weekly*, 14. 367–76.

RAY, T. (1969), *The politics of the barrios of Venezuela*, University of California Press.

REDCLIFT, M. R. (1973), 'Squatter settlements in Latin American cities: the response from government', *Journal of Development Studies*, 10. 92–109.

REED, R. R. (1976), 'Indigenous urbanism in South-East Asia', in Yeung and Lo (eds.) (1976), 14–27.

REMPEL, H. (1981), *Rural–urban labor migration and urban unemployment in Kenya*, International Institute for Applied Systems Analysis.

RENAUD, B. (1981), *National urbanization policy in developing countries*, Oxford University Press.

RICHARDSON, H. W. (1973), *Economics of urban size*, Saxon House.

—— (1976), 'The argument for very large cities reconsidered: a comment', *Urban Studies*, 13. 307–10.

—— (1977), 'City size and national spatial strategies in developing countries', World Bank Staff Working Paper 252.

—— (1980), 'Polarization reversal in developing countries', *Papers of the Regional Science Association*, 45. 76–85.

—— (1981), 'National urban development strategies in developing countries', *Urban Studies*, 18. 267–83.

—— (1989), 'The big, bad city: mega-city myth?', *Third World Planning Review*, 11. 355–72.

—— and SCHWARTZ, G. (1988), 'Economic development, population and primacy', *Regional Studies*, 22. 467–75.

RIOFRIO, G. (1978), *Se busca terreno para próxima barriada: espacios disponibles en Lima, 1940, 1978, 1990*, DESCO, Lima.

ROBERTS, B. R. (1978), *Cities of peasants: the political economy of urbanization in the Third World*, Edward Arnold/Sage.

—— (1989), 'Employment structure, life-cycle, and life-chances: formal and informal sectors in Guadalajara', in A. Portes, M. Castells, and L. A. Benton (eds.), *The informal economy: studies in advanced and less developed countries*, Johns Hopkins University Press, 41–59.

ROBINSON, G., and SALIH, K. B. (1971), 'The spread of development around Kuala Lumpur', *Regional Studies*, 5. 303–14.

ROBLEDO, J. E. (1985), *El drama de la vivienda en Colombia*, El Ancora Editores, Bogotá.

ROCCA, C. A. (1970), 'Productivity in Brazilian manufacturing', in Bergsman (1970), 22–41.

RODWIN, L. L. (1961), 'Metropolitan policy for developing areas', in W. Isard and J. H. Cumberland (eds.), *Regional economic planning*, OECD.

—— (1973), 'Choosing regions for development', in C. J. Friedrich and S. E. Harris (eds.), *Public policy*, Graduate School of Public Administration, Harvard University, 141–62.

RODWIN, L. L. and associates (1969), *Planning urban growth and regional development: the experience of the Guayana program in Venezuela*, MIT Press.

ROFMAN, A. B. (1974), *Dependencia, estructura de poder y formación regional en América Latina*, Editorial Suramericana, Buenos Aires.

ROLLWAGEN, J. R. (1971), 'Region of origin and rural–urban migration in Mexico: some general comments and a case study of entrepreneurial migration from the West', *International Migration Review*, 5. 277–338.

ROMERO, E. (1949), *Historia económica del Perú*, Editorial Suramericana, Buenos Aires.

RONDINELLI, D. (1983), *Secondary cities in developing countries: policies for diffusing urbanization*, Sage.

—— and RUDDLE, K. (1976), *Urban functions in rural development: an analysis of integrated spatial development policy*, USAID.

ROSENBERG, M. B., and MALLOY, J. M. (1978), 'Indirect participation versus social equity in the evolution of Latin American social security policy', in J. A. Booth and M. A. Seligson (eds.), *Political participation in Latin America*, 1: *Citizen and State*, Holmes & Meier, 157–71.

ROSENSTEIN-RODAN, P. N. (1943), 'Problems of industrialization of eastern and south-eastern Europe', *Economic Journal*, 53. 202–11.

ROSENZWEIG, M. R. (1988), 'Labor markets in low-income countries', in H. Chenery and T.N. Srinivasan (eds.), *Handbook of development economics*, vol. 1, Handbooks in Economics 9, North-Holland, 713–62.

ROSING, K. E. (1966), 'A rejection of the Zipf model (rank-size rule) in relation to city size', *Professional Geographer*, 18. 75–81.

ROSSER, C. (1972), 'Housing and planned urban change: the Calcutta experience', in Dwyer (ed.) (1972), 179–90.

ROSTOW, W. W. (1960), *The stages of economic growth: a non-communist manifesto*, Cambridge University Press.

ROUCH, J. (1956), 'Migrations au Ghana (Gold Coast): enquête 1953–1955', *Journal de la société des africanistes*, 26. 33–196.

ROXBOROUGH, I. (1979), *Theories of underdevelopment*, Macmillan.

—— (1984), *Unions and politics in Mexico: the case of the automobile industry*, Cambridge University Press.

RUGH, A. B. (1984), *Family in contemporary Egypt*, Syracuse University Press.

RULE, S. P. (1989), 'The emergence of a racially mixed residential suburb in Johannesburg: demise of the apartheid city?', *Geographical Journal*, 55. 196–203.

RUSSELL, C. A., MILLER, J. A., and HILDNER, R. E. (1974), 'The urban guerrilla in Latin America: a select bibliography', *Latin American Research Review*, 9 (1). 37–79.

RYAN, W. (1971), *Blaming the victim*, Pantheon.

SABERWAL, S. (1986), *India: the roots of crisis*, Oxford University Press.

SABOT, R. H. (1979), *Economic development and urban migration: Tanzania 1900–1971*, Clarendon Press.

SACHS, C. (1983), 'The growth of squatter settlements in Sao Paulo: a study of the perverse effects of the state's housing policies', *Social Science Information*, 4/5. 749–75.

SADA, P. O. (1972), 'Residential land-use in Lagos: an inquiry into the relevance of traditional models', *African Urban Notes*, 7. 3–25.

SALAFF, J. W. (1981), *Working daughters of Hong Kong: filial piety or power in the family?* Cambridge University Press.

SALISBURY, R. F. (1970), *Vunamami: economic transformation in a traditional society*, University of California Press.

SANCHEZ-LEON, A. *et al.* (1979), *Tugurización en Lima metropolitana*, DESCO, Lima.

SANDBROOK, R. (1982), *The politics of basic needs: aspects of assaulting poverty in Africa*, Heinemann.

—— and ARN, J. (1977), *The labouring poor and urban class formation: the case of Greater Accra*, Occasional Monograph, Center for Developing-Area Studies, McGill University.

SANDERSON, S. S. (ed.) (1985), *The Americas in the New International Division of Labor*, Holmes and Meier.

SANYAL, B. (1981), 'A critical look at the housing subsidies in Zambia', *Development and Change*, 12. 409–40.

SARGENT, C. S., Jr. (1972), 'Toward a dynamic model of urban morphology', *Economic Geography*, 48. 357–74.

SARIN, M. (1982), *Urban planning in the Third World: the Chandigarh experience*, Mansell.

SAUL, J. S. (1975), 'The "labour aristocracy" thesis reconsidered, in R. Sandbrook and R. Cohen (eds.), *The development of an African working class: studies in class formation and action*, Longman, 303–10.

SAUNDERS, P. (1980), *Urban politics: a sociological interpretation*, Penguin. (1st edn. 1979.)

SAWERS, L. (1989), 'Urban primacy in Tanzania', *Economic Development and Cultural Change*, 37. 841–59.

SAXENA, D. P. (1977), *Rururban migration in India: causes and consequences*, Popular Prakashan.

SCHILDKROUT, E. (1982), 'Dependence and autonomy: the economic activities of secluded Hausa women in Kano, Nigeria', in E. Bay (ed.), *Women and work in Africa*, Westview Press, 55–83.

SCHMITZ, H. (1984), 'Industrialisation strategies in less-developed countries: some lessons of historical experience', *Journal of Development Studies*, 21. 1–21.

SCHNORE, L. F. (1965), 'On the spatial structure in the two Americas', in P. Hauser and L. F. Schnore (eds.), *The study of urbanization*, John Wiley, 347–98.

SCHWERDTFEGER, F. (1972), 'Urban settlement patterns in northern Nigeria', in P. J. Ucko, R. Tringham and G. W. Dimbleby (eds.), *Man, settlement and urbanism*, Duckworth, 547–56.

SCOBIE, J. R. (1964), *Argentina: a city and a nation*, Oxford University Press.

SEERS, D., SCHAFFER, B., and KILJUMEN, M. L. (eds.) (1979), *Underdeveloped Europe: studies in core-periphery relations*, Harvester Press.

SEGALEN, M. (1986), *Historical anthropology of the family*, Themes in the Social Sciences, Cambridge University Press. (First pub. (1981), *Sociologie de la famille*, Armand Colin.)

SELDEN, M. (1988), *The political economy of Chinese socialism*, M. E. Sharpe.

SEN DOU CHANG (1968), 'The million city of mainland China', *Pacific Viewpoint*, 9. 128–53.

SENGUPTA, A. K. (1977), 'Trade unions, politics and the state: a case from West Bengal', *Contributions to Indian Sociology*, NS 11. 45–68.

SHAHEED, Z. A. (1979), 'Union leaders, worker organization and strikes: Karachi 1969–72', *Development and Change*, 10. 181–204.

SHARMA, U. (1986), *Women's work, class, and the urban household: a study of Shimla, North India*, Tavistock.

SHAW, R. P. (1975), *Migration theory and fact: a review and bibliography of current literature*, Bibliography Series 5, Regional Science Research Institute, Philadelphia.

—— (1976), *Land tenure and the rural exodus in Chile, Colombia, Costa Rica and Peru*, Latin American Monographs, 2nd Series 19, University Presses of Florida.

SHAWCROSS, W. (1979), *Sideshow: Kissinger, Nixon and the destruction of Cambodia*, André Deutsch.

SHIBLI, K. (1974), 'Metropolitan planning in Karachi: a case study', in Jakobson and Prakash (eds.) (1974), 109–29.

SIMMIE, J. M. (1974), *Citizens in conflict: the sociology of town planning*, Hutchinson.

SIMMONS, A. B. (1978), 'Slowing metropolitan city growth in Asia: a review of policies, programmes and results', paper presented to the ECLA/CELADE Seminar on Population Redistribution, Aug. 1978.

—— DIAZ-BRIQUETS, S., and LAQUIAN, A. A. (1977), *Social change and internal migration: a review of research findings from Africa, Asia, and Latin America*, International Development Research Centre, Ottawa.

SIMON, D. (1989), 'Crisis and change in South Africa: implications for the apartheid city', *Transactions of Institute of British Geographers*, 14. 189–206.

SINCLAIR, U. (1906), *The jungle*, Heinemann.

SINGER, P. (1973), 'Urbanización, dependencia y marginalidad en América Latina', in Castells (ed.) (1973), 287–312.

—— (1975), 'Urbanization and development: the case of São Paulo', in Hardoy (ed.) (1975), 435–56.

SINGH, A. M. (1976), *Neighbourhood and social networks in urban India*, Marwah, New Delhi.

SJOBERG, G. (1960), *The pre-industrial city*, Free Press.

—— (1963), 'The rise and fall of cities: a theoretical perspective', *International Journal of Comparative Sociology*, 4. 107–20.

SKINNER, E. P. (1965), 'Labor migration among the Mossi of Upper Volta', in H. Kuper (ed.), *Urbanization and migration in West Africa*, University of California Press, 60–84.

SKINNER, R. (1982), 'Self-help, community organization and politics: Villa El Salvador, Lima', in Ward (ed.) (1982), 209–29.

SKLAIR, L. (1985), 'Shenzhen: a Chinese "development zone" in global perspective', *Development and Change*, 16. 571–602.

—— (1987), 'Does Mexico's maquiladora programme represent a genuine development strategy?', paper presented to Center of US–Mexican Studies Research Fellows' Seminar, April 1987.

SKLAR, R. L. (1963), *Nigerian political parties: power in an emergent African nation*, Princeton University Press.

SKOCPOL, T. (1979), *States and social revolutions: a comparative analysis of France, Russia, and China*, Cambridge University Press.

SLATER, D. (1975), 'Underdevelopment and spatial inequality', *Progress in Planning*, 4. 97–167.

—— (1982), 'State and territory in post revolutionary Cuba—some critical reflections on the development of spatial policy', *International Journal of Urban and Regional Research*, 6. 1–34.

SMIT, D., OLIVIER, J. J., and BOOYSEN, J. J. (1982), 'Urbanization in the homelands', Smith (ed.) (1982), 90–105.

SMITH, ADAM (1776), *An inquiry into the nature and causes of the wealth of nations*, Whitestone.

SMITH, D. M. (ed.) (1982), *Living under apartheid: aspects of urbanization and social change in South Africa*, George Allen and Unwin.

SONGRE, A., SAWADOGO, J.-M., and SANAGOH, G. (1974), 'Réalités et effets de l'émigration massive des Voltaïques dans le contexte de l'Afrique Occidentale', in S. Amin (ed.), *Modern migrations in Western Africa*, Oxford University Press, 384–402. (Abridged and rev. English version: A. Songre (1973), 'Mass emigration from Upper Volta: the facts and implications', *International Labour Review*, 108, 209–25.)

SOUSSAN, J. (1982), 'Urban planning and the community: an example from the Third World', *Environment and Planning A*, 14. 901–16.

SOUZA, P. R., and TOKMAN, V. (1976), 'The informal urban sector in Latin America', *International Labour Review*, 114. 355–65.

SOVANI, N. V. (1964), 'The analysis of "over-urbanization" ', *Economic Development and Cultural Change*, 2. 113–22.

SPENCE, J. (1979), *Search for justice: neighborhood courts in Allende's Chile*, Westview Press.

SQUIRE, L. (1981), *Employment policy in developing countries: a survey of issues and evidence*, Oxford University Press.

SROLE, L. (1978), 'The city versus town and country: new evidence on an ancient bias, 1975', in L. Srole and A. Kassen Fischer (eds.), *Mental health in the metropolis: the midtown Manhattan study*, New York University Press, rev. ed., 433–59.

STACEY, J. (1983), *Patriarchy and socialist revolution in China*, University of California Press.

STANDING, G. (ed.) (1984), *Labour circulation and the labour process*, Croom Helm.

STEIN, S. J., AND STEIN, B. H. (1970), *The colonial heritage in Latin America*, Oxford University Press.

STEWART, C. T. (1958), 'The size and spacing of cities', *Geographical Review*, 48. 222–45.

STEWART, F. (1981), 'International technology transfer: issues and policy options', in P. P. Streeten and R. Jolly (eds.), *Recent issues in world development: a collection of survey articles*, Pergamon, 67–110.

STÖHR, W. B. (1975), *Regional development in Latin America: experience and prospects*, Mouton.

—— and TAYLOR, D. R. F. (eds.) (1981), *Development from above or below? The dialectics of regional planning in developing countries*, John Wiley.

—— and TÖDTLING, F. (1979), 'Spatial equity: some anti-theses to current regional development doctrine', in H. Folmer and J. Oosterhaven (eds.), *Spatial inequalities and regional development*, Martinus Nijhoff Publishing, 133–60.

STREN, R. (1975), 'Urban policy and performance in Kenya and Tanzania', *Journal of Modern African Studies*, 13. 267–94.

—— (1982), 'Underdevelopment, urban squatting and the state bureaucracy: a case study of Tanzania', *Canadian Journal of African Studies*, 16. 67–91.

STRETTON, A. (1985), 'Circular migration, segmented labour markets and efficiency', in Standing (ed.) (1985), 290–312.

STUCKEY, B. (1975), 'Spatial analysis and economic development', *Development and Change*, 6. 98–101.

SUSMAN, P. (1987), 'Spatial equality and socialist transformation in Cuba,' in D. Forbes and N. Thrift (eds.), *The socialist Third World: urban development and territorial planning*, Blackwell, 250–81.

SVENSON, G. (1977), *El desarrollo económico departmental 1960–1975*, Inandes, Bogotá.

TEMPLE, F. T., and TEMPLE, N. W. (1980), 'The politics of public housing in Nairobi', in M. S. Grindle (ed.), *Politics and policy implementation in the Third World*, Princeton University Press, 224–49.

TERNENT, J. A. S. (1976), 'Urban concentration and dispersal: urban policies in Latin America', in Gilbert (ed.) (1976*a*), 169–96.

Thailand, National Statistical Office (1986), *Statistical Handbook of Thailand*, 1985.

THOMAS, H. (1977), *The Cuban Revolution*, Harper & Row. (1st edn. (1977), *Cuba: the pursuit of freedom*.)

THOMPSON, R. (1979), 'City planning in China', in Maxwell (ed.) (1979), 299–312.

THRIFT, N. (1986), 'The geography of international economic disorder', in R. J. Johnston and P. J. Taylor (eds.), *A world in crisis: geographical perspectives*, Blackwell, 12–67.

TIMBERG, T. A. (1978), *The Marwaris: from traders to industrialists*, Vikas Publishing House, Delhi.

TIMBERLAKE, M. (ed.) (1984), *Urbanization in the world-economy*, Academic Press.

TIPPLE, A. G. (1976), 'Self-help housing policies in a Zambian mining town', *Urban Studies*, 13. 167–9.

—— (1988), 'The development of housing policy in Kumasi, Ghana, 1901–1981', Centre for Architectural Research and Development Overseas, University of Newcastle upon Tyne.

—— and WILLIS, K. G. (1989), 'Who owns, who rents: tenure choice in a West African city', University of Newcastle upon Tyne, mimeo.

TOKMAN, V. E. (1978), 'An exploration into the nature of informal–formal sector relationships', *World Development*, 6. 1065–75.

TOWNROE, P. M. (1979), 'Employment decentralization: policy instruments for large cities in less developed countries', *Progress in Planning*, 10. 85–154.

—— and KEEN, D. (1984), 'Polarization reversal in the state of São Paulo, Brazil', *Regional Studies*, 10. 45–54.

TRAGER, L. (1988), *The city connection: migration and family interdependence in the Philippines*, University of Michigan Press.

TRAVIESO, F. (1972), *Ciudad, región y subdesarrollo*, Fondo Editorial Común, Caracas.

TROWBRIDGE, J. W. (1973), *Urbanization in Jamaica*, Ford Foundation International Urbanization Survey, New York.

TURNER, J. F. C. (1967), 'Barriers and channels for housing development in

modernizing countries', *Journal of the American Institute of Planners*, 33. 167–81.
—— (1968), 'Housing priorities, settlement patterns and urban development in modernizing countries', *Journal of the American Institute of Planners*, 34. 354–63.
—— (1969), 'Uncontrolled urban settlements: problems and policies', in Breese (ed.) (1969), 507–31.
—— (1972), 'Housing as a verb', in Turner and Fichter (eds.) (1972), 148–75.
—— (1976), *Housing by people*, Marion Boyars.
—— and Fichter, R. (eds.) (1972), *Freedom to build*, Collier Macmillan.
TURNER, R. (ed.) (1962), *India's urban future*, University of California Press.
TUROK, B., and MAXEY, K. (1985), 'Southern Africa in crisis', in P. C. W. Gutkind and I. Wallerstein (eds.), *Political economy of contemporary Africa*, Sage, 2nd edn. 243–78. (1st edn. 1976.)
UDALL, A. T., and SINCLAIR, S. (1982), 'The "luxury unemployment" hypothesis: a review of recent evidence', *World Development*, 10. 49–62.
UNCRD (United Nations Centre for Regional Development) (1976), *Growth pole strategy and regional development planning in Asia*, UNCRD.
UNECLA (United Nations Economic Commission for Latin America) (1971), *Income Distribution in Latin America*, New York.
UNECLAC (Economic Commission for Latin America and the Caribbean) (1989), 'Preliminary Overview of the Latin American Economy 1989', *Notas sobre la economía y el desarrollo*, 485–6.
UNESCO (1957), *Urbanization in Asia and the Far East*, UNESCO.
UNIKEL, L. (1976), *El desarrollo urbano de México: diagnóstico e implicaciones futuras*, El Colegio de México, Mexico City.
United Nations (1969), *Growth of the world's urban and rural population 1920–2000*, United Nations.
—— (1980), *Patterns of urban and rural population growth*, Population Studies 68, United Nations.
—— (1984), *1982 Demographic Yearbook*, United Nations.
—— (1985), *Compendium of human settlement statistics*, 1982–84, New York.
—— (1986), *1984 Demographic Yearbook*, United Nations.
—— (1987), *1985 Demographic Yearbook*, United Nations.
—— (1988), *1986 Demographic Yearbook*, United Nations.
—— (1989a), *1987 Demographic Yearbook*, United Nations.
—— (1989b), *1989 World survey on the role of women in development*, United Nations.
—— (1990), *1988 Demographic Yearbook*, United Nations.
—— (1991), *World urbanization prospects 1990*, United Nations.
UNRISD (United Nations Research Institute for Social Development) (1971), *Regional development: experiences and prospects*, vol. 1: *South and Southeast Asia*, Mouton.
URQUIDI, V. L., and CARRILLO, M. M. (1985), 'Desarrollo económico e interacción en la frontera norte de México', *Comercio exterior*, 35. 1060–70.
VALENTINE, C. A. (1968), *Culture and poverty: critique and counter-proposals*, University of Chicago Press.
VALENZUELA, J., and VERNEZ, G. (1974), 'Construcción popular y estructura del mercado de vivienda: el caso de Bogotá', *Revista interamericana de planificación*, 8. 88–104.

VALLADARES, L. do P. (1978*a*), *Passa-se uma casa: análise do programa de remoção de favelas do Rio de Janeiro*, Biblioteca de Ciencias Sociais, Zahar Editores, Rio de Janeiro.
—— (1978*b*), 'Working the system: squatter response to resettlement in Rio de Janeiro', *International Journal of Urban and Regional Research*, 2. 12–25.
VAN HUYCK, A. P. (1968), 'The housing threshold for lowest income groups: the case of India', in J. D. Herbert and A. P. Van Huyck (eds.), *Urban planning in the developing countries*, Praeger, 64–108.
VAPÑARSKY, C. A. (1969), 'On rank-size distribution of cities: an ecological approach', *Economic Development and Cultural Change*, 17. 584–95.
VARLEY, A. (1985), 'Urbanisation and agrarian law: the case of Mexico City', *Bulletin of Latin American Research*, 4. 1–16.
VATUK, S. (1972), *Kinship and urbanization: white collar migrants in North India*, University of California Press.
VELLINGA, M. (1989), 'Power and independence: the struggle for identity and integrity in urban social movements', in F. Schuurman and T. van Naerssen (eds.), *Urban social movements in the Third World*, Routledge, 151–76.
VELSEN, J. van (1960), 'Labor migration as a positive factor in the continuity of Tonga tribal society', *Economic Development and Cultural Change*, 8. 265–78.
VERNEZ, G. (1973), *The residential movements of low-income families: the case of Bogotá, Colombia*, Rand Institute.
VINING, D. R. (1982), 'Recent dispersal from the world's industrial core regions', in T. Kawashima (ed.), *Urbanization processes: experiences of eastern and western countries*, Pergamon, 171–92.
VIOLICH, F. (1944), *Cities of Latin America*, Reinhold.
WALKER, R., and STORPER, M. (1981), 'Capital and industrial location', *Progress in Human Geography*, 5. 473–509.
WALLERSTEIN, I. (1974), *The modern world-system: capitalist agriculture and the origins of the European world economy in the sixteenth century*, Academic Press.
—— (1980), *The modern world-system, 2: mercantilism and the consolidation of the European world-economy, 1600–1750*, Academic Press.
—— (1989), *The modern world system, 3: the second era of the great expansion of the capitalist world-economy, 1730–1840s*, Academic Press.
WALTERS, A. A. (1978), 'The value of urban land', in World Bank (1978), 65–98.
WALTON, J. (1978), 'Guadalajara: creating the divided city', in W. A. Cornelius and R. V. Kemper (eds.) (1978), *Metropolitan Latin America: the challenge and the response*, Latin American Urban Research, 6, Sage, 25–50.
—— (1982), 'The international economy and peripheral urbanization', in N. I. Fainstein and S. S. Fainstein (eds.), *Urban policy under capitalism*, Sage Urban Affairs Annual Reviews 22, Sage, 119–36.
—— (ed.) (1985), *Capital and labor in an industrializing world*, Sage.
—— (1989), 'Debt, protest, and the state in Latin America', in S. Eckstein (ed.), *Power and popular protest: Latin American social movements*, University of California Press, 299–328.
—— and MASOTTI, L. H. (eds.) (1976), *The city in comparative perspective*, Halsted Press.
—— and RAGIN, C. (1989), 'Austerity and dissent: social bases of popular

struggle in Latin America', in W. L. Canak (ed.), *Lost promises: debt, austerity, and development in Latin America*, Westview Press, 216–32.
WARD, B. (1964), 'Creating man's future: goals for a world of plenty', *Saturday Review*, 9 Aug. 1964, 27–9 and 191–3.
WARD, P. M. (1976*a*), 'The squatter settlement as slum or housing solution: the evidence from Mexico City', *Land Economics*, 52. 330–46.
—— (1976*b*), 'Intra-city migration to squatter settlements in Mexico City', *Geoforum*, 7. 369–83.
—— (1978), 'Self-help housing in Mexico City: social and economic determinants of success', *Town Planning Review*, 49. 38–50.
—— (ed.) (1982), *Self-help housing: a critique*, Mansell.
—— (1986), *Welfare politics in Mexico: papering over the cracks*, Allen and Unwin.
WARREN, B. (1973), 'Imperialism and capitalist industrialization', *New Left Review*, 81. 3–44.
WEBB, S. D. (1984), 'Rural–urban differences in mental health', in H. Freeman (ed.), *Mental health and the environment*, Churchill Livingstone, 226–49.
WEE, A. (1972), 'Some social implications of rehousing programmes in Singapore', in Dwyer (ed.) (1972), 216–30.
WEGELIN, E. A. (1977), *Urban low-income housing and development*, Martinus Nijhoff.
WEINER, M. (1978), *Sons of the soil: migration and ethnic conflict in India*, Princeton University Press.
WEISNER, T. S. (1972), 'One family, two households: rural–urban ties in Kenya', Ph.D. dissertation, Harvard University.
—— (1976), 'Kariobangi: the case history of a squatter resettlement scheme in Kenya', in W. Arens (ed.), *A century of change in East and Central Africa*, Mouton, 77–97.
—— and ABBOTT, S. (1977), 'Women, modernity, and stress: three contrasting contexts for change in East Africa', *Journal of Anthropological Research*, 33. 421–51.
WELCH, C. E. (1977), 'Obstacles to "peasant war" in Africa', *African Studies Review*, 20. 121–30.
WELLMAN, B. (1988), 'Structural analysis: from method and metaphor to theory and substance', in B. Wellman and S. D. Berkowitz (eds.), *Social structures: a network approach*, Cambridge University Press, 19–61.
WELLS, J. (1983), 'Industrial accumulation and living standards in the long-run: the São Paulo industrial working class, 1930–1975', *Journal of Development Studies*, 19, parts II and III. 145–69; 297–328.
WHEATLEY, P. (1967), *City as symbol*, Inaugural Lecture, University College London.
—— (1970), 'The significance of traditional Yoruba urbanism', *Comparative Studies in Society and History*, 12. 393–423.
WHITE, A. (1975), 'Squatter settlements, politics and class conflict', Occasional Paper 17, Institute of Latin American Studies, University of Glasgow.
WHITE, G. (1988), 'State and market in China's labour reforms', *Journal of Development Studies*, 24 (4). 180–202.
WHYTE, M. K., and PARISH, W. L. (1984), *Urban life in contemporary China*, University of Chicago Press.

WHYTE, W. F. (1981), *Street corner society: the social structural of an Italian slum*, University of Chicago Press, 3rd edn. (1st edn. 1943).

WILKIE, J. W., and HABER, S. (eds.) (1981), *Statistical Abstract of Latin America*, 22. UCLA Latin America Center Publications.

—— and PERKAL, A. (eds.) (1984), *Statistical Abstract of Latin America*, 23. UCLA Latin America Center Publications.

WILLIAMSON, J. G. (1965), 'Regional inequality and the process of national development: a description of the patterns', *Economic Development and Cultural Change*, 13. 3–45.

WILSHER, P., and RIGHTER, R. (1975), *The exploding cities*, Deutsch.

WILSON, F. (1972*a*), *Labour in the South African gold mines, 1911–1969*, African Studies 6, Cambridge University Press.

—— (1972*b*), *Migrant labour in South Africa*, Johannesburg, South African Council of Churches and Spro-cas.

WINSTON, G. (1979), 'The appeal of inappropriate technologies: self-inflicted wages, ethnic pride and corruption', *World Development*, 7. 835–45.

WIRSING, R. G. (1973), 'Associational "micro-arenas" in Indian urban politics', *Asian Survey*, 13. 408–20.

—— (1976), 'Strategies of political bargaining in Indian city politics', in D. B. Rosenthal (ed.), *The city in Indian politics*, Thomson Press (India), 192–212.

WIRTH, L. (1938), 'Urbanism as a way of life', *American Journal of Sociology*, 44. 1–24.

WISEMAN, J. (1986), 'Urban riots in West Africa, 1977–85', *Journal of Modern African Studies*, 24. 509–18.

WOLF, E. R., and MINTZ, S. W. (1957), 'Haciendas and plantations in Middle America and the Antilles', *Social and Economic Studies*, 6. 380–412.

WOLF-PHILLIPS, L. (1979), 'Why Third World?', *Third World Quarterly*, 1. 105–16.

WOMACK, J. (1969), *Zapata and the Mexican Revolution*, Thames & Hudson.

World Bank (1978), 'Urban land policy issues and opportunities', vol. 1, World Bank Staff Working Paper 283.

—— (1979), *World Development Report 1979*, World Bank.

—— (1980), *World Development Report 1980*, World Bank.

—— (1989), *World Development Report 1989*, World Bank.

YAP, L. Y. L. (1977), 'The attraction of cities: a review of the migration literature', *Journal of Development Economics*, 4. 239–64.

YEUNG, Y. M., and LO, C. P. (eds.) (1976), *Changing South-East Asian cities: readings on urbanization*, Oxford University Press.

YOUNG, C. (1976), *The politics of cultural pluralism*, University of Wisconsin Press.

YOUNG, K. (1982), 'The creation of a relative surplus population: a case study from Mexico', in L. Benería (ed.), *Women and development: the sexual division of labor in rural societies*, Praeger Special Studies, Praeger, 149–77.

ZAPATA, F. S. (1975), 'Action syndicale et comportement politique des mineurs chiliens de Chuquicamata', *Sociologie du travail*, 17. 225–42. (English trans. (1979), 'Trade-union action and political behaviour of the Chilean miners of Chuquicamata', in R. Cohen, P. C. W. Gutkind, and P. Brazier (eds.), *Peasants and proletarians: the struggles of Third World workers*, Monthly Review Press, 460–81.)

ZIPF, G. K. (1941), *National unity and disunity*, Principia Press.

ZOLA, E. (1877), *L'assommoir*, Bibliothèque Charpentier.

Name Index

Subject index

(Cities are listed by country)